Energy Science, Engineering and Technology

THIN FILM SOLAR CELLS

CURRENT STATUS AND FUTURE TRENDS

Energy Science, Engineering and Technology

Additional books in this series can be found on Nova's website under the Series tab.

Additional E-books in this series can be found on Nova's website under the E-book tab.

Energy Science, Engineering and Technology

THIN FILM SOLAR CELLS

CURRENT STATUS AND FUTURE TRENDS

**ALESSIO BOSIO AND ALESSANDRO ROMEO
EDITORS**

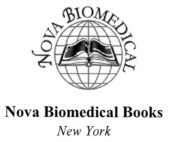

Nova Biomedical Books
New York

Copyright © 2010 by Nova Science Publishers, Inc.

All rights reserved. No part of this book may be reproduced, stored in a retrieval system or transmitted in any form or by any means: electronic, electrostatic, magnetic, tape, mechanical photocopying, recording or otherwise without the written permission of the Publisher.

For permission to use material from this book please contact us:
Telephone 631-231-7269; Fax 631-231-8175
Web Site: http://www.novapublishers.com

NOTICE TO THE READER

The Publisher has taken reasonable care in the preparation of this book, but makes no expressed or implied warranty of any kind and assumes no responsibility for any errors or omissions. No liability is assumed for incidental or consequential damages in connection with or arising out of information contained in this book. The Publisher shall not be liable for any special, consequential, or exemplary damages resulting, in whole or in part, from the readers' use of, or reliance upon, this material. Any parts of this book based on government reports are so indicated and copyright is claimed for those parts to the extent applicable to compilations of such works.

Independent verification should be sought for any data, advice or recommendations contained in this book. In addition, no responsibility is assumed by the publisher for any injury and/or damage to persons or property arising from any methods, products, instructions, ideas or otherwise contained in this publication.

This publication is designed to provide accurate and authoritative information with regard to the subject matter covered herein. It is sold with the clear understanding that the Publisher is not engaged in rendering legal or any other professional services. If legal or any other expert assistance is required, the services of a competent person should be sought. FROM A DECLARATION OF PARTICIPANTS JOINTLY ADOPTED BY A COMMITTEE OF THE AMERICAN BAR ASSOCIATION AND A COMMITTEE OF PUBLISHERS.

LIBRARY OF CONGRESS CATALOGING-IN-PUBLICATION DATA

Thin film solar cells : current status and future trends / editors, Alessio Bosio, Alessandro Romeo.
    p. cm. -- (Energy science, engineering and technology)
Includes index.
ISBN 978-1-61668-326-9 (hardcover)
1. Solar cells. 2. Thin films. I. Bosio, Alessio. II. Romeo, Alessandro. III. Romeo, Alessandro.
TK2960.T444 2011b
621.31'244--dc23
                            2011035651

Published by Nova Science Publishers, Inc. † New York

Contents

Preface		v
Chapter 1	**Status and Perspectives of Thin Film Photovoltaics** *Arnulf Jäger-Waldau*	1
Chapter 2	**Introduction to Inorganic Thin Film Solar Cells** *Francesco Roca, Alessio Bosio and Alessandro Romeo*	25
Chapter 3	**Co-Evaporation of CIGS and Alternative Buffer Layers for CIGS Devices** *Marika Edoff and Charlotte Platzer-Björkman*	59
Chapter 4	**CIGS Absorber Layers Prepared by Sputtering Based Methods** *António Ferreira da Cunha and Pedro Manuel Parracho Salomé*	89
Chapter 5	**Electrical Characterization of Cu(In,Ga)Se2 - Based Thin Film Photovoltaic Devices** *Pawel Zabierowski*	103
Chapter 6	**CdTe solar Cells by Low Temperature Processes** *Alessandro Romeo and Alessio Bosio*	135
Chapter 7	**Polycrystalline CdTe Thin Films Solar Cells** *Alessio Bosio, Alessandro Romeo and Nicola Romeo*	159
Chapter 8	**Thin Film Silicon Solar Cells** *Paola Delli Veneri and Lucia Vittoria Mercaldo*	199
Chapter 9	**Organic and Hybrid Solar Cells** *Thomas M. Brown, Andrea Reale and Aldo Di Carlo*	245
Index		283

PREFACE

The need of more and more energy supply due to increased demand from emerging countries such as India, China and Brazil and the contemporary necessity to preserve the environment has increased the interest to the development of new technologies that make use of solar energy.

In particular photovoltaic solar energy, the direct conversion of solar energy into electricity by means of semiconducting materials, had a very strong development in the last 30 years.

The most important parameter that characterizes a photovoltaic device is the ratio between its conversion efficiency and its cost.

A value less than 0.5 $/Wp is considered competitive with the electricity obtained from fossil energy sources.

Despite the strong development of photovoltaic devices in the last years, the best result so far obtained is around 1 $/Wp for CdTe/CdS thin film modules fabricated by "First Solar".

In any case, in order to reach the goal of 0.5 $/Wp or less, the only way is that of further developing thin film photovoltaics.

In this book, the most recent results concerning thin film solar cells, covering all kinds of relevant materials, are presented by qualified experts in the field. The book is particular important for researchers who are already engaged or intend to start a research in this area.

It could be also useful for the industries who want to invest their money in the new thin film photovoltaic technologies.

Professor Nicola Romeo

In: Thin Film Solar Cells: Current Status and Future Trends   ISBN 978-1-61668-326-9
Editors: Alessio Bosio and Alessandro Romeo   © 2010 Nova Science Publishers, Inc.

Chapter 1

STATUS AND PERSPECTIVES OF THIN FILM PHOTOVOLTAICS

Arnulf Jäger-Waldau
European Commission, Joint Research Centre, Renewable Energy Unit, JRC - Via Enrico Fermi I - 21020 Ispra, Italy.

ABSTRACT

Since a number of years photovoltaics continues to be one of the fastest growing industries with growth rates well beyond 40% per annum. This growth is driven not only by the progress in materials and processing technology, but by market support programmes in a growing number of countries around the world as well. The spiking fossil energy prices in 2008 and the European gas crises in 2007 and early 2009 showed the necessity to diversify energy supplies for the sake of energy security and highlighted the benefits of local renewable energy resources like solar energy. The very high growth is being attained by an increase in manufacturing capacities based on the technology of crystalline silicon, single junction devices, but in the last few years despite the already very high growth rates of the whole industry, thin film photovoltaics is growing at an even quicker pace increasing its market share from 2005 to 2009 from 6 to over 15%. Consistent with the time needed for any major change in energy infrastructure, another 10 to 15 years of sustained and aggressive growth will be required for photovoltaic solar electricity to substitute a significant share of conventional energy sources. The question is whether a switch will be possible with the current technologies alone or whether this growth will only be possible with the continuous introduction of new technologies. It leads us to the search for new developments with respect to material use and consumption, device design and production technologies as well as new concepts to increase overall efficiency. This paper analyses the current status of thin film solar cells and their outlook for future developments.

Keywords: renewable energies, photovoltaic, thin film solar cells, technological development, policy options, market development

1. INTRODUCTION

Photovoltaics has enjoyed extraordinary growth during the last few years with overall growth rates around 40% per annum, indicating that further increase of production facilities are an attractive investment. The World Energy Outlook released by the International Energy Agency (IEA) in November 2009 indicates that more and more renewable energy and in particular solar energy will be needed and used in the coming decades to ensure stable and secure energy supply [1]. This development has to be seen in light of the last gas crises in Europe in 2007 and early 2009, when many people realised how dependent our energy supply relies on a few key suppliers. In addition, the spiking oil price in 2008 pointed out, that economic growth in the BRIC[1] countries leads to an increasing demand for oil, which is higher than the growth rate in oil production. The speculative aspects fuelled by the fact that daily oil demand was higher than the actual production accelerated the price spike. The current world-wide recession is easing the price pressure on this side, but as soon as the economic recovery will demand more oil similar developments will be observed again.

These developments as well as the various support schemes in place for the introduction of renewable energies and Photovoltaic solar electricity are one of the reasons that despite the current constraints in the financing sector, investments in Photovoltaic solar cell/module production capacities are still attracting large amounts of capital.

Production data for the global cell production[2] in 2009 vary between 10.5 GW and 12 GW (Fig. 1), which is again an increase of 40% to 50% compared to 2008. The significant uncertainty in the data for 2009 is due to difficult market situation, which was characterised by a declining market environment in the first half of 2009 and an exceptional boom in the second half of 2009. In addition, the fact that some companies report shipment figures, whereas others report production figures add to the uncertainty. In addition, the difficult economic conditions lead to a decreased willingness to report confidential company data. Nevertheless, the figures show a significant growth of the production as well as an increasing silicon supply situation. Since 2006 the production of Thin Film solar modules grew faster than the Photovoltaic industry as a whole.

---

[1] BRIC: Brazil, Russia, India, China
[2] Solar cell production capacities mean:
- In the case of wafer silicon based solar cells only the cells
- In the case of thin films, the complete integrated module
- Only those companies which actually produce the active circuit (solar cell) are counted.
- Companies which purchase these circuits and make cells are not counted.

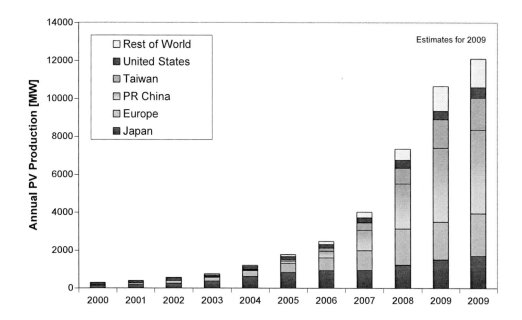

Figure 1. World-wide PV Production from 2000 to 2009
Data Source: PV News [2], Photon International [3] and own analysis

If all these ambitious plans can be realised by 2012, China will each have about 28% of the worldwide production capacity of 48 GW followed by Europe (22%), Japan and Taiwan with 15% each (Figure 2) [5,6].

Announcements of completion of a capacity increase frequently refer to the installation of the equipment only. It does not mean that the production line is really fully operational. This means, especially with new technologies, that there can be some time delay between installation of the production line and real sales of solar cells. In addition, the production capacities are not equal to sales and therefore, there is always a noticeable difference between the two figures, which cannot be avoided.

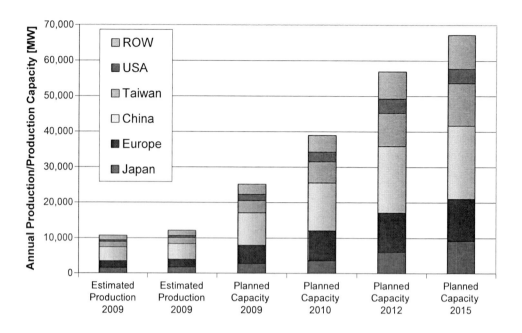

Figure 2. World-wide PV Production 2009 with future planned production capacity increases.

Last but not least it should be mentioned that all these ambitious plans to increase production capacities at such a rapid pace depend on the hope that markets will grow accordingly. This however is at the moment the biggest uncertainty as can be seen with the market estimates for 2009 which vary between 7.2 GW and 7.6 GW. Also, most markets are still dependent on public support in the form of feed-in tariffs, investment subsidies or tax-breaks. Already now, Electricity production from photovoltaic solar systems has shown that it can be cheaper than peak prices in the electricity exchange. But only if markets and competition will continue to grow, prices of the photovoltaic systems will decrease and make electricity from PV systems price-competitive for consumers. In order to achieve the price reductions and reach grid-parity for electricity generated from photovoltaic systems, public support in one ore the other way will be necessary for the next decade.

2. THIN FILM TECHNOLOGIES AND RESEARCH NEEDS

Equally competitive technologies are thin film silicon (amorphous, amorpous/micromorph), Cadmium Telluride (CdTe) and Chalcopyrite (Cu(In,Ga)(S,Se)$_2$) thin films. Crystalline silicon thin films on glass or foreign substrates are additional technology options which are already or planned to be transferred from laboratory to production[3]. Dye-cells are another technology getting ready to enter the market. The growth of these technologies is accelerated by the positive development of the PV market as a whole. It is interesting to note

---

[3] High concentration solar cells are not covered here due to the fact that they are manly based on Gallium Arsenide (GaAs) wafer substrates.

that not only new players are entering into thin film production, but also established silicon wafer based PV cell manufacturers diversify into thin film photovoltaics.

2.1. Technology Overview

In order to give an overview of the technologies discussed, this chapter will list the different technologies under investigation and in production with general descriptions rather than techno-scientific details, which are better championed by respective specialists.

2.1.1. Amorphous and Thin Film Silicon

Within thin film silicon the various technologies can be ordered according to the respective grain size: none (amorphous silicon – a-Si), smaller than 0.1 µm (nanocrystalline – n-Si), between 0.1 and 50 µm (polycrystalline – pc-Si), 100 to 1000 µm (multicrystalline – mc-Si) and single crystals (c-Si). The term TF-Si (thin film silicon) is used here to describe silicon material between 0.7 µm and 200 µm. Another classification often referred to is:
1) amorphous silicon;
2) microcrystalline (µ-Si) and low temperature thin film crystalline silicon (LT-f-Si) with deposition temperatures < 600°C;
3) high temperature crystalline silicon (HT-f-Si) or the deposition temperatures > 600°C.

Both categorisations are frequently used and quoted according to their use in the literature.

2.1.2. CdTe

Cadmium Telluride has two features that make it appear like an ideal candidate for thin film solar cells. It can be deposited with various deposition methods which all result in reasonable quality and a direct energy gap at $E_g = 1.45$ EV - within the ideal range for solar energy conversion.

When CdTe is deposited onto substrates above 449°C it condenses stoichiometrically as the stable phase in this regime [7]. These films are in general p-type with carrier concentrations of $p < 10^{15}$ cm^{-3} due to a slight cadmium deficiency. The most common CdTe solar cell structure is the n-CdS/p-CdTe hetero-junction, where the CdS is deposited on a transparent conductive oxide (TCO) coated glass. An important feature of this type of solar cell is that CdS and CdTe can be deposited with the same deposition technologies. CdTE also allows bandgap engineering if additional elements like Mercury (Hg) or Manganese (Mn) are added. With increasing mercury content the badgap of $Cd_{1-x}Hg_xTe$ decreases, whereas the increase of the manganese concentration increases the bandgap in $Cd_{1-x}Mn_xTe$. The proof of concept for a two terminal tandem solar cell based on CdMnTe and CdHgTe absorbers was shown by Alvin Compaan in 2004 [8].

The standard techniques to deposit p-type CdTe of good crystalline quality and high electron mobility are: sublimation/condensation (S); close spaced sublimation (CSS) a modification of the first process; chemical spraying (CS); screen printing (SP); chemical vapour deposition (CVD); sputtering; and electro-deposition (ED). The top efficiency achieved at NREL was 16.5 ± 0.5% [9] prepared by CSS (CdTe) and chemical bath deposition (CdS).

2.1.3. Chalcopyrites

Chalcopyrites with $Cu(In,Ga)(S,Se)_2$ are an interesting material system, which offers the possibility to tailor the device's band gap between 1.01 eV for $CuInSe_2$ to 1.68 eV for $CuGaSe_2$ or 2.4 eV for $CuGaS_2$ by substituting indium with gallium or selenium with sulphur. This renders chalcopyrites not only an interesting material for single junction devices, but also opens the route to a possible tandem structure device made from the same class of materials. The use of chalcopyrite solar cells in concentrator applications is being investigated by NREL and other research groups like the Tokyo Institute of Technology.

Significant progress has not only been made in the basic understanding of the material properties of these devices, but also in the field of large area production of monolithically interconnected modules. The highest efficiency for small area devices was realised at NREL with 19.9% efficiency [10].

2.1.4. Dye sensitised and Organic Solar Cells

Dye-sensitized solar cells (DSSc, DSC or DYSC) are based on a semiconductor formed between a photo-sensitized anode and an electrolyte, a *photoelectrochemical* system. This cell type was invented by Michael Grätzel and Brian O'Regan at the École Polytechnique Fédérale de Lausanne in 1991 and are also known as Grätzel cells [11].

Although the stability and light-conversion mechanisms are currently inadequately understood, it can be said that (1) there is no expected limitation on material, (2) stable 10%-efficient modules are certainly within reach, and (3) the energy-payback period should be significantly shorter than other PV technologies. Demonstrated levels of efficiency and degradation have inspired investment, and several companies are working toward commercialising this technology. However, further advance in the fundamental understanding of the factors that govern cell performance and stability is essential.

There are three main types of organic photovoltaic technologies:

1) **Molecular OPV**: These devices are in general fabricated by sublimating successive layers of electron and hole conducting materials under vacuum. Common materials include PTCBI, PTCDA, Me-PTCDI, Pe-PTCDI, H_2Pc, MPc where M stands for (Zn, Cu), TPyP, TPD, CBP, C60, and PCBM.

2) **Polymer OPV**: These photovoltaic devices are typically synthesised by processing blends of two conjugated polymers or a conjugated polymer with a molecular sensitiser. Most common materials are PPV - Poly(p-phenylene vinylene), polyfluorenes, or polythiophenes.

3) **Hybrid OPV**: These devices use both organic and inorganic materials, e.g. polymer-nanocrystal blended active layers, including the use of quantum dots.

2.2. Research Needs

Thin film solar cell technologies have made remarkable progress in efficiency, reliablity and production over the last few years. Nevertheless, all roadmaps to reach the goal of 1

€/Wp include as major topics: improvement of production yields, reduction of material costs through economy of scale and further improvement of cell and module efficiency.

The main scientific and technical issues (not exhaustive) for the thin film technologies can be summarised as follows:
- In a-Si:H the light induced degradation of the electronic properties or Staebler-Wronski effect (SWE) results in a "stabilised" efficiency of the devices, which is considerably lower than the initial one. Designs of higher efficiency levels would be possible (20 – 30% gain possible) if the SWE could be minimised or even eliminated.
- Due to the wide band gap nature of a-Si:H, the absorption in the red part of the light spectra is rather low. Multi-junction concepts with narrow band gap materials could overcome this problem. Materials under investigation are a-Si:Ge and μ-Si (15 – 25% gain possible).
- a-Si:H is a highly defective material with a low hole mobility, which requires hydrogen passivation of the intrinsically high concentration of dangling bonds. Although it has the advantage of being relatively easy to deposit from gases which allow easy H-incorporation on large areas, an a-Si:H material with high hole mobility would avoid a number of technical problems.
- TF-Si is a rather complex structure and carriers have to pass through imperfect crystallites of different grain size, grain boundaries as well as amorphous parts. In order to improve device performance, it is necessary to increase basic understanding of the limiting factors for mobility and lifetime as well as the effectiveness of grain boundary passivation or the role of different inter-crystalline defects.
- CdTe solar cells exhibit strong fluctuations in parameters of nominally identical devices. For example, it is typical to observe noticeable (~10%) experimental differences between cells ~1 cm apart on the same substrate.
- The formation of good ohmic contacts of high stability is still a major issue in the CdTe cell technology. A number of materials are being investigated, but the physical problems are that metals of very high work function are needed and CdTe is difficult to be p-doped due to a strong tendency towards self compensation.
- The hetero-junction formation between CdTe/CdS is the critical part of the solar cell. Especially the inter-diffusion of CdS and CdTe influences the cell behaviour and is determined by the activation process.
- Environmental, safety, and health (ES&H) continues to be an important aspect of the CdTe technology development and should be constantly updated and studied.
- Bandgap engineering of $Cd_{1-x}Y_xTe$ could be used for the design and manufacturing of tandem structures based on CdTe derived absorber materials.
- Theoretic modelling and understanding of loss mechanisms in chalcopyrite devices.
- Investigation of the different interfaces of chalcopyrite devices and improvement of the understanding of their microstructure and chemical composition. Back contact: focus on corrosion stability and transparent back contacts for tandem applications.
- Development of an augmented fundamental science and engineering basis for $Cu(InGa)Se_2$ materials, devices, and processing requirements. There is a particular need for more fundamental understanding of wide band-gap alloys and devices as

these should lead to improved module performance and could enable the development of thin film based tandem cells.
- Investigation of the role of the "buffer" layer and search for an optimal one.
- Investigation of different transparent conductive oxides with possibilities of n- and p-doping.
- Commercially limited availability of rare earth metals like indium, tellurium etc.
- Issues including device sensitivity to water vapour and enhancing processing approaches to improve commercial module efficiency of all different kind of thin film modules.
- The primary challenge for OPV is to increase the efficiency and reliability. The limitations to efficiency are generally understood, but a rigorous fundamental understanding is lacking. Issues related to device degradation, such as photo-oxidation, interfacial instability and de-lamination, inter-diffusion, and morphology changes are poorly understood. Development of more complex device designs, such as multi-junction devices or inclusion of more exotic third-generation mechanisms into the OPV design, may be necessary to push efficiencies to competitive levels or to enable substantially higher efficiencies.
- The long-range goal of OPV is large-scale power generation. But as the technology develops, the potential for low-cost and flexible form-factors may enable other applications in the short term.

3. RESEARCH SITUATION OF THIN FILM SOLAR CELLS

This chapter deals with the different research activities for thin film solar cells and tries to analyse, why funding organisations give different attention to this technology.

3.1. Thin Film Activities in the Japanese NEDO PV Programme

In Japan, the Independent Governmental Entity New Energy Development Organisation (NEDO) is responsible for the Research Programme for Renewable Energies. The current programmes for Photovoltaics in the frame of Energy and Environment Technologies Development Projects has three main pillars [12]:
- New Energy Technology Development
- Introduction and Dissemination of New Energy and Energy Conservation
- International Projects

One of the dominant priorities, besides the future increase in PV production, is obviously the cost reduction of solar cells and PV systems. The "PV Roadmap towards 2030" (Figure 3), which was drafted by NEDO, METI, PVTEC[4] and JPEA[5] depicts the planned targets.

---

[4] Phtovoltaic Power Generation Technology Research Association
[5] Japan Photovoltaic Energy Association

Within the New Energy Technology Development Programme there are projects on Photovoltaic technology specific issues, problems of grid-connected systems, as well as public solicitation. In the following only the Thin Film related projects are listed.

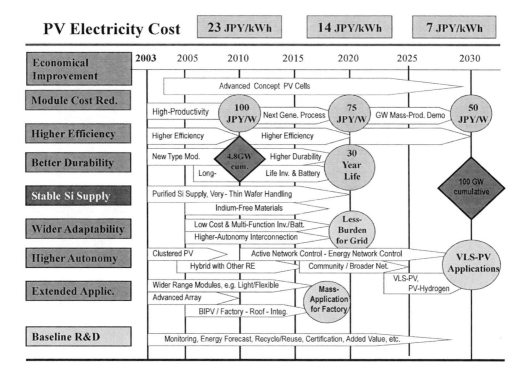

Figure 3. Japanese roadmap for PV R&D and market implementation.

- **Research and Development of Next-generation PV System Technologies**
  FY2006 - FY2009

To play an important role in energy generation in the future, the cost-effectiveness, performance, function, applicability, and usability of Photovoltaic systems must be drastically improved to facilitate the promotion and dissemination of solar power generation. Given this, medium- to long-term innovative technological development efforts beyond simple extensions of currently available technologies are underway. More specifically, the following research and development themes are being undertaken:
- **Technologies to improve the efficiency of CIS thin-film solar cells and elemental technologies to form solar cells on lightweight substrates.**
  Target efficiencies:
  18% for sub-module area of 100 cm^2
  16% for sub-module area of 900 cm^2
  16% for sub-module area of 100 cm^2 on a light-weight substrate
- **Technologies to enable higher productivity and to improve the efficiency of thin-film silicon solar cells.**
  High Productivity Targets:
  1) μc-Si thin films with large area (4m2)
     deposition rate > 2.5 nm/s and single junction cell efficiency > 8%

2) μc-Si thin films 100 cm2 substrates
deposition rate > 10 nm/s and single junction cell efficiency > 8%
3) Thin film silicon etching rate: 20 nm/s
High Efficiency:
15% for module area of 1000 cm^2 with (film deposition rate: 2.5 nm/s)

- **Technologies to enable highly efficient, modular, and durable dye-sensitized solar cells.**
  High efficiency of 15% for small area (1cm^2) cells
  Durability of modules wit target efficiency of 8% (900 cm^2)

- **Technologies and associated processes to produce highly efficient next-generation ultra-thin crystalline silicon solar cells.**
  Development of production technology for crystalline silicon solar cells with a
  Monocrystalline: 100-μm substrate thickness, 125 × 125 mm^2 and 21% efficiency
  Polycrystalline: 100-μm substrate thickness, 150 × 150 mm^2 and 18% efficiency

- **Technologies to improve the efficiency and durability of organic thin-film solar cells.**
  Target efficiency of 7% for small area (1 cm^2) cells
  Relative efficiency degradation ≤ 10% after 100 hours of exposure to air and direct light

- **Search for next-generation technologies that would enable significant cost reductions, improved performance, and extend the usable life of solar power generation systems.**

♦ **Research and Development on Innovative Solar Cells**
FY2008 - FY2014 (peer review after 3rd year)

The objective of this project is to improve drastically the conversion efficiency of solar cells using new and innovative concepts. Tokyo University and AIST Tsukuba in collaboration with the Tokyo Institute of Technology were selected in July 2008 as Centres of excellence (CoE) to carry out the tasks. The following research topics were selected and are open for international collaboration:

- **Post-silicon Solar Cells for Ultra-high Efficiencies**
  (1) Super high-efficiency concentrator multi-junction solar cells
  (2) High efficiency quantum structure tandem solar cells and their manufacturing technologies
  (3) Ultra-high efficiency solar cells based on quantum dots and super lattice
  (4) Ultra-high efficiency multiple junction solar cells with hybrid materials

- **Thin Film Full Spectrum Solar Cells with low concentration ratios**
  (1) Band-gap control of nano dots/ multi-exiton/ band-gap engineering of strained Ge/ novel Si-based and amorphous alloy thin films/ thin film materials design
  (2) Si-based thin film concentrators/ wide band-gap Si based thin films/ multi-cell interface junction/ Chalcopyrite based thin film concentrators on metal substrates/ optical design/ CdTe thin film concentrators
  (3) Surface plasmons/ p-type TCO/ full-spectrum TCO/ grapheme transparent conductive film

- **Exploring Novel Thin Film Multi-junction Solar Cells with Highly-ordered Structure**
  (1) Highly-ordered plane poly-silane/ ordered nano-crystalline Si-materials/ Ge-based narrow band-gap materials/ heterojunction devices
  (2) Wide band-gap chalcogenide-based materials/ solar cells using novel wide band-gap material/ Oxynitride-based wide band-gap materials/ Oxide-based wide band-gap materials/ CIGSSe-based tandem-type solar cells
  (3) Novel concept solar cells using nano-Si, nano-carbon and single-crystalline organic semiconductors/ novel concept solar cells using correlated materials/ novel concept solar cells using nano-materials with controlled structure
  (4) Mechanical stacking-techniques/ highly efficient light-trapping techniques/ improved transparent conduction oxides films using preparation techniques for improved glass substrates

♦ **Research and Development of Common Fundamental Technologies for Photovoltaic Generation Systems**
FY2006 - FY2009

To facilitate the dissemination of Photovoltaic generation systems in the future, it is essential to develop and incorporate commonly-used fundamental technologies and to reduce the cost of solar cells. For this purpose, the following research and development activities are currently ongoing:

- **Development of new solar cell evaluation technologies**
  To increase the number of installations, methods to evaluate the performance and reliability of solar cell modules and solar generation systems are being developed.
- **Development of Photovoltaic environmental technologies**
  Studies are being conducted under a variety of environmental conditions and guidelines for Photovoltaic (PV) generation systems. The development of technologies related to solar cell recycling and the development of life-cycle assessment (LCA) evaluation methods for PV generation are also being carried out.
- **Study on Photovoltaic generation technology development trends**
  Research and development trends, future development directions, and the analysis and evaluation of the state of PV generation abroad are being tracked.

3.2. Thin Film Research in the United States

The National Renewable Energy Laboratory (NREL) supports the U.S. Department of Energy's (DOE) Office of Energy Efficiency and Renewable Energy (EERE) "Solar Energy Technologies Programme". NREL's National Centre for Photovoltaics (NCPV) conducts research to support the U.S. Department of Energy's goal to reduce the average cost of all grid-connected PV systems from 6.25 $/W to 3.30 $/W for end users.

DOE's "Solar Energy Technologies Programme" runs from 2008 to 2012 [13] and covers both Photovoltaic Electricity generation and Concentrated Solar Thermal Electricity. The vision of the Solar Energy Technologies Programme is that:

- Inexpensive solar energy will become available for all Americans,

- Millions of homes and commercial buildings across the nation will use solar technology to provide all or much of their energy needs, and
- Solar energy will constitute a significant portion of the Nation's energy production.

3.2.1. Photovoltaic Technology Roadmaps

Ten Photovoltaic technology roadmaps were developed in 2007 by staff at NREL, Sandia National Laboratories, DOE, and experts from universities and private industry [14]. This work was done, in part, to support activities within the Solar America Initiative. These technology roadmaps summarise the current status and future goals for the specific technologies. The Roadmaps for Intermediate-Band PV, Multiple-Exciton-Generation PV and Nano-Architecture PV are still in a draft stage.

3.2.1.1. Film-Silicon PV

The roadmap highlights the following pathways to advance the technology to advance:
1. For a-SI:H, cells dramatically reduce area-costs and moderately increase efficiency.
2. Despite the fact that c-Si film technologies still lack the efficiencies of wafer-based silicon, a high-risk/high-payoff pathway uses c-Si films fabricated on one of several candidates of inexpensive substrates such as glass, glass-ceramics, metallurgical-grade Si, or stainless steel. This wafer-replacement approach has the potential to raise efficiencies to levels competitive with polysilicon-wafer technology, while maintaining the low-cost structure of a-SI:H thin-film manufacturing.

Table 1. a-Si-Based Thin-Film Technology Performance [13]

Parameter	Status (2007) (costs are estimated)	Future Goal (2015)
Production volume	100 MW/yr	>5 GW/yr
Capital equipment cost	1–2 $/W @ plant capacity	0.7 $/W @ plant capacity
Substrate cost	12–20 $/m^2	4 $/m^2
Module manufacturing cost for a-Si	125–200 $/m^2	0.45–0.70 $/W or 70 $/m^2 @ 10%–15% efficiency
Stabilised efficiency, best a-Si lab cells	13%	15%
Stabilised efficiency, commercial a-Si modules	5%–8%	10%–13%
Reliability of a-Si panels	~1%/yr degradation	1%/yr degradation

3.2.1.2. CdTe PV

Enhanced open-circuit voltage (V_{oc}) with some improvements from short-circuit current density (J_{sc}) in thin-film CdTe devices will most likely be the pathway to higher cell and module efficiency. Factors limiting fill factor (FF) have to be analysed and evaluated to improve solar cell and module performance. The CdTe deposition processes have the distinct advantage of rapidly transferring the material needed to compose the cells, but it also limits the ability to introduce and control constituents to modify the electro-optical properties of the materials.

The reliability of the current glass-glass encapsulated thin-film CdTe modules appears to be comparable to conventional Si-based technology. As new technologies are added to boost efficiency, tests are needed to ensure that reliability is not sacrificed. Environmental, safety, and health (ES&H) continues to be an important aspect of the technology development and should be constantly updated and studied. Efforts should be made to increase public awareness of the perceived cadmium issue.

Table 2. c-Si Film Technology Performance [13]

Parameter	Status (2007) (costs are estimated)	Future Goal (2015)
Production volume	<1 MW/yr	1 GW/yr
Capital equipment cost	2–3 $/W @ plant capacity	0.7 $/W @ plant capacity
Substrate cost	26 $/m^2	10 $/m^2
Module manufacturing cost for waferless silicon	Not available	0.50 $/W or 65 $/m^2 @ 13% efficiency
Efficiency, best supported-film c-Si lab cells	10%	16%–18%
Efficiency, best supported-film c-Si modules	5%–6%	13%–16%
Reliability of Si-film panels	? ? %/yr degradation	1%/yr degradation

Table 3. Cadmium Telluride Technology Performance [13]

Parameter	Status (2007)	Future Goal (2015)
Commercial module efficiency	>9%	13%
Champion device efficiency	16.5%	18%–20%
Module cost	1.21 $/W	0.70 $/W
$/W installed system cost	4–5 $/W	2 $/W
LCOE	18–22 ¢/kWh	7–8 ¢/kWh
Overall process yield	90%	95%
Identify relevant degradation mechanisms and develop appropriate ALTs for device and mini-modules	1.2% per year	0.75% per year

3.2.1.3. CIGS PV

A primary challenge for CIGS is to provide the science and technology needed to close the gap in efficiency between the entry-level prototype products and champion devices. A second challenge is to discover and qualify new materials and device schemes that can enhance performance, absorber band-gap and voltage, material usage, stability, yield, and process simplicity.

Issues including device sensitivity to water vapour, the commercially limited availability of indium, and enhancing processing approaches to improve commercial module efficiency, are all significant challenges for CIGS. Building-integrated products may provide an entry channel for the technologies, taking advantage of the demonstrated capability to manufacture flexible modules and the potential to conform the film PV to building-material geometries.

Overall, the following issues need to be addressed:
- Enhance module efficiency,
- Improve module manufacturing processes,
- Discover alternative approaches and new materials
- Assess and interact.

Table 4. CIGS Technology Performance [13]

Parameter	Status (2007)	Future Goal (2015)
Commercial module efficiency	5%–11%	10%–15%
Champion device efficiency	19.5%	21%–23%
Module cost	Not established, estimated < 2 $/W	~ 1 $/W
$/W installed system cost	5–12 $/W	3 $/W
Reliability goal	0% to 6% annual degradation in pilot arrays	<1% annual power loss for commercial product
Overall process yield	Not available	> 95%
New manufacturing methods	Pilot Flexible "roll-to-roll" manufacturing (initially packaged as a glass to glass laminate)	Develop new encapsulation schemes and appropriate accelerated life testing for flexible and rigid modules
Deposition rate and cell thickness	5 µm/h, 1.25–3 µm CIGS absorber thickness	30–40 µm/h <1 µm CIGS absorber thickness

Table 5. OPV Technology Performance [13]

Parameter	Status (2007)	Future Goal (2015)
Champion device efficiency	5.2%	12%
Cell degradation	< 5% per 1000 h, research-scale	< 2% per 1000 h, module
Material figure-of-merit efficiency. Identification of candidate materials whose fundamental properties, such as optical absorption, band structure, and carrier mobility, allow for high theoretically attainable efficiencies.	Some material sets with improved figure-of-merit efficiencies exist.	Identification and synthesis of multiple donor-acceptor materials that meet all the fundamental requirements to achieve the Shockley-Queisser limit.

3.2.1.4. Organic PV

The primary challenge for OPV is to increase the efficiency and reliability. The limitations to efficiency are generally understood, but a rigorous fundamental understanding is lacking. Issues related to device degradation, such as photo-oxidation, interfacial instability and de-lamination, inter-diffusion, and morphology changes are poorly understood. Development of more complex device designs, such as multi-junction devices or inclusion of more exotic third-generation mechanisms into the OPV design, may be necessary to push efficiencies to competitive levels or to enable substantially higher efficiencies. The long-range goal of OPV is large-scale power generation. But as the technology develops, the potential for low-cost and flexible form-factors may enable other applications in the short term.

3.2.1.5. Sensitised Solar Cells

Although the stability and light-conversion mechanisms are currently inadequately understood, it can be said that (1) there is no expected limitation on material, (2) stable 10%-efficient modules are certainly within reach, and (3) the energy-payback period should be significantly shorter than other PV technologies. Demonstrated levels of efficiency and degradation have inspired investment, and several companies are working toward commercialising this technology. To reach the 2015 targets, further advance in the fundamental understanding of the factors that govern cell performance and stability is essential.

Table 6. Sensitised Solar Cells Technology Performance [13]

Parameter	Status (2007)	Future Goal (2015)
Champion device efficiency	11%	16%
Laboratory cell degradation	<5% after stress at 80°C for 1000h in dark or after light-soaking for 1000 h @ 1 sun at 60°C	<5% after stress at 85°C for 3000 h in the dark or after light-soaking for 3000 h @ 1 sun at 60°C
Module efficiency	5–7%	10%
Outdoor module degradation	<15% in 4 yrs	<15% in 10 yrs
Identification of key degradation mechanisms	Degradation mechanisms are controversial	Primary degradation mechanisms identified

3.2.1.6. Intermediate-Band PV

This roadmap addresses intermediate-band (IB) solar cell technology which currently is in a concept stage. The main challenge is to experimentally prove the concept, because so far, an increase in cell efficiency due to the presence of an IB has not been experimentally demonstrated. The reason for this is that a suitable materials system, with the required properties, has yet to be discovered.

3.2.1.7. Multiple-Exciton-Generation PV

This roadmap addresses the development of solar cells based on inorganic semiconductor nanocrystals (NCs) – such as spherical quantum dots (QDs), quantum rods (QRs), or quantum wires (QWs) – focusing on their potential to improve upon bulk semiconductor cell efficiencies by efficient multiple-exciton generation (MEG). The generation of multiple excitons (i.e., electron-hole pairs) for each absorbed photon of sufficient energy raises the thermodynamically attainable power conversion efficiency of a single-junction Photovoltaic (PV) solar cell from 33.7% to 44.4%. Semiconductor NCs are produced at much lower temperatures than their bulk counterparts, enabling significantly lower production cost. Several possible implementations of semiconductor NC-based solar cell devices may be realised.

The current state of this technology is in the fundamental and exploratory research phase, and is focused on:

- Pursuing an experimental and theoretical understanding of the MEG mechanism— i.e., how NCs enhance charge-carrier pair production for high photon energies.
- Materials selection and characterisation.
- Efficient charge separation for photocurrent collection from MEG-active NCs.

Table 7. Intermediate-Band PV [14]

Parameter	Status (2007)	Future Goal (2015)
Understand the material design needed to implement the IB concept.	The importance of choosing a materials system with the IB at optimal energy is understood, but a method to avoid harmful non-radiative recombination is lacking.	By **2010**, identify the material requirements needed to demonstrate added efficiency from excitation through the IB.
Identify a materials system: new compound (e.g., GaPTi); new alloy (e.g., ZnMnOTe); quantum dot array, with the required properties to demonstrate an IB cell with an efficiency greater than the present record efficiency for a single-junction solar cell.	Not accomplished	Demonstrate an IB cell with an efficiency that exceeds the present record efficiency for a single-junction solar cell (~25%).
Champion device efficiency (1 sun)	Not accomplished	>25%
Champion device efficiency (under concentration)	Not accomplished	>30%
Cost target (assuming a single-junction single-crystal IB cell with an efficiency of 40% under concentration can be fabricated with a similar cost to a single-junction crystalline Si cell.)	Not accomplished	5–7 ¢/kWh
Cost target (assuming a single-junction IB cell with an efficiency of >20% can be fabricated by a low-cost route similar to CIGS thin films.)	Not accomplished	7–10 ¢/kWh

Table 8. Multiple-Exciton-Generation PV [14]

Parameter	Status (2007)	Future Goal (2015)
MEG quantum yield at hv = 2.5 x E_g	105% – 110%	180%
IPCE at hv > $2E_g$	45%	>100%
Champion NC solar cell efficiency	1%–3%	25%
AM1.5 photocurrent density at Voc = 1.0 eV	~1 mA/cm^2	36 mA/cm^2
Carrier mobility (DC value for coupled NC array)	~1 cm^2/(V·s)	100 cm^2/(V·s)

3.2.1.8. Nano-Architecture PV

This roadmap addresses nano-architecture solar cells that use nanowires, nanotubes, and nanocrystals, including single-component, core-shell, embedded nanowires or nanocrystals, either as absorbers or transporters.

These technologies are mostly in the stage of concept proposal or proof-of-principle device demonstration, although few have reached the stage of offering decent efficiency (although still not comparable to the more mature technologies, e.g., Si, CdTe, and CIGS).

Table 9. Nano-Architecture PV [14]

Parameter	Status (2007)	Future Goal (2015)
Concept; Proof-of-principle device	Either with only concept proposals or proof-of-principle devices	By 2010, the material systems that could, in principle, offer the desirable material properties should be identified; the proof-of-principle solar cells should be demonstrated
Materials; Device structures; Efficiency	Materials might not have the desirable properties; Device structures are not optimised; Efficiency < 3%	By 2015, the most-promising device structures and materials should be identified; the target efficiency of 15% should be achieved in the laboratory; the compatibility with thin-film and/or CPV technologies should be assessed

3.3. Thin Film Research in the European Union and the Research Framework Programme

The European research activities in photovoltaics are funded from a variety of different sources. A lot of activities are done in Universities often by small groups and it is very difficult to determine the funding level for such activities. To determine the level of regional or state funding would clearly be beyond the scope of this article and I will therefore concentrate on the funding on the European Union level.

In addition to the 27 national programmes for market implementation, research and development, the European Union has been funding research (DG RTD) and demonstration projects (DG TREN) with the Research Framework Programmes since 1980. Compared to the combined national budgets, the EU budget is rather small, but it plays an important role in creating a European Photovoltaic Research Area. This is of particular interest and importance, as research for Photovoltaics in a number of Member States is closely linked to EU funds. A large number of research institutions from small University groups to large research centres, covering everything from basic material research to industry process optimisation, are involved and contribute to the progress of Photovoltaics. In the following, only activities on the European level are listed, as the national or regional activities are too manifold to be covered in such a report.

The European Commission's Research and Development activities are organised in multi-annual Framework Programmes (FP), which until recently had a duration of 4 years. For the first time, the 7th EC Framework Programme for Research, Technological Development has a duration of 7 years and runs from 2007 to 2013. Support for Photovoltaic Research Projects started in 1980. Descriptions of EC funded projects can be found at the CORDIS web site (http://cordis.europa.eu/guidance/services_en.html)

In addition to the direct project funding there is a instrument, known as 'Article 169', a reference to the treaty establishing the Framework Programmes. This Article 169 instrument allows the Commission to support the opening and joining of national research programmes of Member States.

During the 6th Framework Programme, the PV Technology Platform was established [15]. The aim of the Platform is to mobilise all the actors sharing a long-term European vision for Photovoltaics. The Platform developed the European Strategic Research Agenda for PV for the next decade(s) and gives recommendations for its implementation to ensure that Europe maintains industrial leadership [16].

In the first call for projects, which closed 2007, the following thin film related research projects were selected:

- **HETSI**: Heterojunction solar cells based on a-Si c-Si

  The project aims to design, develop and test novel a-Si/c-Si hetero-junction solar cell structure concepts with high efficiency.

  The project covers all aspects of the value chain, from upstream research of layer growth and deposition, to module process and cell interconnection, down to upscaling and cost assessment of hetero-junction concept.

  The project started on 1 February 2008 and has a duration of 36 months.

  Coordinator: Commissariat à l'Energie Atomique (CEA), France

- **HIGH-EF:** Large grained, low stress multi-crystalline silicon thin film solar cells on glass by a novel combined diode laser and solid phase crystallisation process

  The project will develop a unique process allowing for high solar cell efficiencies (potential for >10%) by large, low defective grains and low stress levels in the material at competitive production costs. This process is based on a combination of melt-mediated crystallisation of an amorphous silicon (a-Si) seed layer (<500 nm thickness) and epitaxial thickening (to >2 µm) of the seed layer by a solid phase crystallisation (SPC) process.

  The project started on 1 January 2008 and has a duration of 36 months.

  Coordinator: Institute of Photonic Technology e.V., Germany

- **ROBUST DCS**: Dye Sensitised Solar Cells (DSC)
  ROBUST DSC aims to develop materials and manufacturing procedures for *Dye Sensitised Solar Cells* (DSC) with long lifetime and increased module efficiencies (7% target). The project intends to accelerate the exploitation of the DSC technology in the energy supply market. The approach focuses on the development of large area, robust, 7% efficient DSC modules using scalable, reproducible and commercially viable fabrication procedures.
  The project started on 1 February 2008 and has a duration of 36 months.
  Coordinator: Energy Research Centre of the Netherlands (ECN), The Netherlands
- **ULTIMATE**: Ultra Thin Solar Cells for Module Assembly – Tough and Efficient
  The main objective of the project is to demonstrate the production feasibility of PV modules with substantially thinner solar cells (100 μm) than today.
  Duration of Project: 36 months.

The second call for projects was launched on 3 September 2008 and the Call specified the following topics in the area of Photovoltaics (ENERGY 2.1):

♦ Photovoltaics is the most capital-intensive renewable source of electricity. Research will include the development and demonstration of new processes for Photovoltaic manufacturing, including the manufacturing of equipment for the PV industry, new Photovoltaic-based building elements complying with existing standards and codes and the demonstration of the multiple additional benefits of Photovoltaic electricity. Longer term strategies for next generation Photovoltaics (both high-efficiency and low-cost routes) will also be supported.

- **Topic ENERGY.2009.2.1.1: Efficiency and material issues for thin-film Photovoltaics**
  Content/scope: Thin-film Photovoltaics has an inherent low-cost potential because its manufacture requires only small amounts of active materials and it is suited to fully-integrated processing and high throughputs. Research is needed to improve device quality and module efficiency, and to develop a better understanding of the relationship between the deposition processes and parameters, the electrical and optical properties of the deposited materials, and the device properties that result. Key issues to be addressed are improvement of understanding of electronic properties of materials and their interfaces, improvement of the quality and stability of transparent conductive oxides (TCOs), and development of advanced methods for optical confinement. Results should be transferred to production lines by the end of the project.
  Funding scheme: Collaborative project. Application Deadline 25 November 2008.
  Expected impact: Accelerated market development of cost-effective and more efficient thin-film Photovoltaics.
  Other information: In order to maximise industrial relevance and impact of the research effort, the active participation of SMEs represents an added value to this topic. This will be reflected in the evaluation. The active participation of relevant Chinese partners could add to the scientific and/or technological excellence of the project and/or lead to an increased impact of the research to be undertaken; this will also be considered by the evaluators.

- **Topic ENERGY.2009.2.1.2: Solar Photovoltaics: Manufacturing and product issues for thin-film Photovoltaics**
  Content/scope: Demonstration of standard production equipment and better processes to reduce materials and energy use, achieve higher throughputs and yields, increase recycling rates and improve both the environmental profile and the overall economics of thin-film Photovoltaics. Quality assurance procedures, in-line monitoring techniques, integration and automation of production and processing steps are also needed to improve production yield and module efficiency and reduce production costs. Equipment manufacturers will play a leading role in this development. Knowledge gained in relevant industries outside PV should be also exploited.
  Funding scheme: Collaborative project. Application Deadline 29 April 2009.
  Expected impact: Improved productivity parameters (e.g. process yield, throughput) and lower costs leading to accelerated market development and market uptake of cost-effective and more environmentally friendly thin-film Photovoltaics.
  Other information: This topic is coordinated with the parallel research work. The active participation of key industrial partners and technology suppliers is essential to achieve the full impact of the project. This will be considered in the evaluation. The guidelines for demonstration projects figure in the guide for applicants. The industrial partners should include a realistic and convincing market deployment plan with clear roles, tasks and responsibilities of defined partners if the project is successful.
  Up to two projects may be funded.

4. THIN FILM PRODUCTION AND FUTURE DEVELOPMENTS

In 2005 production of Thin-Film solar modules reached for the first time more than 100 MW per annum. Since then the *Compound Annual Growth Rate* (CAGR) of Thin-Film solar module production was even beyond that of the overall industry increasing the market share of Thin-Film products from 6% in 2005 to 10% in 2007 and 16 – 20 % in 2009. . Thin-Film shipments in 2009 more than doubled again compared to 2008.

More than 150 companies are involved in the thin film solar cell production process ranging from R&D activities to major manufacturing plants. The first 100 MW Thin Film factories became operational in 2007 and the announcements of new production capacities accelerated again in 2008. If all expansion plans are realised in time, thin-film production capacity could be 20.2 GW or 36% of the total 56.9 GW in 2012 and 23.0 GW or 34% in 2015 of a total of 67.2 GW (Fig. 4). [2, 4, 5, 6]. The first Thin Film factories with GW production capacity are already under construction for various thin film technologies.

In August 2007, when the first survey was made for a presentation at the 22nd EUPVSEC in Milano, it was found that in 2006 just 21 companies had produced thin film solar modules which were available in the market with quantities above 1 MW. Between 2007 and 2009 at least 20 new companies started their ramp-up face and delivered first modules. Compared to this it is interesting to note, that about 82 companies had made announced their time frames for build up or increase of production capacities in August 2007 whereas at the beginning of 2010 this number increased to more than 150 companies. This shows that a lot of thin-film newcomers are starting production lines and it has to be seen if the ambitious time frames can be kept.

Until a few years ago it was not possible to purchase complete Thin Film manufacturing lines with guaranteed deposition and manufacturing processes. This led to substantial higher initial investment costs and risks compared to the wafer silicon based solar cell lines as every potential manufacturer had to specify the respective production equipment and ramp it up individually with limited support from equipment suppliers.

When major semiconductor equipment supply companies like Applied Materials, Oerlikon and ULVAC started to offer complete production lines for thin film solar cells this together with the then existing silicon shortage and the high demand for solar cells changed the investment behaviour of investors and potential manufacturers.

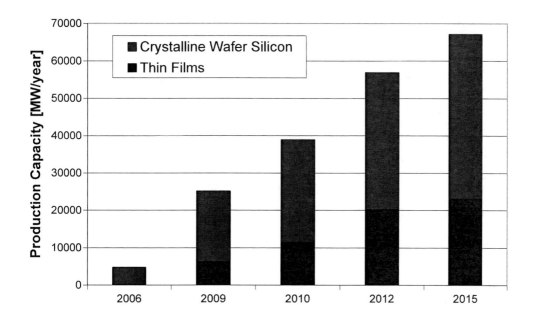

Figure 4. Actual and planned PV Production capacities of Thin Film and Crystalline Silicon based solar modules.

The fact that the majority of the supply companies offering complete "turn-key" production lines do this for amorphous silicon is reflected in the fact that the majority of production capacity increase is in this technology (Figure 5). The option to expand the amorphous silicon technology to micromorph tandems with the help of their equipment supplier is a strategy followed by most of the companies.

The increase of production capacities in the non silicon based thin film technologies is still driven mainly by companies holding the technology knowledge and working together with selected supply companies. However, the situation is changing rapidly. The first equipment suppliers like Centrotherm already offer complete production lines and in the near future others will follow. This will help potential customers with a lower technology experience to advance.

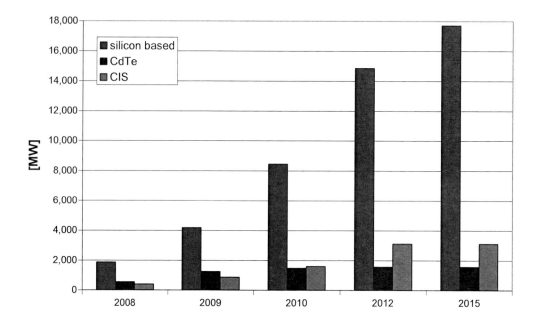

Figure 5. Actual and planned Thin Film Photovoltaic Production capacities of the different technologies.

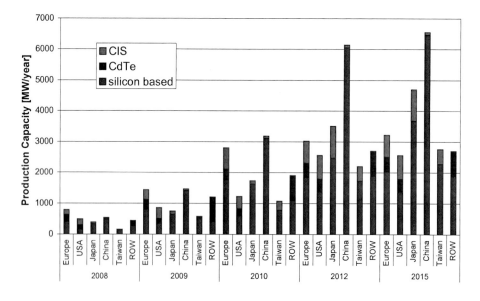

Figure 6. Current and planned Thin Film PV Production capacities. Regional distribution of the different technologies.

The technology as well as the company distribution varies significantly from region to region (Fig. 6). 49 companies are located in Europe, 43 in China, 26 in the US, 18 in Taiwan,

9 in Japan and 16 elsewhere. More than 100 companies are silicon based. 30 companies will use Cu(In,Ga)(Se,S)$_2$ as absorber material for their thin-film solar modules, whereas 9 companies will use CdTe and 8 companies go for Dye & other materials.

This regional distribution reflects on the one hand side the scientific knowledge base concerning the different thin film technologies but also the investment options and availability of human resources. This last issue is very crucial one for the future development of Photovoltaics and Thin Film Photovoltaics in particular. All this new thin film factories will not only need qualified operators and process engineers but scientists with a good knowledge of the respective materials as well in order to improve the guaranteed start-up processes and secure the competitiveness of the companies for the years to come. Already now the number of scientists in the filed of thin film photovoltaics is limited and the number of students aiming to continue their studies for a Ph.D. or post doc in research institutions is decreasing rapidly due to the attractive offers from industry. On the one hand side this is a very good development but efforts have to be made to secure the scientific knowledge base and to work not only on short term research and development issues but to solve fundamental long term problems as well.

In August 2007 there were just 7 companies which offered "turn-key" thin film production lines, exclusively for amorphous silicon based solar cells sometimes with the option to move to the micromorph options. This number has increased to more than 20 and the range of product covers the different thin film technologies: amorphous, amorphous and micromorph, CIS, CdTe and Dye cells.

5. CONCLUSION

The increase of conventional energy prices has increased the investment attention for renewable energies and in particular photovoltaics significantly. Since 2006 the investments and growth in thin film photovoltaics has surpassed the already high growth rates of the whole photovoltaics industry due to the then existing constraints in the silicon supply and hopes of faster cost reductions with thin film technologies.

Thin film solar cells offer the possibility of reducing the manufacturing costs considerably, however, considering the increasing maturity of wafer based production technologies and observed learning curves, newcomers have to enter the game at already very competitive levels. In addition, the entry ticket, i.e. factory size for thin film manufacturers into the market is becoming more and more expensive the more the market grows.

Thin film technologies still need a lot of research over a wide range of issues, ranging from improvement of the understanding of basic material properties to advanced production technologies and the possible market perspectives. To tackle these problems, a long term vision for photovoltaics and long term research is needed. To meet these challenges the European Photovoltaic Technology Platform is finalising the Implementation Plan of the Strategic Research Agenda for Photovoltaics.

However, there is no "winning technology" and a viable variety of technology options has to be ensured. To focus on any single technology option now could be a road block in the future. Funding structures should take into account that different technologies are at different development stages and need different support measures.

In order to realise high production volumes for PV we must now look towards already available high throughput, high yield production technologies analysing if and how they can be utilised for PV in the future. This is especially important for thin film solar cell materials which have only a limited backing by other industries, such as that provided by the microelectronic industry, in the development of production technologies. In addition, there are a number of research issues common to all thin film technologies which have to be solved. No single solar cell technology can neither satisfy the world-wide demand nor satisfy all the different wishes consumers have for the appearance or performance of PV systems.

REFERENCES

[1] International Energy Agency, *World Energy Outlook 2008*, International Energy Agency, Paris, 2009, ISBN 978-92-64-06130-9 .
[2] PV News 2009, published by The Prometheus Institute, ISSN 0739-4829.
[3] Photon International, March 2008
[4] Paula Mints, *The PV industry's black swan, Phtovoltaics World, March 2010*
[5] Company Web-sites and Press Releases
[6] Data received from Companies during personal visits.
[7] Zanio K., Cadmium Telluride: Material Preparation, Physics, Defects and Application in Semiconductors and Semimetals Vol.13, Academic Press 1978.
[8] Compaan A., The Status of and Challenges in CdTe Thin-Film Solar-Cell Technology, Spring 2004 MRS Symp. O Amorphous and Nanocrystalline Silicon Science and Technology, MRS Symp. Proceedings 808.
[9] Wu X., Keane J.C., Dhere R.G., DeHart C., Albin D.S., Duda A., Gessert T.A., Asher S., Levi D.H. and Sheldon P., *Proceedings of 17th European Photovoltaic Solar Energy Conference,* 22- 26 October 2001, Munich, Germany, pp 995-1000.
[10] Repins I., Contreras M.A., Egaas B., DeHart C., Scharf J., Perkins C.L., To B., and Noufi R., 19.9% Efficient ZnO/CdS/CuInGaSe2 Solar Cell with 81.2% Fill Factor, Prog. Photovolt: Res. Appl. 2008; 16:235–239.
[11] O'Regan B., Grätzel M., A low-cost, high-efficiency solar cell based on dye-sensitized colloidal TiO_2 films, *Nature 353* (6346): 737–740.
[12] NEDO Brochure, Energy and Environment Technologies, December 2007 http://www.nedo.go.jp/kankobutsu/pamphlets/kouhou/2007gaiyo_e/87_140.pdf
[13] U.S. Department of Energy, April 2008, Solar Energy Technologies Programme (Solar Programme): *2008-2012 Multi-Year Programme Plan*
[14] DOE, Solar America Initiative, http://www1.eere.energy.gov/solar/solar_america/publications.html# technology_roadmaps
[15] PV Technology Platform; http://www.eupvplatform.org/
[16] PV Technology Platform: *Strategic Research Agenda for Photovoltaic Solar Energy Conversion Technology*, June 2007, Luxembourg: Office for Official Publications of the European Communities, ISBN 978-92-79-05523-2.

In: Thin Film Solar Cells: Current Status and Future Trends   ISBN 978-1-61668-326-9
Editors: Alessio Bosio and Alessandro Romeo   © 2010 Nova Science Publishers, Inc.

Chapter 2

INTRODUCTION TO INORGANIC THIN FILM SOLAR CELLS

Francesco Roca[1,], Alessio Bosio[2] and Alessandro Romeo[3]*

[1]Enea Portici Research Centre, Italy;
[2]Department of Physics University of Parma, Italy;
[3] Faculty of Mathematical, Physical and Natural Science University of Verona, Italy.

ABSTRACT

Photovoltaics stumbling block has always been its cost but it has held the promise of providing clean electricity and competitive rates. The cost is declined by a factor of nearly 150-160 times since the invention in 1954 of the modern solar cell based on crystalline silicon technology. More than 90% of the current production uses 1st generation PV wafer based cSi (1st G PV) a technology with the ability to continue to reduce its cost at its historic rate. The direct production costs for crystalline silicon modules are expected to be around 1 €/Wp in 2013, below 0.75 €/Wp in 2020 and lower in the long term as indicated in table 1 [5].

Thin-film deposited directly on large area substrates, such as glass panels (square meter-sized and bigger) or foils (several hundred meters long) in roll-to-roll application, recognized as 2nd generation approach (II G) is always looked at as the "younger cousin" of the silicon technology, as pointed out by L.L Kazmerski [3]. It has an inherent low-cost potential because it requires only a small amount of expensive photo-active materials and its manufacture is suited to fully integrated processing and high throughput with a very low energy pay back time (<1.0 year or less). It poised to take over the energy production responsibilities of its older relative but it never quite fulfilling its expectations or potential except during last year when its market share moved from 6 to over 10%. During 2005- 2008– period (chapter 1 this book) the higher growth rates of the whole PV industry.

---

[*] Correspondence concerning this article should be addressed to: Francesco Roca, email: francesco.roca @enea.it. Address: ENEA Portici Research Centre località Granatello.- 80055 Portici (NA), Italy.

The fantastic boom of thin film technology, during last years, can suggest further development principally during next years mainly due to the application of innovative concept to conventional materials and new class of thin film material coming from nanotechnologies, photonics, optical metamaterials, plasmonics and new semiconducting organic and inorganic sciences, all them recognized as 3rd generation approach (3rd G PV) to overcome efficiency limitation at low cost [15].

Within the next 20 years, it is reasonable to expect that 2nd G PV technologies cost reductions and the implementation of some new technologies and introduction of high efficiency 3rd G PV concepts can lead to long term fully cost-competitive solar energy based on thin film approach.

First, second and third generation PV are mainly based on inorganic approach. They cover a very wide area of material science and only a short overview can be outlined in this introduction leaving to the reader to analyse thoroughly the introduced concepts in the remaining part of the book as well by consulting a very wide available bibliography partially indicated in the references of each chapter.

1. INTRODUCTION

Solar energy represents the largest energy input into the terrestrial system. Although photovoltaics is not the sole answer to the myriad of energy challenges offered by the clean and economical use of sun spectrum, this renewable energy option can make an important contribution to the economy of each country.

Despite its relatively low power density, the solar electricity market is booming. By the end of 2007, the cumulative installed capacity of solar photovoltaic (PV) systems around the world had reached more than 9,200 MW. This compares with a figure of 1,200 MW at the end of 2000. Installations of PV cells and modules around the world have been growing at an average annual rate of more than 35% since 1998. The EPIA/Greenpeace Advanced Scenario [1] shows that by the year 2030, PV systems could be generating approximately 2,600 TWh of electricity around the world to satisfy the electricity needs of almost 14% of the world's population

Photovoltaics remain one of the most dynamic sectors worldwide with a level of expansion during last year up to 40-50% /per year. The target of achieving by 2010 already a worldwide annual production over then 5.5 GWp has been anticipated to 2008. The 2020 goal of 100GWp installed modules which correspond to the 3% of electricity in Europe by photovoltaics are very realistic and with sequential grid parity on peak demand for PV, retail prices and increase in cost of bulk power from 6% up to 10% can be envisaged and we could also prepare for 12% corresponding to 400 GWp in Europe on or before 2020 [2].

The typical southern Europe electricity generation costs moved from 2.0 €/KWh (1980) to < 0.30 €/KWh (2006) with the typical commercial flat-plate module conversion efficiencies improved by 8% to 15% [Table 1]- The typical energy pay-back time in southern Europe changed by >10 years (1980) to ~2years (2006) and perspective for its striking reduction are foreboded mainly due to reduce the thickness of wafer and overall to the improved use of thin film approach.

Photovoltaic market is now supported by incentive schemes (feed-in tariff) but the increased production encourages the support policies in order to sustain a virtuous cycle that can stimulate the market and in turn can ensure further reduction cost.

Table 1. Expected development of PV technology over the coming decades by ref. [5]

	1980	Today	2015	2030	Long term potential
Typical turn-key system price (2006 €/W$_p$, excl. VAT)	>30.0	5.0	2.5	1.0	0.5
Typical Electricity Generation cost southern Europe (2006 €/kWh)	>2.0	0.30	0.15 (competitive with retail electricity)	0.06	0.03
Typical commercial flat-plate module efficiency	up to 8%	up to 15%	up to 20%	up to 25%	Up to 40%
Typical commercial concentrator module efficiency	(~10%)	up to 25%	up to 30%	up to 40%	up to 60%
Typical system energy pay-back time southern Europe (years)	>10	2.0	1.0	0.5	0.25

"Flat plate" refers to standard module for use under natural sunlight, "concentrator" refers to system that concentrate sunlight (and, by necessity, tracks the sun across the sky)

Wafer-based crystalline silicon has dominated the photovoltaic industry since the dawn of the new solar PV started in 1954. More than 90% of the current PV production uses wafer based crystalline silicon technology (mono, multi-crystalline silicon and aSi-cSi heterojunction). This is a well-established product, which achieves sufficient efficiency for at least 20 years of lifetime convincing track-record in reliability. Its cost has decreased by ~20% for each doubling of cumulative installed and this means that the driving forces are based on market size and technology improvements.

The last ones are mainly based on the improved efficiency achieved by the development of new cost-effective crystal growth requesting less energy- for preparation which found in multi-crystalline silicon a very useful solution. Further enhancement in efficiency was obtained through innovative and effective passivation schemes (Si$_x$N$_y$). Low kerf loss sawing and the reduction of thickness of the wafer, from 400 μm in 1990 to 200 μm in 2006 and with target of 150 μm or less on or before 2013 should assure technology needs to less the 7-5 g/Wp on or before 2013. By having in mind just >15-13 g/Wp few year before and 9-10 g/Wp today the reduction in thickness of cSi will produce the first and relevant cost reduction. Increased wafer size, from 100 cm^2 to 240 cm^2 will also assure reduced cost production and mainly higher throughput.

But cSi technology is based on silicon. What about its availability? Silicon feedstock is acknowledged as a bottleneck in keeping up the booming PV industry. In fact during last years the availability for the PV industry was mainly limited due to the needs of the semiconductor industry. The products realized by the last one represent a much larger economic value per unit area than solar cells production, and the electronic industry easily affords to pay higher prices to the feedstock manufacturers then photovoltaic industry [8]. As effect of booming in 2007, solar companies started to use more silicon than the semiconductor industry- about 30,000 of 48,900 metric tons of silicon produced worldwide and the high growth rates of PV industry has not been followed at beginning by the silicon producers.

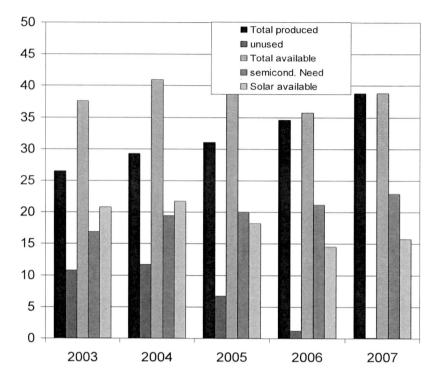

Figure 1. Development of silicon feedstock production and unused feedstock in kton/a and availability for the semiconductor and PV industry period 2003-2007. Data elaborated by ref. [8].

Why? Silicon purification plants cost hundreds of millions of dollars to build and take three years just to turn them "on" and to produce the first ingots. After an initial muddled period, according to Travis Bradford president of the Prometheus Institute, as indicated in his report, "Polysilicon: Supply, Demand and Implications for the PV Industry" the "black period" was now behind. [6]. New companies started to recognized PV as a fully fledged industry that provides a stable business segment as opposed to microelectronics industry which is strongly dependent on the demand cycles. As effect the investment at today constitutes a low-risk placement with high expectations for return on investments and cSi producers stated to follow PV market with higher reliance. By 2012, the total silicon-manufacturing capacity could reach more than 261,742 metric tons through the major silicon suppliers investments.- REC, MEMC, Wacker and Hemlock and also highlighted new companies entering the field by China-based LDK Solar, which had unused credit facilities totalling US$785 million as of April 14, 2009. and the company expects to use these credit facilities to fund ongoing business activities, construct a 1,000- and a 15,000-metric-ton plant; DC Chemical, a South Korean firm due to open its 5,000-ton plant this year. Upgraded metallurgical silicon production will also be blended with purer silicon through Elkem investment in Norway, Dow Corning in Brazil and Timminco in Canada [6].

As effect of the renewed enthusiasm, the goal for crystalline (cSi) and multi crystalline (mc-Si) silicon photovoltaic modules to arrive at a market prices of around 1 €/Wp in 2013, under 0.75 €/Wp in 2020 or lower, in the long term started to become a feasible target if R&D effort will be directed towards the issues of greatest strategic concern. On other hand, costs have become increasingly dominated not only by the costs of the starting materials, namely those of the silicon wafer, but strengthened by encapsulant, nominally ethylene vinyl acetate (EVA) resins and Tedlar®-PVF, low-iron glass cover sheet, and the costs of other components. This trend towards dominance by starting material costs is expected to continue as the photovoltaic industry continues to mature. New bottleneck could threaten the development of today PV market and as effect will be cSi alone able to sustain PV market in the long term?

2. WHY THIN FILM SOLAR CELLS?

Thin film science has grown world-wide into a major research area. It is no understatement to say in recent years the importance of coatings and the synthesis of new materials for industry have resulted in a tremendous increase of innovative thin film processing technologies fundamentally changed both condensed solid state physics and everyday life.

Well-established thin-film technologies are used to realize integrated circuits in all electronic components, especially solid-state devices and microelectronic integrated circuits, multi-layer configuration realise selective optical filter, electronic component, LCDs, solid state lasers, solar cells, several kind of different thin layer are used to change the optical electrical and mechanical surface nature of materials and products for application as optical coatings, magnetic film and data storage, antistatic coatings, hard surface coatings, and novel effects in thin films technology continue to be discovered and explored by both solid-state physicists and optical physicists-engineers in order to develop new products utilized in current or in specialized application including bio-medical and space application.

Thin-film for PV modules recognized as 2nd PV generation directly deposited on large area substrates, such as glass panels or foils are cheaper to manufacture owing to their reduced quantity of material, less energy and handling costs because suited to fully integrated processing and high throughputs.

Thin film solar cells offer a wide variety of choices in terms of device design, fabrication methods and a great variety of substrates (flexible or rigid, metal or insulator) can be used for the deposition of different layers (contact, buffer, absorber, reflector, etc.) using several techniques (LPCVD and plasma-based CVD, sputtering, sublimation, etc) last but not least such versatility allows tailoring and engineering of the layers in order to match solar spectrum and to improve device performance. Such versatility also allows tailoring and engineering of the layers in order to match solar spectrum and to improve device performance.

At present, the market share of thin-film PV within total PV production is below 10%, but might grow to 20% by 2010 and beyond 30% in the long term.[5]. In fact the global production capacity of thin films is expected to reach 1 GWp/year in 2010 and 2 GWp/ year in 2012. It is being installed mainly in Japan, USA and Europe (chapter 1 this book). During next years, the research and the development in innovative simple materials and devices, as well simple manufacturing will lead to completely different processes to overcome the costs again. It is expected increasingly that in next future the cost of PV module will be also dominated for the 2nd generation PV by the cost of the glass sheet on which the cells are deposited or by procedures and process used to encapsulate the material. Further aspect will also related to the quantity of pure materials (metals and TCO) that the technology needs for each Wp produced, because bottlenecks are not the photo-converter material only.

This will impose a new approach by which more power from a given investment in material is possible by increasing energy-conversion efficiency. This leads to the notion of new generation of both high-efficiency and thin-film introduced in the previous paragraph recognized as a 3rd PV generation by M.A. Green [14, 15]

As effect the challenges that need to be addressed to make solar energy systems viable and competitive on a large scale cannot put on the basis of cost alone, the low toxicity and the availability of materials today but it should also involve mainly in medium-long term the development and research of new class of material and process, will be able to significantly increase efficiency at low cost.

3. THIN FILM DEPOSITION PROCESS

There exists a huge variety of thin film deposition processes currently utilized in laboratories as well in the electronic and mechanical industries which originate from purely physical or purely chemical processes or physical-chemical process [12]. The more important thin film processes are based on liquid phase chemical techniques, gas phase chemical processes, glow discharge processes and evaporation methods as summarized in following paragraphs

3.a. Chemical Deposition

In chemical deposition techniques a fluid precursor undergoes a chemical change at a solid surface, producing a solid thin or thick layer. The fluid usually surrounds the solid object, and deposition happens with little regard to direction. As effect thin films from chemical deposition techniques tend to be conformal, rather than directional. Chemical deposition techniques are categorized by the phase of the precursors, the process conditions and the agents by which chemical reactions are initiated (e.g., activation process).

Classified by operating pressure:
- *Atmospheric pressure CVD* (APCVD) - CVD processes at atmospheric pressure.
- *Low-pressure CVD* (LPCVD) - CVD processes at sub atmospheric pressures. Reduced pressures tend to reduce unwanted gas-phase reactions and improve film uniformity across the wafer. Most modern CVD processes are either LPCVD or *Ultrahigh vacuum CVD* (UHVCVD) - CVD processes at a very low pressure, typically below 10^{-6} Pa (~10^{-8} torr). Note that in other fields, a lower division between high and ultra-high vacuum is common, often 10^{-7} Pa.

Classified by physical characteristics of vapour/liquid:

- *Plating* relies on liquid precursors, often a solution of water with a salt of the metal to be deposited. The most commercially and diffused plating process are based on electroplating were an external potential is applied (as example CdS electroplating) but some plating processes are driven entirely by reagents in the solution (*electroless plating*). Further applications take advantage –from the potential produced by illuminated solar cell itself where only one side electrical connection is requested. In electrolytic anodization an oxide is formed from the substrate, the anode reacts with negative ions from the electrolyte in solution and become oxidize forming an oxide or a hydrate oxide coating. *Electrophoretic coatings* are based on deposition of a thin film from a dispersion of colloidal particles onto a conductive substrate. The dispersion in a conductive liquid dissociates into negative charged colloidal particle and positive ions (cations), or the reverse. On application of an electric field between the opposite substrate electrode (anode) and substrate (cathode) the colloidal particle migrate to the substrate.
- *Chemical solution deposition (CSD)* uses a liquid precursor, usually a solution of organo-metallic powders dissolved in an organic solvent and applied through spray, dipping or spinning techniques. In *Direct liquid injection* CVD (DLI-CVD) the precursors are in liquid form (liquid or solid dissolved in a convenient solvent) and are injected in a vaporization chamber towards injectors (typically similar to car injectors). This is a relatively inexpensive, simple thin film process that is able to produce stoichiometrically accurate crystalline phases. The simplicity of process was also applied to several class of coatings including indium tin oxide (ITO) [10] or copper, or CdS
- *Chemical vapour deposition (CVD)* generally uses a gas-phase precursor, often a halide or hydride of the element to be deposited. CVD involves the dissociation and/or chemical reactions of gaseous reactants in a activated (heat, light, plasma) environment, followed by the formation of a stable solid product. The deposition involves homogeneous gas phase reactions, which occur in the gas phase, and/or heterogeneous chemical reactions which occur on/near the substrate.
  - *Metalorganic chemical vapor deposition* (MOCVD) - CVD processes based on metalorganic precursors. Often is also named metalorganic vapour phase epitaxy (MOVPE). It has established itself as the method of choice for mass production of modern compound semiconductor devices like InGaN multi-quantum well (MQW) structures are at the heart of today modern blue-green and white LED emitters and III-V very high efficiency multijunction solar cells.
  - *Plasma-CVD* uses an ionized vapour, or plasma, as a precursor relies on electromagnetic means (electric current, microwave excitation), rather than a chemical reaction, to produce a plasma. It is divided in several class of plasma CVD depending from the excitation regime and the geometry of plasma reactor
    - *Plasma-Enhanced CVD* (PECVD) - CVD processes that utilize plasma to enhance chemical reaction rates of the precursors. PECVD processing allows deposition at lower temperatures, which is often critical in the manufacture of semiconductors solar cells.
    - *Remote plasma-enhanced CVD (RPECVD)* - Similar to PECVD except that plasma discharge region is put outside the deposition area. Removing the wafer from the plasma region allows reduced bombardment of the substrate

and it results in an improved control of reaction mechanisms, thereby yielding films of higher purity.
- *Microwave plasma-assisted CVD* (MW-CVD) Using microwaves as the gas phase excitation allows to efficiently synthesize a wide variety of materials
- *Hot wire CVD* (HWCVD) - also known as catalytic CVD (Cat-CVD) or hot filament CVD (HFCVD). Uses a hot filament to chemically decompose the source gases.

3.b. Physical Vapor Deposition (PVD)

It includes a variety of methods and vacuum deposition techniques to grow thin films by the condensation of a vaporized form of the material onto various surfaces The coating method involves purely physical processes such as high temperature vacuum evaporation or plasma sputter bombardment rather than involving a chemical reaction at the surface to be coated as in chemical vapour deposition, but it is quite normal understand that some chemical effect are also involved. The material to be deposited is placed in an energetic, entropic environment, so that particles of material escape its surface. Facing this source is a cooler surface which draws energy from these particles as they arrive, allowing them to form a solid layer. The whole system is kept in a vacuum deposition chamber, to allow the particles to travel as freely as possible. Since particles tend to follow a straight path, films deposited by physical means are commonly *directional*, rather than *conformal*,

Variants of PVD include, in order of increasing novelty:
- *Thermal Evaporative deposition*: in which a vapour is generated by a boiling or subliming source material heated typically by electrically resistive heating (boat) under low-medium vacuum. The vapour is transported from source to the substrate where it condenses.
- *Electron beam* physical vapour deposition: in which the material to be deposited is heated to a high vapour pressure by a high energy electron bombardment in "high" vacuum. The beam is usually bent through an angle of 270° in order to ensure that the gun filament is not directly exposed to the evaporant flux.
- *Sputter deposition*: in which a glow plasma discharge (usually localized around the "target" by a magnet) bombards the material sputtering some away as a vapour. The target can be kept at a relatively low temperature, since the process is not one of evaporation class. It is especially useful for compounds or mixtures, where different components would otherwise tend to evaporate at different rates. Sputtering step coverage is more or less conformal, but always better in uniformity then evaporation. It is also widely used in optical media.
- *Cathodic Arc Deposition* (arc-PVD): in which a high power arc directed at the target material blasts away some into a vapour which is a kind of ion beam deposition where an electrical arc is created that literally blasts ions from the cathode. The arc has an extremely high power density resulting in a high level of ionization (30-100%), multiply charged ions, neutral particles, clusters and macro-particles (droplets).
- *Laser Induced deposition*: utilizes high power laser beam ablates material from the target into a vapour phase. Pulses of focused laser light vaporize the surface of the

target material and convert it to plasma; this plasma usually reverts to a gas before it reaches the substrate. In photo-enhanced CVD (PH-CVD) application, the reactants are activated in phase vapour by high energy radiation, usually short-wave ultraviolet radiation.

3.c. Other Deposition Processes

Some methods fall outside these two categories indicated above, relying on a mixture of chemical and physical process.
- *In reactive sputtering*, a small amount of reactive gas such as oxygen or nitrogen is mixed with the plasma-forming gas based on a noble gas (Ar, Kr,).- After the material is sputtered from the target, it reacts with this gas, so that the deposited film is a different material, i.e. an oxide or nitride of the target material.
- *In molecular beam epitaxy (MBE)*, slow streams of an element can be directed at the substrate, so that material deposits atomic layer by atomic layer in a controlled condition by repeatedly applying a layer of one element (i.e., gallium), then a layer of the other (i.e., As), so that the process is chemical, as well as physical. The beam of material can be generated by either physical means (that is, by a furnace) or by a chemical reaction (chemical beam epitaxy).

4. THIN FILM GROWTH KINETICS

Thin film growth involves always a very complicated kinetics starting from the precursor production by chemical and/or the physical process up to condensation of the species on the substrate [11,12,13] and different effects are produced on main physical and chemical properties indicated in the following list [12]:
i. *Electrical properties*: conductivity/resistively, dielectric constant, dielectric strength, dielectric loss, stability under voltage bias, polarization, permittivity, radiation hardness and stability
ii. *Morphology:* amorphous, nano and micro crystalline; polycrystalline and crystalline orientation, structural defect density, planarity, conformal or step coverage, microstructure and porosity.
iii. *Chemical properties*: composition, impurities, thermodynamic stability, reaction with ambient and substrate, corrosion and erosion resistance, hygroscopicity, impurity barrier against diffusion, toxicity and carcinogenicity.
iv. *Thermal properties*: expansion coefficient, thermal conductivity, dependence of all chemical/physical properties by temperature, stability against thermal treatment.
v. *Mechanical properties*: intrinsic, residual and composite stress, anisotropy, adhesion, hardness, density, fracture, ductility, elasticity.
vi. *Optical properties*: refractive index, absorption, birefringence, spectral characteristics.
vii. *Magnetic properties*; permeability, coercive forces, saturation flux density.

The indicated structural, chemical, metallurgical and physical properties of the grown thin film are not only dependent by the chemical nature of the material but are also strongly dependent by a large number of deposition and process parameters.

Just by considering thin silicon as an example, the nature of grown material changes totally by amorphous to nano-micro and polycrystalline morphology in dependence of the particular chemical or physical technique used to grow the material.- such as sputtering, PECVD, electron beam, and it also radically changes if by using the same PECVD process in silane plasma if it is changed the hydrogen content in plasma and/or substrate temperature. In fact, as will clearly indicated in chapter 8 the increase in substrate temperature and in hydrogen content during the plasma process increase always the crystallinity of the realized thin film. Analogy about the effect of changes in process parameters on the final physical and chemical properties can also be applied to CdTe, CIGS, TCO or any other class of PV thin film.

A deep understanding of the growth kinetics of thin material is fundamental not only for research application but it makes process adjustments in commercial production efficient to find the better compromise between costs and performances

The kinetic of thin film growth is indicated in the schematic diagram of figure 2.

i. Precursors and chemical species are produced by the source in direct or remote configuration.
ii. Depending from the specific deposition technique used, precursor and chemical species partially interact in vapour phase and after a path in vapour phase they arrive to the substrate in conformal or directional condition.
iii. The material can also arrive with/without additional energy with a rate can vary by several orders of magnitude.
iv. The material starts random nucleation and growth processes based on the individually condensing/reacting atomic/ionic/molecular species on the substrate. Any atom arriving to the surface has a probability to sticking to surface can range from 0 to 1 with a process is influenced by surface roughness, substrate surface energy and it is also often thickness dependent.
v. The atoms that arrive on the surface can be frozen in the arriving position or can move on the surface with a process well know as surface diffusion. The atoms continue to move until they reach a position that minimizes the total energy.
vi. As more atoms are arriving, they undergo similar process (sticking and surface diffusion) and some of them interact by forming nucleated particles (nucleation process).
vii. Initially the nucleated particles are metastable clusters but once the nuclei reach a critical size they will become stable and fixed (brinding-coalescence process).
viii. Two mechanisms are considered for the further growth of layer by layer.
   a. With Van der Merwe mechanism, atoms cover the whole surface before start to grow the second layer and so on.
   b. With the Volmer-Weber mechanism the nuclei grow as hemispheres and a thin film is formed when hemisphere diameters is wide enough to assure each hemisphere touch another.
   c. When a growth process can be considered a combination of both indicated mechanism by which it is first covered the surface by a first layer and after start the hemisphere growing, it is referred as Stabski-Krastanow mechanism.

ix. Radical and chemical species can chemically interact and can diffuse onto the substrate.

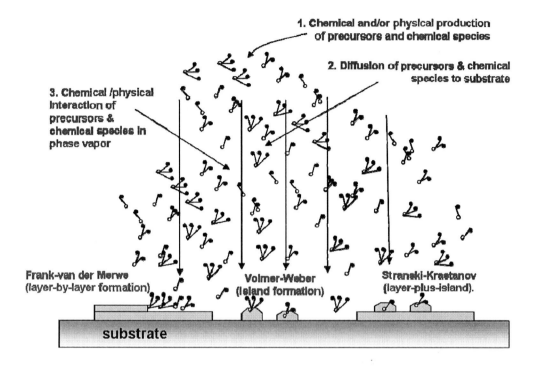

Figure 2. Thin film growth sequence.

- Growth material can assume amorphous/nanocrystalline highly oriented and/or epitaxial morphology depending on deposition parameters and substrate condition.
- Based on solubility conditions and or non equilibrium growth doping and alloying with compatible/incompatible materials can/cannot be obtained.
- Change in optical and electrical properties can be tuned in desired manner Graded bandgap and lattice matching can be realized by changing the deposition parameters during the growth.
- The density of interface and surface defects strongly depends by interaction with impinging radical and chemical species and by the surface cleaning and passivation procedures.
- The adsorption and insertion of the radicals onto the substrate surface; can also induce chemical reaction inside the substrates mainly at the interface.
- Surface, interface and grain boundaries can be passivated or damaged during the growth process. Surfaces and interfaces can be modified by the growth process itself to provide or to avoid interlayer diffusion or to achieve desired roughness through enhanced optical reflectance/transmission characteristics, haze and optical trapping effects or to avoid it depending from the specific requested process.

5. THIN FILM SOLAR CELL BASED ON EXISTING MATERIALS

Based on the utilization of a wide class of deposition techniques, thin-film solar cells are deposited directly on large area substrates, such as square metre-sized glassily substrates or several hundred meters long superstrate foils.
- Two main configurations are used: substrate configuration, light passing before a transparent substrate, typically glassily substrates, and after it reaches photovoltaic layers and TCO.
- superstrate configuration, light passing before TCO and photovoltaic grown material and after it reaches the stratum or layer lying underneath.

The first one (substrate configuration) always needs of an additional encapsulation layer and/or glass.

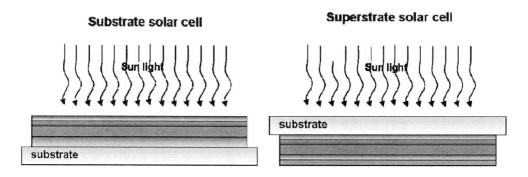

Figure 3. Thin film solar cell in substrate and superstrate configuration, respectively.

There are three major existing inorganic thin-film technologies of worldwide, very high interest [5]:
i) amorphous/microcrystalline silicon (TFSi-13% efficiency)
ii) the polycrystalline semiconductors CdTe (16.7% efficiency)
iii) CIGSS (the abbreviation of $Cu(In,Ga)(S,Se)_2$ - 19.9% efficiency).

The existing thin film cells have almost similar structures (figure 4). The typical Thin film solar cells in general consist of several layers of different materials in thin-film form: TCO, window layer (p or n-type), absorber layer (i or p-type) and metal contact layer. Each of them has different physical and chemical properties and their electrical/optical properties as well interfacial properties affect the overall performance of the device.

All these technologies have been manufactured at pilot scale and are being or have been transferred to high volume production. First Solar, Inc., announced in March 2009 that it has produced 1 gigawatt (GW) since beginning commercial production in early 2002. Its advanced manufacturing process transforms at today a piece of glass into a complete CdTe solar module in less than 2.5 hours [22]. Further initiatives are in progress on CdTe in several countries including Italy by Arendi company.

Oerlikon Solar offers field proven equipment and end-to-end manufacturing lines for the mass production of thin film silicon solar modules, engineered to reduce manufacturing costs while maximizing productivity, with extensive experience in both amorphous and Micromorph® tandem technology [23]. The Applied Material SunFab line is a fully integrated thin film production line capable of producing both single junction (amorphous silicon) and tandem junction (amorphous/microcrystalline silicon stack) panels [23].

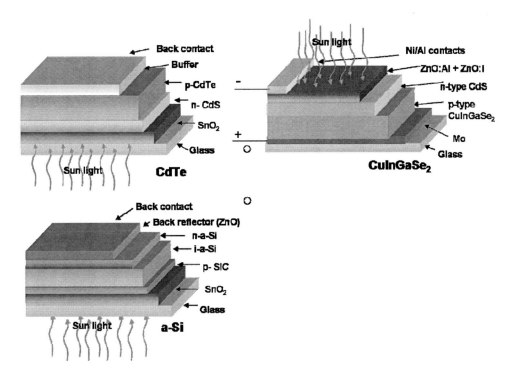

Figure 4. Typical: (a)superstrate CdTe solar cell ; (b) substrate Cu(InGa)Se$_2$ solar cell; and (c) superstrate p-i-n a-Si solar cell.

The SunFab line provides customers with the first and only industry able to realize 5.7m^2 panels, the world's largest, which can reduce production cost as well as enable installation efficiencies that can significantly lower balance of system.

Global Solar achieved 10% average solar cell efficiency the first CIGS company to reach this milestone in a production environment with a full-scale production of CIGS (Copper Indium Gallium diSelenide) PV cells on a flexible substrate [24]

A critical understanding of the behaviour of the individual layer alone is essential for designing of a thin film device but the effects at interface between the different layers in terms of band bending and electron affinity/work function, recombination, expansion coefficient, diffusion coefficient, as well mechanical aspect like adhesion and thermal expansion coefficient are very relevant for the better knowledge of the key issue by which can be realized the enhancement of the performance of the single device.

Furthermore interface causes also reflection/transmission/scattering, as well chemical interdiffusion and exchanges of chemical species.

The key steps directed at one specific thin film technology could be applied to another, increasing its usefulness. It was therefore not a coincidence that the more relevant integrated project (IP) funded by the European Community through 6 FP ATHLET- Advanced Thin Film Technologies for Cost Effective Photovoltaics joined a wide consortium formed by several European Public and private Research Organization and Companies on the development, assessment and consolidation of photovoltaic thin film technology based on

Thin Film Silicon large-area modules on glass, Chalcopyrite specific heterojunctions and Transparent conductive Oxide.[16].

The common aspect concerning efficiency and material issues are summarized below [5].
General aspects:
i. Reliable, cost-effective production equipment for all technologies.
ii. Low cost packaging solutions both for rigid and flexible modules.
iii. Low cost transparent conductive oxides.
iv. Reliability of products: advanced module testing, and improved module performance assessment.
v. Handling of scrap modules, including re-use of materials.
vi. Developing replacements for scarce substances such as indium.

Specific aspects of technology concern:
i. Better fundamental understanding of growth process in order to enhance deposition rate and to improve optical and electronic properties of the existing three families of thin film materials (TFS, CdTe & CIGS) and their interfaces.
ii. Alternative absorber materials with different band gaps for wide spectrum absorption and application in multijunction.
iii. Simplified and efficient cells interconnection in order to produce reliable modules in comparison with wafer-based technologies.
iv. Development of large-area deposition equipment and process technology, for low-cost and high-volume production.
v. Reducing the fabrication and installation costs so that these systems can be deployed at a large scale on the ground and in build-integrated application. This last application should be also based on flexible and lightweight modules produced using thin polymer or metal substrates by means of roll-coating techniques.
vi. Use of low quantity of materials stable and durable against ageing.
vii. Development of transparent Conductive Oxide having improved optical performance (transmittance> 85-90%) and electrical properties (resistivity $< 8 \times 10^{-4}$ Ω cm) and wide availability.

We leave to the reader to extend its interest in deepening through the excellent analysis sketched in chapter 1 with its detailed overview for each thin film solar cells technologies, to the chapter 4 & 5 for all arguments related to CIGS growth and alternative buffer layers indicated there, chapter 6&7 for the technological improvement observed on Cadmium Telluride including new proposed process schemes environmental friendly and last but not least Thin Silicon solar cells (chapter 8) can be considered the next approach to silicon technology.

We will focus through following paragraphs on thin film material like thin poly-Si, Transparent Conductive Oxides, dielectric and low conductive layers very useful and relevant for the photovoltaic technological applications.

6. POLYCRYSTALLINE THIN FILM SILICON SOLAR CELL

By the end of the 21st century, PV will probably be based on materials different from what we know today, but the current photovoltaic market is by far dominated by silicon.

About 84% of the solar-cell PV world production for terrestrial applications has its origin in monocrystalline- (c-Si) and multicrystalline-silicon (mc-Si) wafer technology [4]. Most of the remaining 16 % market is dominated by thin-silicon (amorphous, microcrystalline, nanocrystalline, etc.) and by hybrid amorphous-crystalline solar cells (microcrystalline Si on low-cost substrates, a-Si on CZ slides). In the short term (until 2011-2013) PV technology has to rely on the development of silicon technology in order to sustain the rapid market growth, the continuous decrease of PV module prices and on the efforts for closing the gap between laboratory efficiency and efficiency in production

For a long time, silicon wafer and thin-film silicon technologies have evolved as competing options, i.e. as if the solutions to the technical problems could have come out from only one of these two research lines. In wafer silicon production technology the key point concerns the reduction of wafer thickness and the economy of scale triggered by the rapid growth of wafer silicon production capacities. The prime goal to lower costs leads to multicrystalline instead of monocrystalline silicon, thinner active layers (~10^1 instead of ~10^2 μm) and a more efficient use of raw materials.

Thin silicon is also forced in the same direction from the opposite side by improving the optoelectronic properties and stability of the material, its growth rate and consistently of making the cell active-layer thicker. This leading force result in requests for an increased crystallinity, a thicker active layer (~10^1 instead of ~1 μm), and higher growth rates.

A rough view at the guidelines of both tendencies leads to conclude that the key properties of silicon material in next-generation photovoltaics would be: i) Medium crystallinity (Multi-or polycrystalline silicon) ii) Medium active-layer thickness (>3μm; <50 μm) iii) High fabrication throughput, either by epitaxy fast solidification of melted silicon, fast film growth from gaseous silicon sources or similar processes [25]. Several efforts are in progress based on techniques such as molecular-beam graphoepitaxial growth (MBGE), solid-phase crystallisation (SPC), zone-melting re-crystallisation (ZMR), plasma-spray silicon growth (PSSG), liquid-phase epitaxy (LPE) molecular-beam epitaxy (MBE), hot-wire CVD, VHF-PECVD or other suitable techniques [25].

The thin-film approaches are grouped in three main class [18] according to substrate temperature used for the deposition of thin film poly-silicon: i) high-temperature substrates (HTS) low temperature substrates (LTS) and layer transfer process (LTP).

- *HTS high-temperature substrates* approach requires typically process up to 1000°C on low cost substrates such as low-cost metallic grade Si, graphite, ceramics. Some approaches apply higher process temperature up to 1420 °C to re-crystallize materials in order to enhance grain size through process like Zone Melt Re-crystallization.
- *LTS low temperature substrates* approach requires temperatures lower then 550°C in order to use glassily substrate metals or in perspective plastic substrates. The main techniques used are the following: Solid Phase Crystallization (<600°C), PECVD-Plasma Enhanced Chemical Vapour Deposition and VHF-PECVD Very High Frequency Plasma Enhanced Chemical Vapour Deposition (<300°C).
- *LTP Layer Transfer* Approach use a special surface conditioning of the growth substrate in order to permit the transfer of the reusable growth substrate to a low cost device carrier.

A large variety of substrates, deposition techniques and cell architectures are currently under development in order to assure potential for industrialization.

Whereas it is unclear whether this kind of material will be obtained from the evolution of wafer technology, or from thin-film technology the mutual convergence of these two research lines is more and more obvious.

Thin film technology is intrinsically compatible with the predicted material in order to remove the detrimental high temperature process of conventional diffusion technology (700°C-1100°C) to avoid degradation of minority-carrier lifetimes and thermal damage of substrate and thin silicon layers. These requirements, added to other factors such as the search for automatic module-assembly techniques, the need to lower costs by simplifying the technology and spending less energy, or the tendency to produce large-area devices, have led to the development of new approaches that could merge together both technologies.

Some promising results could come from Rapid Thermal Processing (RTP) where heating and cooling rates can de dramatically increased up to 100 K/s instead of 15-20 K/min, but silicon-heterojunction solar cells (SHJ) basically made of a crystalline-silicon (mono or multi-crystalline) wafer or ribbon absorber and one or two thin-film-silicon emitter(s) grown through Plasma- Enhanced Chemical Vapour Deposition (PECVD) or similar techniques represents a realistic relatively simple technology to fabricate solar cells with very high efficiency.

Also it is an excellent example of technological convergence for both thin-film and wafer technologies. Key features of the silicon-heterojunction technology are:
  i) a very simple low-temperature fabrication process.
  ii) it eliminates the degradation of bulk properties (diffusion of impurities, defect formation).

No wrapping problems on thin c-Si or multi-Si wafers. Work at Sanyo Electric during the early 1990s showed that solid phase crystallisation (SPC) at ~600 °C of a relatively thick (~5 µm) PECVD-deposited a-Si film gives 9.2% conversion efficiency on a metal substrate [18]. The cells featured a p^+-doped a-Si heterojunction emitter on a n^+-n polycrystalline silicon (pc-Si) structure crystallised by SPC. Despite these excellent results, this thin-film approach seems to have been abandoned at Sanyo around 1996, but new perspective will sure now opened by Silicon Genesis, a leader in process and technology, which engineered 20µm substrates. The 125 mm square monocrystalline silicon (mono c-Si) foils were found to be robust and highly flexible and this achievement represents an important milestone in the development of kerf-free wafering technology.

In the late 1990s, Pacific Solar Pty Ltd, a spin-off company of the University of New South Wales (UNSW) in Australia, successfully transferred the PECVD-based SPC approach to borosilicate glass sheets (Borofloat33 from Schott AG, Germany). Major breakthroughs have been achieved at Pacific Solar in the following years in the areas of light trapping (novel glass texture and cell metallisation and interconnection (point contacts) The best efficiency obtained so far with this so-called CSG (Crystalline Silicon on Glass) technology is 10.5%, realised in 2007 with a 94-cm^2, 20-cell mini-module with a fill factor (FF) of 72.1%, a J_{sc} of 29.5 mA/cm^2, and an average cell V_{oc} of 492 mV [27]. The J_{sc} is remarkably high for a silicon film thickness of merely 2.2 µm, confirming that CSG devices feature excellent light trapping properties, but no further progress were noted during last years. Both surfaces of the glass superstrate are textured with a dip coating process in order to achieve light trapping, The Si-coated glass sheets are heated to 600 °C in a batch oven for several hours to achieve solid-phase crystallisation. Crystallographic defects are reduced by briefly heating the samples (~1 min) to over 900 °C, using a Rapid Thermal Anneal (RTA) process. Most of the remaining defects are passivated by exposure to atomic hydrogen by plasma. Effort were put on the ground in 2004 by a new company in Germany, CSG Solar AG established a CSG factory with a rated capacity of 10 MWp/year. Silicon depositions were conducted in a KAI-1200 PECVD tool from Oerlikon Solar, using 1.4m^2 glass sheets. The sale of large-area CSG modules started in late 2006. In mid-2007 the module efficiencies were in the 6–7% range and were improving steadily, and a second KAI- tool installed in 2007, doubled the rated factory capacity to 20 MWp/year.

Remarkable results achieved to University of Stuttgart on thin Si film transfer process of a 45 μm thick layer with and efficiency of 16.7%, on a cell of around 4 cm^2 and with a V_{oc} = 645 mV, J_{sc} = 33 mA/cm2 and FF 78.2 %.

7. TRANSPARENT CONDUCTIVE OXIDE

An excellent overview of transparent conductive oxide (TCO) is given by ref [33]. TCO thin films are extensively used in optoelectronic devices as:
i) transparent electrodes for photovoltaic cells
ii) transparent electrodes for flat panel displays
iii) low emissivity windows
iv) light emitting diodes
v) semiconductor lasers, solar cells and liquid crystal displays due to their high conductivity and high transparency in a wide range of wavelength.

TCO are characterized by a resistivity that could be as low as 10^{-4} Ω cm and by an extinction coefficient k in the visible part of the light spectrum that could be lower than 10^{-4} due to their wide optical band gap (Eg) that could be greater than 3 eV.

In intrinsic and stoichiometric oxides this extraordinary combination of conductivity and transparency is usually impossible; it is normally achieved by introducing appropriate dopants or by non-stoichiometric composition. In the first '900 it was discovered that CdO films have these characteristics. During the last years, the dominant TCOs have been tin oxide (SnO_2), indium oxide (In_2O_3), indium tin oxide (ITO), and zinc oxide (ZnO). All of these materials have been mass-produced in very large volumes over a long period of time but new class of material started to be developed during last time. New TCO materials consisting of multi-component oxides have been developed, for example, combinations of binary compounds such as ZnO, CdO, In_2O_3, and SnO_2, Impurity-Doped ZnO by using Group III elements id. ZnO:Al (AZO); ZnO:In (IZO)) or Group IV element (Si, Ge,Ti, Zr) or ternary compounds such as Zn_2SnO_4; $MgIn_2O_4$: $CdSb_2O_6$:Y; $ZnSnO_3$, $GaInO_3$; $Zn_2In_2O_5$.

Despite the huge volume of experience in the field, there remain many unanswered questions at both applied and fundamental levels. Although multi-component oxides newly developed as TCO materials are suitable for specialized applications, up to now resistivity lower than in ITO films have not yet been reported for the new class of TCO materials, and in PV thin film commercial application SnO_2:F continues to dominate the market.

Since the utility of TCO thin films depends on their electrical and optical properties, these two parameters should be considered together with chemical and physical stability, abrasion resistance, electron work function, and adaptability with substrate and other components of a given device, as appropriate for the application. Economics of the deposition method and the availability of the raw materials are also significant factors in choosing the most appropriate TCO material. The final choice is usually made by maximizing the performance of the TCO thin film by considering all important parameters and minimizing the operating cost. TCO material selection only based on maximizing the conductivity and the transparency can be flawed.

In recent times, the high price and scarceness of Indium needed for ITO, the most popular TCO, stimulate R&D to find a replacement. Its resistivity should be on the order of 10^{-4} Ω cm or less, with an absorption coefficient smaller than 10^{-4} cm^{-1} in the near-ultraviolet and visible range, and with an optical band gap > 3eV. A 100 nm thick ITO film with these parameters will have optical transmission of at least 90% and a sheet resistance of 10 Ω/cm. For the moment, AZO and ZnO:Ga (GZO) oxides are promising options for ITO thin-film transparent electrode applications. From this point of view the best candidate is AZO, which could have a low resistivity, on the order of 10^{-4} Ω.cm, and its origin elements are inexpensive and non-toxic. Another aim of the recent attempts to develop novel materials is to deposit p-type TCO films. Most of the TCOs are n-type semiconductors, but p-type materials are necessary in the field of solid lasers. Such p-type TCOs comprise: ZnO:Mg, ZnO:N, NiO, NiO:Li, $CuAlO_2$, Cu_2SrO_2, and $CuGaO_2$ thin films although these materials have not yet found an application in actual devices.

Since the development of new TCOs is mostly imposed by the requirements of specific devices, low resistivity and low optical absorption are always important characteristics. There are at least two suitable strategies for developing advanced TCOs that could satisfy the requirements. The main strategy is to dope known binary TCOs with other elements, which can increase the density of the charge carriers. As shown in Table 2, more than 20 different doped binary TCOs were produced and characterized, of which ITO was preferred, while AZO and GZO come close to it in their electrical and optical performance.

Table 2. TCO Compounds and Dopants

TCO	*Dopant*
SnO_2	Sb, F, As, Nb, Ta
ZnO	Al, Ga, B, In, Y, Sc, F, V, Si, Ge, Ti, Zr, Hf, Mg, As, H
In_2O_3	Sn, Mo, Ta, W, Zr, F, Ge, Si, Nb, Hf, Mg
CdO	In, Sn
Ta_2O	
$GaInO_3$	Sn, Ge
$CdSb_2O_3$	Y

The above described dopant ions should have appropriate valence to be an effective donor when replacing the native metallic ion. This strategy to increase the conductivity without degrading the transparency was paralleled by a more elaborate approach in which phase-segregated two-binary and ternary TCOs were synthesized. The phase-segregated two-binary systems include $ZnO-SnO2$, $CdO-SnO_2$, and $ZnO-In_2O_3$. Despite expectations, the electrical and optical properties of two-binary TCO were lower than that of ITO.

The phase diagram of ternary TCOs could be schematically presented by a three-dimensional or four-dimensional phase combination of the most common ternary TCO materials based on known binary TCO compounds.

As a result, the ternary TCO compounds could be formed by combining ZnO, CdO, SnO$_2$, InO$_{1.5}$ and GaO$_{1.5}$ to obtain Zn$_2$SnO$_4$, ZnSnO$_3$, CdSnO$_4$, ZnGa$_2$O$_4$, GaInO$_3$, Zn$_2$In$_2$O$_5$, Zn$_3$In$_2$ and Zn$_4$In$_2$O$_7$. However, since Cd oxides are highly hygroscopic and, as a consequence, they are toxic, the utilization of these TCOs is limited, though they have adequate electrical and optical properties. Other binary TCOs were synthesized from known binary TCOs and also from non-TCO compounds, such as In$_6$WO$_{12}$ and the p-type CuAlO$_2$.

All the TCOs discussed above are n-type semiconductors. In addition, p-type doped TCOs were also developed and could find interesting future applications, in particular in the new optoelectronic field. Fabricating un-doped or doped p-type TCOs was found to be more difficult than the n-type. It has been reported that is possible to form acceptor levels in ZnO, doping with N, P and As. The difficulty in producing p-type oxide was hypothesized to result from the strong localization of holes at oxygen 2p levels or due to the ionicity of the metallic atoms. Oxygen 2p levels are far lower than the valence orbit of metallic atoms, leading to the formation of a deep acceptor level. Hence, these holes are localized and require sufficiently high energy to overcome a large barrier height in order to migrate within the crystal lattice, resulting in poor hole-mobility and conductivity. Following this hypothesis, an effort was made to grow p-type TCO based on "Chemical Modulation of the Valence Band (CMVB)", where the oxide composition and structure were expected to delocalize the holes in the valence band. In recent times several groups of p-type TCOs were synthesized, e.g., CuM^{iii}O$_2$, AgM^{iii}O$_2$ where M^{iii} is a trivalent ion. Compared with the n-type TCOs, these TCO have relatively lower conductivities, on the order of 1 S/cm, and lower transmission, < 80%.

Growing p-ZnO was an important milestone in ''Transparent Electronics'', allowing fabrication of wide band gap p-n homo-junctions, which is a key structure in this field.

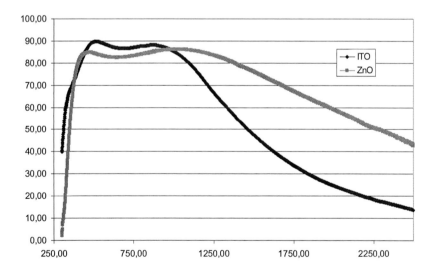

Figure 5. Trasmittance of ZnO (ENEA) and ITO (commercial product) growth on glass. For both layer thickness is 100 nm.

The exigency to produce n-type TCOs with higher conductivity and better transmission, without the use of In, inspired R&D to discover some non-conventional TCOs. Novel transparent conductors were proposed using oxides of Mg, Ca, Sc and Al which also exhibited the desired optical and electronic characteristics; however, they have not been considered as candidates for achieving good electrical conductivity because of the difficulty of efficiently generating carriers in these wide band gap materials. The suggested approach was to increase the mobility rather than the carrier density. If this purpose will be achieved, the optical properties would not degrade at higher conductivity. Recently, new thin film geometries were also explored searching TCO films with lower resistivity. It was demonstrated that higher conductivity could be obtained by doping modulation, which spatially separates the conduction electrons and their parent impurity atoms (ions) and in this manner the effect of ionized and impurity scattering on the electron motion is reduced. By using a zone confining process to deposit TCO it is possible to obtain an ITO layer with $\rho = 4.4 \cdot 10^{-5}$ Ω·cm and $\mu = 103$ cm^2/V·s. A theoretical outline of a method to engineer high mobility TCOs is based on the high mobility transistor structure. The film should consist of alternating thin layers of two semiconductors. One layer provides a high density of carriers, while the second is a high mobility material. Electrons are supplied by the former and transported in the latter, attenuating the scattering due to ionized impurity. The model assumes that the electrons move into the high mobility material in response to differences in electron affinity. However, the success of the proposed TCO design depends upon controlling the layer thickness at nano dimensions, (e.g. ~5 nm). In addition, this approach depends on having materials of excellent quality and with compatible crystal structure in order to avoid problems related to interface defects.

Further arguments are related to the problems and necessity of TCO texturization during the growth or with post-deposition treatment, but we leave to the reader to deepen the argument.

In summary, AZO, GZO and FTO are at present the only TCOs with electrical conductivity close to that of ITO, and with appropriate high optical transmission in the near-UV, VIS and NIR. The objective of producing TCO materials with optical and electrical characteristics equivalent to the ITO ones has not yet been achieved.

8. DIELECTRIC AND HIGH RESISTIVELY COATING

The active part of a Si solar cell is often regarded as the Si wafer n-p junction. However, surfaces and interfaces play an important role in optical or electronic performance of a solar cell, thereby influencing both the generation and collection of photocarriers.

The dielectric and high resistively coatings constitute a very wide area of PV technology. Their use in Si solar cell technology is very profitable for several applications. It concerns i) antireflection treatment of the front surface ii) reduction of interface recombination iii) application as a device component, needed for the operation of the device iv) profitable layer growth during solar cells fabrication v) shield against chemical and physical treatments.

The field of dielectric and high resistively coatings represent a wider area then what could be covered by an introduction to material science. Thin layers of SiO_2 are often used for passivation of solar cell surfaces, resulting in surface recombination properties similar to those in metal oxide semiconductor (MOS) devices. Likewise, the materials used for antireflection coatings include SiO_2, SiN, TiO_2, Ta_2O_5, ZnS, MgF, and SiN. In this paragraph we will focus on silicon nitride and amorphous Silicon only because both materials are considered among the more relevant thin film dielectric/high resistively coating used in photovoltaics.

8.1. Silicon Nitride

One of the most important manufacturing steps of crystalline silicon solar cells is the cost-effective fabrication of an antireflection (AR) coating which should not only reduce optical losses but simultaneously provide a reasonable degree of surface passivation.

During last years it has become clear that Silicon Nitride is considered the best candidate for cost-effective antireflective treatment on multi-crystalline (mc) silicon material because the plasma growth of the material produce and excellent hydrogen passivation of bulk defects and/or grain boundaries. In fact, conventional SiO_2 or TiO_2 fabricated by atmospheric pressure are not able to produce the same effect. For example, TiO_2 provides no electronic surface passivation because the process cannot assure the presence of hydrogen required for the mc-Si bulk passivation and the refractive index of SiO_2 (1.46) is too low for optimal AR performance.

The most important properties of Si_xN_y for PV applications include:
i) excellent reduction of reflection losses
ii) good surface passivation on phosphorus-diffused emitters
iii) a very effective hydrogen passivation of bulk defects and grain boundaries in m-cSi
iv) outstanding surface passivation mainly on low-resistivity Si wafers
v) the compatibility with cost-effective process schemes including fabrication of the contact by screen printing.

The Si_xN_y optical and the electrical properties strongly depend from the selected deposition technique. In fact, when the growth is realized by simple sputtering of Si_3N_4 target or by reactive sputtering of Si in nitrogen ambient, a stoichiometric amorphous silicon nitride is produced but no effect on passivation properties of the thin layer are usually noted when Si_xN_y is realized by nitration of Silicon or by high energy ion implantation of nitrogen and very thin Si_xN_y layer is only produced but it isn't useful for the antireflective effect and it strongly damages cSi surface.

Plasma enhanced techniques employ the growth of silicon nitride at temperatures in the range from 250°C to 400 °C. The material obtained results in production of non-stoichiometric SiN with a variable hydrogen content from few % up to 40% depending from the process condition. Ammonia is the most common oxidant, although pure nitrogen is also used or a combination of both. At low power density and low ammonia concentrations, the gas phase is dominated by SiH_2, SiH_3, and polymerized silanes. In this circumstance nitrogen must be added in the discharge to increase the presence of nitrogen at the surface. Films produced under these conditions have copious Si-H bonds and are relatively poor insulators. Increasing the power density and the ammonia:silane ratio triaminosilane, $Si(NH_2)_3$ is produced in the gas phase, with lesser amounts of hydrogenated species. Films deposited show poor quantity of Si-H bonds and have excellent electrical properties. The process conditions during the growth in PECVD reactor can be easily diagnosed by monitoring the Si_2H_6 signal in the exhaust by mass spectroscopy. As ammonia is increased the amount of Si_2H_6 conversely decreases rapidly.

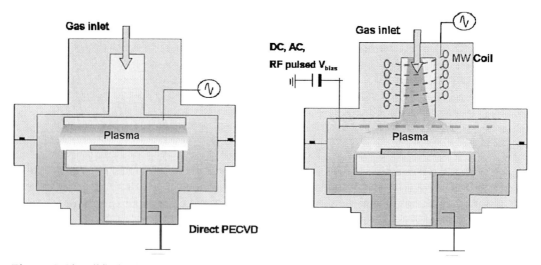

Figure 6. Simplified scheme of a direct and remote PECVD reactor.

There are two fundamentally different excitation regimes for PECVD SixNy deposition methods by direct PECVD and remote PECVD. In direct PECVD reactors all processing gases are excited by an electromagnetic field applied to the two reactor electrodes and the sample are located within the plasma. The electromagnetic field has a frequency of either 13.56 MHz (high-frequency excitation) or in the 10-500kHz range (low-frequency excitation).

In the remote-plasma technique the excitation of the plasma is spatially separated from the sample. A number of different plasma excitation regimes are used ranging from radiofrequency to microwaves. Further used processes concern arc jet, hollow cathode, etc.

In direct plasma excitation the frequency has a strong impact on the electronic properties of the resulting Si/Si_xN_y interfaces. When the excitation of plasma is below the so-called plasma frequency (~4 MHz) ions are able to follow the plasma excitation frequency and therefore they produce a surface damage due to ion bombardment. As effect Si_xN_y obtained at low frequency can be considered an intermediate-quality material for surface passivation on silicon surfaces. Si_xN_y prepared by direct PECVD at high frequency (13.56MHz) or remote plasma provide a better surface passivation and higher stability against UV exposure, moisture and other environmental agents than Si_xN_y prepared at low frequency.

The kinetics produced by impingement of radicals on the surface involve several process steps such as: i) adsorption and insertion of radicals onto the crystalline silicon deposition surface, ii) diffusion of the etching species in the substrate ii) surface dimerization of adsorbed N-H and Si-H groups, or of the other impinging radicals; iv) formation chains and islands of radicals on the surface, through the formation of higher surface hydrides and exchange of hydrogen or other chemical species, v) dangling-bond-mediated dissociation of surface hydrides.

As effect during the PECVD Si_xN_y deposition an interfacial layer is usually grown responsible of interface states at Si-Si_xN_y interface. The control of this intermediate layer is very relevant in order to improve the surface passivation quality.

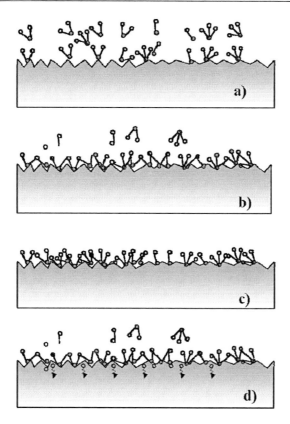

Figure 7. Deposition kinetics of SixNy thin film growth on cSi by RF-VHF PECVD, ECR-CVD, HW-CVD. a) Production, impingement, absorption and insertion of radicals onto the crystalline silicon surface. b) Formation of chains and island of radicals on the surfaces through the formation of surface Si-H, N-H radicals and exchange of hydrogen and other chemical species. c) Nucleation coalescence mechanism. d) Diffusion of the etching species in the substrates.

8.2. Amorphous/Crystalline Heterojunction

In 1974, Sanyo started to investigate amorphous silicon (a-Si)-based solar cell technology and developed various technologies for high conversion efficiency and stable devices. Starting from early '90, Sanyo developed a new artificially constructed junction based on a-Si/c-Si heterojunction structure called ACJ-HIT (Artificially Constructed Junction-Heterojunction with Intrinsic Thin layer). This structure features a very thin intrinsic a-Si layer inserted between a doped a-Si layer and a c-Si substrate. A conversion efficiency of more than 18% has been achieved, since 1992 which was the highest ever value for solar cells in which the junction was fabricated at a low temperature (<200°C).

This simple and novel structure has been attracting a growing amount of attention. This is because:
i) a very simple low-temperature fabrication process eliminates the degradation of bulk properties (diffusion of impurities, defect formation);
ii) no wrapping problems on thin c-Si or multi-Si wafers;
iii) the junction depth can be easily and accurately controlled (~1nm- up to 100 nm or more);

iv) large gap emitter 1,6 eV-2,3 eV to enhance the response in the blue area of solar spectra (window effect);
v) passivation of bulk material and wafer surface;
vi) excellent back surface field;
vii) and enhanced throughput due to reduced process time (only for in line application in production, due to the vacuum process;
viii) opportunities for tandem structures.

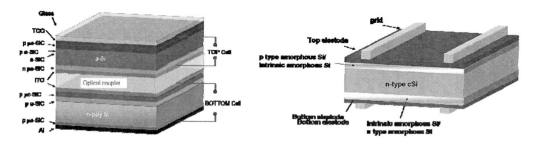

Figure 8. The four-electrical contacts Honeymoon solar cell proposed by prof. Hamakawa and the ACJ-HIT Solar cell proposed by Sanyo.

An interesting solution was proposed at the end of '80 by professor Hamakawa with his Honeymoon solar cell concept [29] when a classical p-i-n junction was proposed to be realized on a p-n artificial constructed heterojunction or similar solar cell.

During the following years Sanyo was able to demonstrate an important cost-reduction potential through the realization of very high efficiency devices (~22-23%) with a high potential for significant improvements.

The solar cells based on SHJ can be considered as interfacial devices. The chemistry and kinetics of the gas surface interaction during Plasma Enhanced Chemical Vapour Deposition (PECVD) and, in general, during plasma treatment is not totally explored. A wide potential is available for further investigation of different process schemes that affect, not only the structural quality of the deposited films, but also surface morphology, roughness, surface reactivity and surface composition.

The deposition conditions can be changed dramatically the electrical and optical properties of the emitter including optical gap, activation energy, band offset, band bending, gap state density, interface state density (Dit(E)), surface roughness. The key issues for improvement in open circuit voltage V_{oc} [30] are the interface state density reduction and the re-optimization of the fabrication condition depending on the interface quality. The high-quality a-Si films should be deposited by a low-damage plasma process and wafer cleaning before loading the CVD reactor is critical. The control of interface properties during a-Si deposition is also important. Even if a-Si layers of same quality are used, different cell properties are obtained due to plasma damage during the deposition.

The development of different plasma process schemes (RF, VHF, MW-PECVD, cat-CVD) based on conventional (SiH_4, NH_3, H_2, N_2,) and new classes of precursor gases (SiF_4, NF_3, other gases) leads to the development of new plasma regime experiments. These options should help to better understand the route by which the kinetics and the effects of impinging radicals on the front and back interface could be changed. As a way of fact, Sanyo was able to increase the conversion efficiency of their devices based on n-type c-Si only after the development of a specific plasma surface conditioning process therefore never indicated in literature.

Notwithstanding this, it is definitive the beneficial role of surface treatment in hydrogen of n-type based devices, but some authors experimented detrimental effects of hydrogen radicals in plasma, always present in such a process, on p-type based SHJ solar cells, due to interfacial acceptor passivation and consequently an increase of interface state density defect was noted. Furthermore, to the knowledge of the authors in all p-type based SHJ solar cells having a conversion efficiency of interest for PV application (>15%) always a conventional high temperature contact was used on the back. No opportunities for a low temperature back surface field are offered to p:Si-H deposition on the back respect n-type based SHJ, where it can be recognized that a very low recombination rate is expected on the back contact due to valence band offset between i-n-type amorphous layer and n-cSi. All indicated considerations explain why the main results on p-type c-Si based SHJ show always lower conversion efficiency in respect to the high efficiency demonstrated by Sanyo on n-type cSi.

As opposite route, full innovation in SHJ technology can be produced by application to multicrystalline silicon and to the new class of predicted thin silicon materials. As effect of the wide extension of knowledge, a number of research groups and R&D braches of companies are actively working on SHJ cells in Europe, USA, Japan and Australia. Despite their high individual level, only during late years interesting applications were seen on devices having a significant conversion efficiency and further efforts need to be address in direction of common objectives.

9. ADVANCED THIN FILM TECHNOLOGIES

The main limitations of photovoltaic are always ascribed to the very high cost and physical efficiency limits. To reach its most competitive long-term position, it has been argued that photovoltaics must push towards ever-increasing energy conversion efficiency with cost effective approaches. This leads to the concept of "third-generation" photovoltaics introduced by Professor M. Green, University of South Wales, Australia. The "third generation" approaches aim to achieve highly efficient photovoltaic devices by using still abundant and nontoxic materials based on "thin film" second generation approach [14,15].

Third generation differs from "first generation" photovoltaics because is approaching advanced technologies to overcome the limiting efficiencies for single band gap devices. In fact solar cells lose energy in a number of ways; optical losses include reflection from interfaces at the surface of a solar cell or a module and photogenerated carrier losses include recombination in the bulk material and as a result of poor interfaces. By far the main losses are due to the nature of the photovoltaic effect itself. In fact the two most important power loss mechanisms in single-bandgap cells arise from the inability to absorb photons with energy less than the bandgap and thermalisation of photon energy exceeding the bandgap, ((figure 9). These two mechanisms alone limit the theoretical maximum efficiency to 31% to 41% for one sun and maximum concentration irradiance respectively by considering the energy difference between the photons being absorbed from the sun at 6000° K and their re-emittance at room temperature at 300° K (the Shockley–Queisser limit calculation).

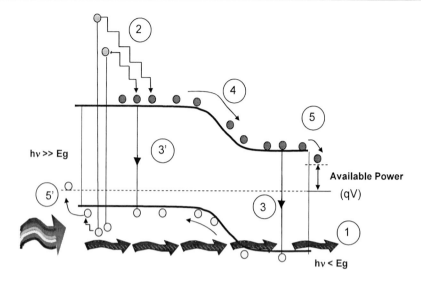

Figure 9. Main loss mechanism in a single junction solar cell ref. [14,15]. Loss processes in a standard solar cell: (1) non-absorption of below bandgap photons; (2) lattice thermalisation loss; (3) recombination loss (4) and (5) junction and contact voltage losses.

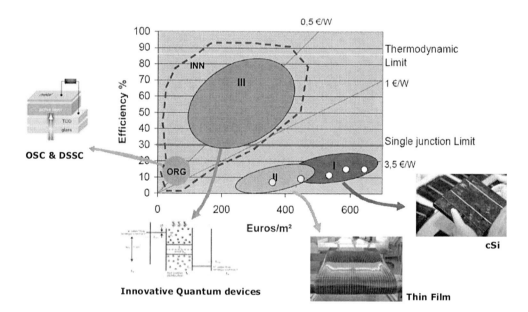

Figure 10. Comparison in terms of efficiency/cost for 1st, 2nd and 3rd photovoltaic generations. The white dots indicate commercial PV modules based on 1st and 2nd generation.

Third generation aims to circumvent the Shockley-Queisser limit for single band gap but with decreased costs to below US$ 0.50/W potentially to US$ 0.20/W or better, by dramatically increasing efficiencies up to 40-50% or more [15].

Several approaches are put on the ground to overcome the conversion efficiency limits which include several options [36].

A lot of exciting ideas are followed by scientist and researchers all around the world, with solutions that have proven very difficult to be demonstrate in principle and often the proposed devices decrease the overall efficiencies instead to improve it, the so called "field of dreams where the research concepts and technology require the PV researchers creativity, patience and funding to prove or develop concepts" (L L. Kazmerski, NREL).

The emergence of 3G PV approaches started to be showed by utilization of GaInP/GaInAs/Ge, gallium indium phosphide, gallium indium arsenide, germanium triple-junction. The cost-expensive III-V technology very useful during last years for space application started to be also considered useful for terrestrial application, through concentration application.

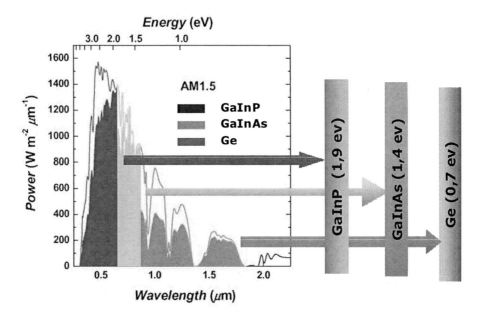

Figure 11. Typical scheme of a III-V stacked triple multi-junction.

Multi-junction devices stack different solar cells with multiple bandgaps tuned to utilise the entire spectrum (figure 11). Light is first incident on the front wide bandgap cell that can produce a relatively high voltage and thereby make better use of high-energy photons, then lower-energy photons pass through to narrow bandgap cell that can absorb the transmitted IR-photons. Theoretical maximum efficiencies of 55.9%, 63.8% and 68.8% are predicted for two-(tandem), three- and four-junction devices [15], but fabrication becomes problematic with the increasing number of interfaces and cells are increased. The final device is realized as a monolithic multi-junction devices in a single growth run with tunnel junctions between each single solar cell. Different compositions of aluminium or indium alloys are used to modify GaAs, InP or GaN bandgaps while maintaining lattice constants (bandgap engineering). Lattice matched (LM) GaInP/ GaInAs/ Ge 3-junction cells have achieved the highest efficiency for a photovoltaic device, at 39.0% at 236 suns, 25°C under the standard AM1.5D. On other hand Ge 3-junction cell leads to excess photogenerated current density in the Ge subcell. Part of this wasted current can be used effectively in the middle cell if its bandgap is lowered, as in lattice-mismatched, or metamorphic (MM) GaInP/GaInAs/Ge 3-

junction cells with a 1.2-1.3 eV GaInAs bandgap. Lattice-mismatched or metamorphic (MM) materials offer still higher potential efficiencies, if the crystal quality can be maintained. It can also be used multi-junction mechanically stacked approach where devices are optically coupled each to the other without internal electrical connection. Researchers at the Fraunhofer Institute for Solar Energy Systems in Freiburg demonstrated efficiency improvement up to world record of 41.1% onto a small 5 mm² MM multi-junction solar cell by concentrating light by a factor of 454 and Spectrolab, a wholly owned subsidiary of the Boeing Company hits with a MM III-V solar cell 41.6% PV cell efficiency world record [34, 35].

On other hand long term approach to 3G is also based on the application of nanotechnology and the utilization of innovative materials with different quantistic behaviour respect to conventional semiconductor.

In fact another parallel approach to tandem/multijunction is the so-called intermediate band solar cell [36]. It is a theoretical concept where the superior theoretical efficiency over single-gap solar cells is obtained by enhancing the photogenerated current, via the two-step absorption of sub-band gap photons. This is achieved through a material with an electrically isolated and partially filled intermediate band located within a wider forbidden gap. A number of attempts, which aim to implement the intermediate band concept, are being followed during these years for the direct engineering of the intermediate band material, and its implementation. It was obtained by means of quantum dots, impurity photovoltaics and highly porous material approach.

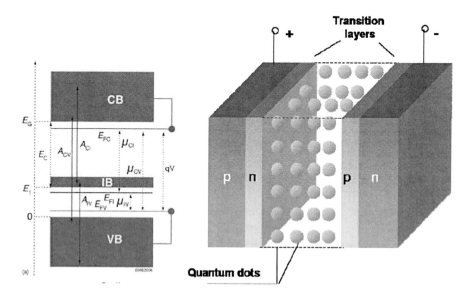

Figure 12. a) the intra-band concept (IB) allows the absorption of sub-bandgap photons. The absorption of photon pumps an electron from the VB to the IB (step 1), while the absorption of photon causes an electronic transition from the IB to the CB (step 2). b) IB-Quantum dot solar cell.

Quantum dots (QDSC) or nanoparticles are semi-conducting crystals of nanometric dimensions showing quantum optical properties that are absent in the bulk material, thanks to the confinement in a region of a few nanometres of electron-hole pairs (called excitons). When quantum dots are ordered in a three-dimensional array, there will be strong electronic coupling between them so that excitons will have a longer life, facilitating the collection and transport of 'hot carriers' to generate electricity at high voltage. In addition, such an array makes possible to generate multiple excitons from the absorption of a single photon.

The tuneable bandgap of quantum dot structures can be considered the first advantage in 3G application. It means that the wavelength at which each layer will absorb or emit radiation can be adjusted. Wider bandgap of a solar cell semiconductor, the more energetic the photons absorbed, and the greater the output voltage. As opposite, a lower bandgap results in the capture of more photons in the red and infrared area resulting in a higher output current at lower output voltage. In contrast to traditional semiconductor materials that are crystalline or rigid, quantum dots can be moulded into a variety of different form, in sheets or three-dimensional arrays. They can easily be combined with organic polymers, dyes, or made into porous films. In a suspended solution colloidal form, they can be processed to create junctions on inexpensive substrates such as plastics, glass or metal sheets.

A quantum well solar cell (QWSC) is a special multiple-band gap device with intermediate properties between heterojunction cells (sum of the current generated in the different materials but voltage controlled by the lowest of the two band gaps) and tandem cells (sum of the voltages but current determined by the worst of the two sub-cells) [37,38]. A QWSC can achieve optimal band-gaps for the highest single-junction efficiencies due to the tuneability of the quantum well thickness and composition. The strain-balanced quantum well solar cell (SB-QWSC) was a concept introduce by Prof. J. Barnham of the Imperial College as a way to extend the spectral range of high efficiency GaAs cells. They demonstrated that the extended spectral range can be achieved and the SB-QWSCs can be grown with zero dislocations in the active region [37].

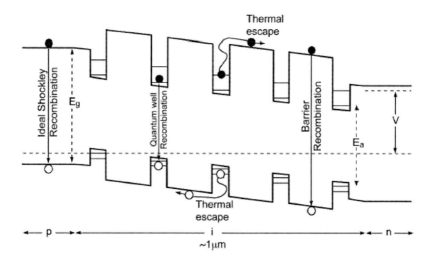

Figure 13. Schematic of a Strained Balanced- Quantum well solar cell (SB-QWSC) with compressively strained [37].

The concept underlying the hot carrier solar cell [41] is to slow the rate of photoexcited carrier cooling, caused by phonon interaction in the lattice to allow the photogenerated photo-carriers to be collected whilst they are still "hot". This will enhance the voltage of the cell. It has been shown that periodic networks of quantum structures, such as semiconductor nanoparticles, incorporated into the absorber material can significantly reduce the scattering between phonon modes. This effect – called the "phonon bottleneck" is used by UNSW to retard thermalization mechanisms. Other approach concerns the development of selective energy contacts able to extract charges through a narrow allowed energy range in order to prevent cold carriers in the contacts cooling the hot carriers to be extracted. Quantum mechanical resonant tunnelling structures are the most likely to satisfy the requirements of selective energy transmission over a small energy range. In thermodynamic terms the carriers are thus collected with a very small entropy increase.

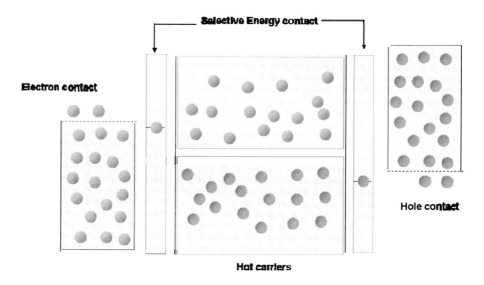

Figure 14. Schematic of an ideal hot carrier solar cell.

This very short overview just outlines the relevance will be on the ground during next year for photovoltaics based always on a thin film approach and we leave to the reader the opportunity to deep the argument on the specific literature [15,36]

10. CONCLUSION

The recent silicon shortage has given thin-film photovoltaic technologies to establish an increased market share presence which is expected to grow during next years. Thin-film PV has a very high potential for cost reduction of photovoltaic energy.

The quality of the materials, the reliability and the cost decrease of manufacturing process can be improved by intensive and effective R&D based on both fundamental science and production technology. The key requirements for the success impose thin-film technology must be based on non-toxic, abundant and reliable materials.

Several option are opened for short-medium terms based on existing materials (high efficiency cSi, TFSi, CIGC, CdTe) and medium-long term (3rd generation) but the solution will not probably come from only one of them and the development in each technology can take profit from the others, but without any doubt the frontier of photovoltaic research will absolutely deal with material science based on thin film approach.

REFERENCES

[1] Solar Generation V, Solar electricity for over one billion people and two million jobs by 2020, GreenPeace International & Epia, Crispin Aubrey Editor, September, 02, 2008, http://www.greenpeace.org/international/press/reports/solar-generation-iv.

[2] Ossenbrink, H., Highlights of the 23rd European Photovoltaic Conference and Exhibition Valencia, Spain 1-5 September 200, JRC EU Ispra Technical Programme Chairman http://www.photovoltaic-conference.com

[3] Kazmerski L.L., Solar photovoltaics R&D at the tipping point:a 2005 technology overview, Journal of Electron Spectroscopy and Related Phenomena, 150 (2006) 105-135.

[4] PV News 2009, published by The Prometheus Institute, ISSN 0739-4829.

[5] A Strategic Research Agenda for Photovoltaic Solar Energy Technology, Luxembourg: Office for Official Publications of the European Communities, 2007 ISBN 978-92-79-05523-2 © European Communities, 2007 PV Technology Platform; http://www.eupvplatform.org

[6] Travis Bradford, Polysilicon: Supply, Demand and Implications for the PV Industry, Prometheus Institute, June, 25, 2008 http://www.greentechmedia.com/research/list/P16

[7] Waldau, A. J., European Photovoltaics in world wide comparison, *Journal of Non-Crystalline Solids 352* (2006) 1922–1927.

[8] W.G.J.H.M. van Sark, G.W. Brandsen, M. Fleuster, M.P. Hekker, Analysis of the silicon market: Will thin film profit?, *Energy Policy 35* (2007) 3121–3125 3122.

[9] K.L. Choy, Chemical vapour deposition of coatings, Progress in Materials Science 48 (2003) 57-170.

[10] F. Tyholdt, A. Ulyashin, M. Mottern, A.T.J. van Helvoort, H. Raeder, ITO films for solar cells made by chemical solution deposition, *Proc. of the 21th European Photovoltaic Solar Energy Conference,* Dresden, Germany, (2006) 311-314.

[11] K. Christiansen, M.Luenenbuerger, R.Schineller, M.Heuken, H.Juergensen, Advances in MOCVD Technology for Research, development and mass production of compound semiconductor devices, *Opto-Electronics Review, 10*(4), 237-242 (2002).

[12] Seshan, K., (2002). *Handbook of Thin-Film Deposition Processes and Techniques - Principles, Methods, Equipment and Applications (2nd Edition).* William Andrew Publishing/Noyes.

[13] Chopra,K. L., Paulson,P. D. and Dutta,V., Thin-Film Solar Cells: an overview, *Prog. Photovolt: Res. Appl. 2004, 1, 69-92.*

[14] Green M.A., Consolidation of Thin-film Photovoltaic Technology: the Coming Decade of Opportunity, *Prog. Photovolt: Res. Appl.* 2006, 14,383–392.

[15] M.A.Green, *Third Generation Photovoltaics: advanced Solar Energy Conversion,* Springer Verlang Berlin, Heidelberg, New York, (2003) ISBN 3-540-40137-7.

[16] 6FP EC ATHLET, Advanced Thin Film Technologies for Cost Effective Photovoltaics project, http://www.ip-athlet.eu/index.html

[17] B. Sopori, Dielectric Films for Si Solar Cell Applications, *Journal of Electronic Materials, Vol. 34*, No. 5, 2005.

[18] R.Brendel, Thin Film Crystalline Silicon Solar cells, Physics and Technology 2003 Wiley-VCH Verlang GmbH ISBN 3-527-40376-0.

[19] G. Aberle, Thin-film solar cells, *Thin Solid Films 517* (2009) 4706–4710.

[20] G. Aberle, Overview on SiN surface passivation of crystalline silicon solar cells, *Solar Energy Materials & Solar Cells*, 65 (2001) 239-248.

[21] G. Prevo, D. M. Kuncicky, O. D. Velev, Engineered deposition of coatings from nano- and micro-particles: A brief review of convective assembly at high volume fraction, Colloids and Surfaces A: Physicochem. Eng. Aspects, 311 (2007) 2.

[22] http://www.firstsolar.com/; http://www.arendi.eu

[23] http://www.oerlikon.com/solar/ http://www.appliedmaterials.com

[24] http://www.globalsolar.com

[25] F. Roca, J. Cárabe & A. Jäger-Waldau, Silicon Heterojunction Cells R&D In Europe, *Proc. 19th European Photovoltaic Solar Energy Conference 7-11 June 2004*, Paris, France

[26] Roca,F; Bruno,G; Losurdo M. et at, Key Issues for the Improvement of the interface and emitter quality in a-Si-:H/c-Si Heterojunction Solar Cells, *Proc. of 21st European Photovoltaic Solar Energy Conference, 4-8 September 2006*, Dresden (Germany)

[27] M. A. Green, K. Emery, Y. Hishikawa, W. Warta, Solar Cell Efficiency Tables (Version 34) *Prog. Photovolt. Res. Appl.*, (2009), 17, 320-326.

[28] Smith, Controlling the plasma chemistry of silicon nitride and oxide deposition from silane, *J. Vac. Sci. Technol., A11* 1843 (1993).

[29] Y. Hamakawa Thin-Film Solar. Cells. Next *Generation Photovoltaics and Its Applications*, Springer Series in. Photonics, Vol.13 Springer-Verlag Berlin Heildelberg (2004).

[30] M.Tanaka, M. Taguchi, T. Matsuyama, T.Sawada, S. Tsuda, S. Nakano, H.Hanafusa, Y. Kuwano "Development of New a-Si/c-Si Heterojunction Solar Cells: ACJ-HIT (Artificially Constructed Junction-Heterojunction with Intrinsic Thin-Layer, *Jpn. J. Appl. Phys. 31* (1992) pp. 3518-3522.

[31] M. Taguchi, A. Terakawa, E. Maruyama M. Tanaka, Obtaining a Higher Voc in HIT Cells, *Prog. Photovol. Res. Appl., 13*, (2005)481-488,

[32] D.M.Bagnall, M.Boreland Photovoltaictechnologies *Energy Policy* 36 (2008) 4390-4396.

[33] C.G.Granqvist, Trasparent conductor ad solar energy material: a panoramic overview, *Solar Energy Material & Solar cells 91* (2007) 1529-1598.

[34] R.R. King et al., 40% efficient metamorphic GaInP/GaInAs/Ge multijunction solar cells, *Applied Physics Letters, 90*, 183516, (2007)

[35] R.R. King et al., Band-Gap-Engineered Architectures for High-Efficiency Multijunction Concentrator Solar Cells, Proc. of 24th European Photovoltaic Solar Energy Conference and Exhibition, Hamburg, Germany, 21-25 Sep. 2009

[36] Next generation Photovoltaics. High efficiency through full spectrum utilization. Istitute of Physics, Publ. Series in Optics and Optoelectronics, Dirac House, Temple Back, Bristol (2004), ISBN 0750309059.

[37] K. W. J. Barnham G. Duggan, A new approach to high-efficiency multi-band-gap solar cells *J. Appl. Phys. 67*, 3490, (1990).

[38] M. Mazzer, et al., Progress in quantum well solar cells *Thin Solid Films* 511 – 512 (2006) 76 -83.

[39] W. H. Weber and J.Lambe, Luminescent greenhouse collector for solar radiation *Applied Optic,s* October 1976, Vol. 15, No. 10.

[40] Goetzberger, W. Greubel, Solar energy conversion with fluorescent collectors. *Appl. Phys. 12*, 123 (1977) 3.

[41] R. T. Ross A.J. Nozik, Efficiency of hot-carrier solar energy converters *J. Appl. Phys. 53*, 3813 (1982).

[42] S. Kolodinski, J. H. Werner, T. Wittchen, and J. Queisser, Quantum efficiencies exceeding unity due to impact ionization in silicon solar cells *Appl. Phys. Lett. 63* (17) 2405 (1993).

Chapter 3

CO-EVAPORATION OF CIGS AND ALTERNATIVE BUFFER LAYERS FOR CIGS DEVICES

Marika Edoff and Charlotte Platzer-Björkman
Solid State Electronics- Uppsala University, Sweden.

ABSTRACT

Co-evaporation is the deposition method leading to the highest efficiencies of $Cu(In,Ga)Se_2$ based thin film solar cells. In this chapter we describe the main co-evaporation processes with focus on routes to high efficiency. Requirements for large area in-line processes are compared to laboratory-based processing. Process control, influence of the back contact and the role of sodium are high-lighted. In addition, alternative buffer layers to replace the standard CdS are treated in detail. Replacement of CdS is not only desirable for environmental reasons, it also has the potential of increasing efficiency further due to reduced absorption in the buffer layer. Properties of the main alternative materials $Zn(S,O,OH)$, $Zn_{1-x}Mg_xO$ and In_2S_3 are treated as well as the commonly observed metastable behaviour of devices with alternative buffer layers.

INTRODUCTION

$Cu(In,Ga)Se_2$ or CIGS, is a semiconductor with high absorption coefficient and direct band gap that can be varied from 1.0 eV for $CuInSe_2$ to 1.67 eV for $CuGaSe_2$. These properties, in combination with a large tolerance to compositional variations and low recombination in grain boundaries are important in order to reach high efficiency using thin, polycrystalline films and fast processing. The CIGS material and devices have been reviewed in several books, see for example [1,2]. The present chapter is divided into two parts. In the first part, we focus on the co-evaporation process while the second part gives an overview of the status of alternative buffer layers.

2. CIGS CO-EVAPORATION

2.1. Background

Since the thin film CIGS work started, many different growth strategies for co-evaporation of CIGS have been tested. Using co-evaporation it is possible to reach up to 16 % efficiency on the cell level, even just by using a straight evaporation profile with constant rates [3]. However, in order to reach the highest efficiencies there are some main advantages coming from using varying evaporation rates.

By varying the Cu flux during the CIGS deposition process, the bulk copper content can exceed stoichiometry in part of the process, which has been observed to lead to improved grain growth. A copper content varying from Cu-rich to Cu-poor or vice versa can also be used actively for process control.

Even if presence of excess copper is desirable during part of the deposition process, device quality CIGS is Cu-poor i.e. with a Cu/(Ga+In) ratio below unity. A copper rich CIGS film with segregated Cu_xSe has a very high conductivity, which makes the material unusable for solar cells. On the other hand, there is a considerable process window on the Cu-poor side of stoichiometric CIGS composition.

One early example of the Cu-rich – Cu-poor co-evaporation process was the so called Boeing process [4]. The most successful co-evaporation so far has been the three-stage process, which is a sequential process which starts with a $(In,Ga)_2Se_3$ growth stage, continues with a deposition of Cu and selenium and ends with another $(In,Ga)_2Se_3$ stage. The best certified small area cell deposited with the three-stage process had an efficiency of 20 % [5].

One inherent feature with co-evaporation is that it is possible to control the in-depth distribution of gallium and indium. By using an intentional variation of the Ga/(Ga+In) ratio in the process, or by using processes that spontaneously lead to this variation, major advantages can be obtained. A high Ga content close to the Mo layer effectively reduces the recombination at the back contact, enabling thinning down of the absorber [6] or improving CIGS devices with long minority carrier lifetimes. Thus already good devices can be improved further.

2.2 Two stage and two stage in-line co-evaporation

In university research material yield and material distribution are non-issues. Basic requirement for sources is a good combination of controllability and a stable flux distribution function. By choosing a long source – substrate distance, a good uniformity across the substrate can be obtained for a variety of source geometries. As soon as scale-up and production issues are addressed, the need for large area homogeneity and material yield call for an in-line approach. Under steady state conditions in the evaporation system, the non-uniformity in the direction of the translation is avoided with the in-line evaporation. Thus only uniformity over the width of the substrate needs to be addressed, for example by using linear sources.

Co-evaporation using an in-line approach, i.e. with substrates passing stationary sources is shown schematically in figure 1. The metal sources are separated by a distance that is dependent on the size of the source. In addition, the flux of material leaving each source will be dependent on the geometry of the source opening, and follow a distribution corresponding to a $\cos^n$ distribution [7], where the exponent, n, is a number between 4 and 7. At some point in the evaporation zone, the flux from each source will reach a maximum. This is valid for all evaporation sources, but will not occur for the same position in the zone. Thus, the in-depth distribution of elements will depend on the source position and of the diffusivity of elements during the evaporation.

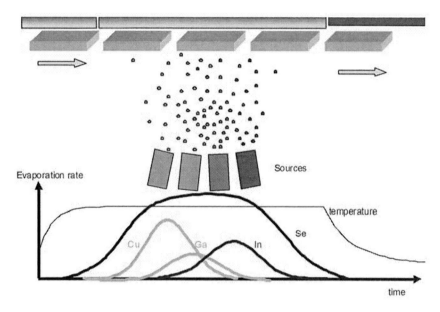

Figure 1. An example of an in-line process and corresponding evaporation rates for all elements.

The source flux distribution as described above is valid for elements with an evaporation temperature which is lower than the temperature of the substrates and walls of the evaporation zone, i.e. Cu, In and Ga. However, for very high evaporation rates, most notably for very high rates of selenium, scattering will tend to smear out the metal evaporation flux distribution.

For Se, the situation is different from that of the metal sources. If the temperature of the Se melt at evaporation temperatures is significantly lower than that of the substrates and the walls, un-reacted selenium will re-evaporate from all hot surfaces. In a hot reactor, Se will be much more uniformly distributed and can more be described by a Se "cloud" in the evaporation zone. It is also likely that the molecular form of Se will change upon re-evaporation from large molecules consisting of chains with eight Se atoms to smaller molecules, which may increase the reactivity of Se. Se thermal cracking at temperatures about the same as the substrate temperatures and hot wall temperatures in an in-line evaporation system was proposed by Kawamura et al [8] as an alternative to rf plasma cracking [9]. In both cases evaporated Se is activated before reaching the evaporation zone.

2.2.1. Simulated in-line in stationary systems

An early version of a simulated in-line profile was published in 1993 by Stolt et al [10]. The basic idea was to transfer the Boeing-like process [4] to an in-line situation. The evaporation profiles for a stationary Cu-rich/Cu-poor process as compared to the in-line process are shown in Figure 2. Group III elements were indium in the first version and In+Ga in a later stage. One basic characteristic for both processes is that the bulk composition of the growing CIGS film is Cu-rich during most of the deposition time and becomes Cu-poor only at the last moments of the evaporation. The substrate temperature is low in the beginning (350 to 450°C) and increased in a later stage of the evaporation to 520°C. Excess Cu from the first stage will be incorporated in the CIGS in the second step. If the sources are placed close to each other, the composition of the growing CIGS film will be close to stoichiometry at all times and the need for Cu diffusion is low.

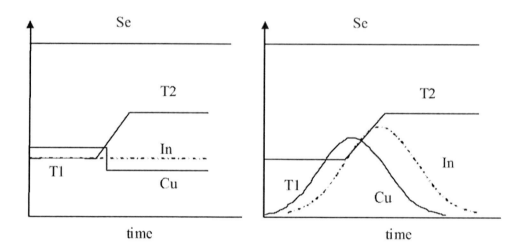

Figure 2. Evaporation profiles of a stationary Cu-rich/Cu-poor process (left) and an in-line simulation (right). The substrate temperatures (denoted T1 and T2) are low (about 350°C) in the first part of the evaporation and high (500-550°C) at the end. Se is evaporated in excess at all times. The bulk composition of the CIGS starts Cu-rich and ends Cu-rich for both evaporation processes.

The in-line version of the Boeing process turned out to be really successful and led to efficiencies as high as 14.8 % for a CuInSe$_2$ thin film solar cell (with less than 1 % Ga), which was world record at that time. A similar process (with about 25 % Ga of the total group III elements) was later used for the world record small area module of 16.6%, which was published in year 2000 [11].

2.2.2. In-line example

An example of a true in-line system for lab use is shown in figure 3.

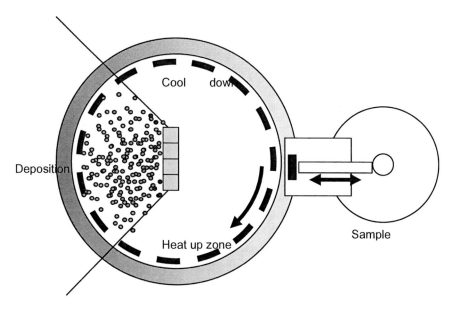

Figure 3. An example of an in-line pilot coevaporation system. The substrates are fed into the main chamber from the load lock through the transfer chamber. The sample holder carousel rotates approximately one rotation per hour. During this hour the substrates experience a heat up phase to 500°C, a deposition phase and a cool down phase. After the completed rotation, the samples are fed out and collected in a load lock chamber.

The in-line system is built as a large flat cylinder, where the substrates are mounted vertically facing inwards onto a moving carousel. A full rotation in the carousel takes about one hour and during this time the substrates undergo a heat-up phase to 500 degrees, followed by an evaporation phase and a cool-down phase. Heating of the substrates is performed with heat lamps mounted behind the substrates. The sources are placed in the middle of the cylinder pointing outwards and positioned such that the bulk composition of the CIGS material is close to stoichiometry at all times. The sources for gallium and indium are separated to have their flux maxima at different positions at the perimeter. Se is evaporated in excess from sources placed in the beginning and at the end of the deposition zone. Many surfaces will act as hot walls and uniformize the Se flux. Scattering of copper, indium and gallium by selenium is commonly observed and will tend to uniformize the CIGS coating as well as reducing the arrival rate of the metals to the substrate.

By placing the sources closely together a composition close to stoichiometry is maintained during the evaporation. Thus in many ways the in-line cylindrical process is similar to the in-line Cu-rich – Cu-poor process described earlier.

The deposition time is in the order of 15-20 minutes and the best result from CIGS deposited in this deposition system is 18.5 % using a Cd-free Zn(O,S) buffer layer [12].

2.3 Two stage and two stage in-line co-evaporation

The three stage CIGS recipe has been successfully used to produce the CIGS for the world record devices [5].

One main characteristic of the three stage process is that copper evaporation is separated from group III elements. With this approach the CIGS is built up by a deposition of (In,Ga)$_2$Se$_3$ film. In the second stage copper is in-diffused until the film is slightly Cu-rich and Cu$_x$Se starts to segregate. In the final stage, the CIGS film is finally made Cu-poor again by deposition of group three elements in absence of Cu.

The crystallographic phase and the texture of the first stage (In,Ga)$_2$Se$_3$ material is believed to have a major impact on the final texture of the Cu(In,Ga)Se$_2$ film [13, 14]. In order to obtain a (204,220) texture, the seed layer should be hexagonal γ-(In,Ga)$_2$Se$_3$ with a (100) orientation of the grains, whereas a (001) oriented c-layer of the same phase will yield (112) orientated grains. The best solar cell devices with efficiencies above 19 % have been fabricated from Cu(In,Ga)Se$_2$ films with a (204,220) texture.

One feature with the three-stage process is that gallium and indium deposited in stage 1 spontaneously will reorganize with more gallium close to the back contact and a grading of the group III elements towards the surface. Another feature is the large grains that form during the second Cu-rich deposition stage. It has been proposed that these large grains may crack during the third stage [15], but other show large grains still after the third stage [16].

An in-depth graded Ga(Ga+In) ratio leads to field assisted diffusion and large high quality grains without grain boundaries parallel to the surface reduce the recombination caused by defects. Both are factors that should be beneficial to increase the diffusivity of electrons towards the junction. However, the results by Jackson et al suggest that the large grains are not a prerequisite for really high efficiencies above 19 %, since their grain sizes are more in the order of 1 µm or less [17].

2.3.1 Multistage in-line

18.3 % efficiency from a multistage in-line co-evaporation process as developed by ZSW (Zentrum für Sonnenenergie und Wasserstoffforschung) is presented in reference [18]. By using an elongated deposition zone and multiple sources, similar grain sizes and efficiencies as with the stationary in-line process have been obtained. ZSW recently presented a device with 19.6 % efficiency also with the multistage in-line process (Press release ZSW 05/2009).

2.4. Control methods for co-evaporation

2.4.1. Flux control

Flux control is in principle a very straight-forward method for controlling the evaporation sources. Several methods are widely used in lab systems, like quartz crystal monitors, QCM, mass spectrometry [19] and atomic absorption spectroscopy, AAS. By placing the control point close to the substrate or further away but at a small angle with respect to the direction of the substrate, a good estimate of the growth process is obtained for QCM and mass spectrometry. By having ports mounted at both sides of the reactor AAS will be able to accurately measure the rates.

In an in-line evaporation system with high material yield the substrates will more or less effectively shield the sources. A measurement which requires a direct view of the source will therefore be difficult to use. In addition, scattered elements will coat all surfaces inside the reactor. Any method using windows will suffer from strongly absorbing coatings that develop during the cause of the process. For in-line deposition systems with up-time in the order of weeks a quartz crystal in a QCM will develop a thick coating and stop working within hours. The answer to some of these problems is to use effective shutters in combination with very short measurement times and to measure the fluxes at regular intervals during the evaporation.

2.4.2. Post process composition control

Controlling the final composition of the CIGS layers is a very useful control method. The most widely used analysis is X-ray Fluorescence Spectroscopy, XRF. Accurately calibrated, the XRF measurement will provide both composition and give a good estimate of the final thickness of the sample. The information is subsequently used for source control. The main drawback with this method is that it gives a slow feed-back and thus requires very stable sources.

2.4.3. In situ composition control by emissivity monitoring

In all evaporation processes using a transition between Cu-rich and Cu-poor bulk composition, or vice versa, the substrates will experience a change in their optical properties. This change can be used for process control, see e.g.[20]. For a constant heating power a change in emissivity will lead to a change in substrate temperature, which can be detected e.g. by thermocouples mounted behind the substrates or by sensors sensitive to changes in IR radiation as in reference [21]. If the temperature is kept constant, the power for substrate heating will decrease when going from Cu-rich to Cu-poor.

The reason for the emissivity change can be found in the ternary phase diagram [22], where single phase $Cu(In,Ga)Se_2$ only is found on the Cu-poor side of the line of stoichiometry. On the Cu-rich side where the Cu/(Ga+In) ratio is above unity, segregation of copper selenide, Cu_xSe, will occur, also at deposition temperatures above 500°C.

The emissivity change will be dependent on the Cu concentration and not significantly on the concentrations of indium and gallium. To a certain extent it will also be dependent on the total thickness of the growing CIGS film.

Using a constant temperature approach, the control parameter is the power to the substrate heater. Figure 4 shows the output power as a function of time for a process following an extreme variation of the Boeing recipe, where the evaporation starts with the deposition of a Cu-rich film, where after the Cu-source is completely turned off [23]. The temperature control is performed with a thermocouple in mechanical contact to the rear side of the substrate. Also shown are the evaporation rates for Cu, In and Ga. As can be seen, the change in power to the substrate heater is abrupt, although it is limited by the thermal dynamics of the system.

In lab scale systems this approach for Cu control is relatively easy to apply. The dynamics are large as compared to the temperature fluctuations in the cold wall system and the time-scale is fast enough. In an in-line system, the application of temperature measurement to monitor emittance changes is more complicated.

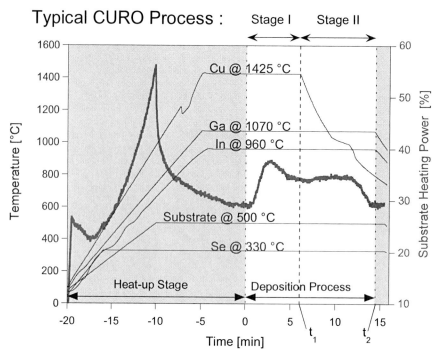

Figure 4. The CURO process scheme, which is an example of an extreme Boeing-type recipe. The power to the substrate heater is recorded and shown as a red curve in the figure. Also shown (in blue) are the source temperatures for all sources as well as the substrate temperature which is constant throughout the process. During stage I of the evaporation, a CIGS layer with excess of Cu is deposited. After stage I, the Cu source is turned off and only In, Ga and Se are evaporated. The box highlighting the end of the red curve shows the transition from Cu-rich to Cu-poor stoichiometry and corresponding decrease in substrate power, caused by reduced emittance of the sample.

For efficient control of Cu content using the end-point detection some basic requirements need to be fulfilled:
- Accurate measurement of temperature on the substrate surface
- Stable background or efficient background subtraction in the calculation for feedback
- Detailed knowledge of the optical properties of the CIGS film in the IR region

2.4.4. In situ compositional control by light scattering

A method which has been recently proposed to monitor and possibly also control CIGS growth is spectroscopic light scattering. It is similar to the emittance change method in the sense that it also measures properties of the sample in-situ during growth. By illuminating the sample during growth with white light, e.g. from a halogen lamp and measure the reflected intensity by a spectrometer, different properties of the sample can be monitored, such as sample thickness in the initial stages of the growth and surface roughness during the following stages. Application of the spectroscopic light scattering method for the three stage process is discussed in references [24] and [25]. The basic idea is that the roughness of the sample will reflect the status of the sample with respect to stoichiometry in the process. Thus, the different stages can be followed by measuring the scattered light intensity at different wavelengths and follow the changes in surface structure.

The method provides a good method for process control provided that
- the radiation from the sources can be subtracted from the SLS signal
- The light source as well as the spectrometer windows are not reduced by covering of selenium and metal selenides
- The fingerprint of the specific process in terms of grain structure is known

2.5. The influence of a copper rich growth stage

The influence of grain size, using a copper rich growth stage is easy to discern from cross section SEM images, see figure 5. It is clear that copper leads not only to improved grain growth conditions, but in some cases also to recrystallization. In the three stage process, large grains are observed after the second stage when the samples have become copper rich. In reference [26], using bilayers of CuGaSe$_2$ and CuInSe$_2$ grown under copper rich conditions, the resulting grains after growing the CuInSe$_2$ layer on top of the CuGaSe$_2$ were larger than after the CuGaSe$_2$ layer alone. This was however only true for films grown on substrates with a barrier for Na. This will be further discussed in section (Influence of Na on CIGS growth).

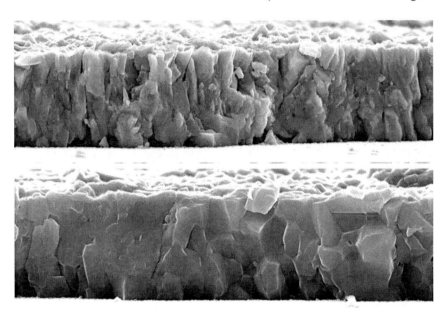

Figure 5. SEM cross section of two positions of the same sample, where on part never was Cu-rich during the CIGS process (upper) and one part was Cu rich during the first part of the deposition.

In processes which are separated in terms of copper and group III elements, the reorganization of the growing film will sometimes lead to unexpected effects. In the extreme version of the Boeing process, where the copper source is completely turned off during stage 2, the source for copper in the third stage is the surplus copper from the earlier stages. This copper will migrate from the grain boundaries, where there is accumulation of Cu$_x$Se to the growing CIGS film, leaving deep crevices between the CIGS grains. The depths of the crevices were found to depend on the Cu excess in the first stage and the time when the Cu source was turned off. As an example Figure 6 shows two examples with different Cu excess in the first stage and thereby different depth of the crevices [23].

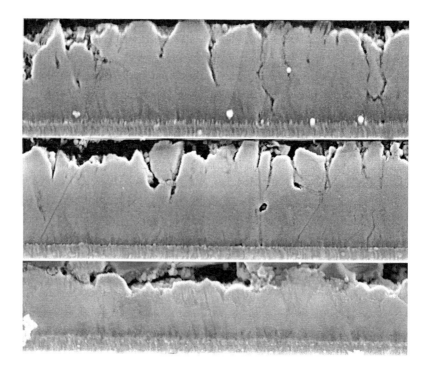

Figure 6. SEM cross sections of a polished sample. The uppermost sample is from a CIGS CURO process (as described in Figure 4), where the sample was very Cu-rich and the Cu source was turned off after 1/3 of the CIGS process. The CIGS from the middle sample was less Cu-rich and the Cu source was turned off after 2/3 of the process. The lower sample is a copper rich CIGS film after stage I, where segregated Cu$_x$Se is visible on the surface of the grains.

2.6. The influence of a copper rich growth stage

CIGS has a very high absorption coefficient. Therefore a thickness of less than 2 μm is needed to absorb all available light. For several reasons it would be interesting to make the CIGS layers thinner:
- A thin CIGS layer takes less material thereby reducing materials cost
- The evaporation time can be reduced and thereby the throughput can be increased in an in-line situation
- Alternatively, for a given process time, growing a thinner layer gives more time for mobility of atoms and reduces the risk for nucleation of new grains instead of growing on top of old grains, which may enhance the CIGS quality as compared to thicker CIGS layers.

A reasonable goal would be to have CIGS layers between 0.5 and 1 μm in thickness. However, already at 1 μm thickness there is considerable loss in efficiency [6]. The main reasons for this efficiency loss are losses in absorption, markedly in the IR region of the spectrum and increased back surface recombination, since a higher fraction of the charge carriers are generated close to the back contact. The losses are thereby found in all parameters. For thinner layers than 1 μm the recombination losses will be even more pronounced and the efficiency decreases rapidly.

In order to reduce back surface recombination it is a viable method to introduce a secondary electric drift field by increasing the Ga/(In+Ga) ratio at the back contact [6, 27]. Higher bandgap at the back contact may, however, lead to increased optical losses, since it reduces the absorption of radiation in the near IR range. Since the absorption coefficient is lower for near IR than for visible and UV light, a higher fraction of IR light will be absorbed further into the CIGS layer or not at all if the layer is too thin or if the Ga concentration increases too early. For this reason the layer with an increase in Ga concentration should be moderate and limited in thickness for thin CIGS absorbers.

Also for CIGS layers with long minority diffusion lengths the beneficial effect of an increased Ga concentration very close to the back contact has been observed [28]. For lower diffusion lengths, however, a continuous Ga grading would be more appropriate.

The strategy for reducing the optical losses for thin CIGS layers is by improving the reflective properties of the back contact, especially in the infrared region, where the losses are larger than in the visible. Materials that have been evaluated for increased reflectivity are nitrides of titanium [29] and zirconium [30].

The electrical properties of alternative nitride back contacts are worse than Mo, thus reducing the voltage and fill factor more than what is gained in the short circuit current by increased optical reflection. By adding a very thin Mo layer good electrical properties can be established, but in order to have the beneficial optical effect of the nitride, this Mo layer needs to be fully selenized before or during the evaporation and thereby transparent in the IR.

Even though the losses can be reduced, especially the electrical losses, the optical losses can not be reduced to zero. In an industrial perspective, the gain in process speed and material cost must be large enough to motivate a reduced efficiency.

2.7. Bandgap grading at the front contact

The so called notch which exists in CIGS samples produced with the three stage process is an example of double Ga grading. The gallium concentration increases both at the back and at the front of the CIGS layer and exhibits a minimum bandgap somewhere in the bulk of the CIGS layer. This structure is spontaneously formed. Since best devices are produced using the three stage process it is easy to draw the conclusion that adding a front Ga-rich layer is beneficial for device performance. Similar gradings as obtained by the three stage process, but using coevaporation of all elements in copper-poor conditions and intentionally varying the gallium and indium relative concentrations were made by Dullweber et al [31] with good cell results. Our experience is that a small beneficial effect can be observed, but that the process window is very narrow. A layer with a thickness well below 100 nm and a Ga/(Ga+In) ratio of the order of 0.6 gave a small beneficial effect on the voltage as compared to reference without the front grading. However, if the thickness of the front layer was increased above 100 nm a strong effect of voltage dependent current collection was observed, which we assigned to a blocking effect at the junction and which deteriorated the current as well as the fill factor. This is in agreement with [27], which came to the same conclusion by modelling.

2.8. Bandgap grading at the front contact

The beneficial effects on Na have been investigated in different ways. Some important results are summarized below
- Na leads to larger grain size for Boeing-like Cu-rich/Cu-poor processes
- Na has less influence on the grain size for three stage process
- Na resides mainly in the grain boundaries
- Na reduces the interdiffusion of gallium and indium.
- Na leads to higher net carrier density and thereby to higher cell voltage

Na in CIGS coevaporation processes is normally provided from a soda-lime glass substrate. It was early found that the beneficial effect of using soda-lime glass substrates could be assigned to the supply of sodium [32]. To have an influence on the CIGS growth, the Na needs to be released from the glass substrate and diffused through the Mo layer in sufficient amounts. For normal soda-lime glass, the Na supply at the CIGS growth temperature is not a limiting factor, but it sets some requirements on the Mo process. If the Mo layer is very dense and/or thick the Na supply may be limited. It is not so easy to discern Na-supply related growth problems from other growth issues, but a rough CIGS surface for a Boeing-like process may be an indication. If there are pinholes in an otherwise Na-tight Mo layer, Na may diffuse in the pinholes and bright areas around the pinhole will be a signal of Na-related problems.

For the three stage process, the influence on the grain size is less important. It is possible to grow large grains without Na [33]. The electrical beneficial effect of Na can be obtained by incorporation of Na in a post-deposition process [34]. Also for Na-free substrates or for soda-lime glass with diffusion barriers for sodium, similar or in some cases even better results can be obtained by using an Na-precursor layer, either deposited before the CIGS process, in the CIGS process or after the CIGS process as described above. The Na layer can consists of various types of Na salts, such as NaF, NaCl, Na_2S or Na_2Se or even thin layers of soda-lime glass. Mostly used is NaF, which has the main advantage of evaporating congruently.

When investigating Na concentration in coevaporated CIGS on soda lime glass, a high level of Na is detected, even up to fractions of percent. This Na is predominantly found in the grain boundaries [35]. For extrinsic dopants in a semiconductor, concentrations at the ppm or ppb levels are normally enough. It is therefore difficult to argue that the grains are completely Na-free.

In experiments with bilayers of CIS and CGS, evaporated using constant evaporation rates, Na has been observed to reduce interdiffusion of indium and gallium [36] and to lead to less recrystallization.

The electrical effect of Na is increased net carrier doping as compared to Na-free CIGS. The increased doping leads in its turn to higher voltage, higher fill factor, but also slightly lower current density has sometimes been observed. All these effects can be explained by higher effective doping density. Since the electrical effect also can be obtained by the post-deposition treatment, it is safe to conclude that the beneficial effect of Na is not from beneficial effects of grain growth only.

3. ALTERNATIVE BUFFER LAYERS

Over the last 20 years there has been substantial effort in trying to find a replacement for the standard CdS buffer layer by chemical bath deposition, CBD. These efforts have been reviewed by Siebentritt [37], Hariskos et al [38] and recently by Naghavi et al [39]. The aim is both to increase the efficiency of the device by reducing absorption losses in the buffer and to avoid toxic cadmium in devices and during processing. There is also potential for improvements of the production flow by replacing the CBD process by a vacuum-based method since all other processing steps (except for scribing of modules) are performed in vacuum. Among the different materials that have been investigated as alternatives, three have emerged as main candidates. These are Zn(S,O,OH), In_2S_3 and (Zn,Mg)O, treated in detail below. For each of these materials, buffer layer properties, band alignment, device performance and industrial efforts are covered. Before turning to these three materials, the role of the buffer layer is briefly discussed followed by an overview of buffer layer deposition methods. The chapter ends with an overview of metastable behaviours of devices with alternative buffer layers.

3.1. Role of the Buffer Layer - why is CBD-CdS so Good?

One advantage of heterojunction solar cells (pn-junction between different semiconductors) as compared to homojunctions (pn-junction with the same semiconductor on both sides) is that recombination in the wide gap semiconductor is small due to the large band gap. On the other hand, the risk for interface recombination is much larger in heterojunctions due to defects and imperfections at the interface. The role of the buffer layer, in combination with the other window layers, is to minimize interface recombination and provide as large total band bending across the junction as possible. In the early days of CIGS research, the device was made with a thick layer of CdS deposited by evaporation. A large improvement was seen when most of the CdS was replaced by ZnO, and evaporation was replaced by chemical bath deposition. The main reason for the improvement using thin CdS was reduced absorption in the window since the band gap of ZnO (3.3 eV) is larger than that of CdS (2.4 eV). The use of CBD instead of evaporation was also clearly beneficial [40]. Attempts to completely remove the CdS and sputter deposit ZnO directly onto CIGS have failed. Several reasons for the role of CBD-CdS have been suggested such as protection of the CIGS against sputter damage, CIGS surface etching in the CBD, cadmium indiffusion, passivation by sulfur, good lattice match and good conduction band alignment. A number of alternative materials and deposition methods have been investigated in order to replace the CdS. Some of the results are shown in table 1.

One option that was early recognized was to increase the band gap of CdS by inclusion of ZnS. Linear increase of the band gap from 2.4 eV for CdS to 3.6 eV for ZnS has been shown [41]. This would also lead to a more optimal lattice match between CIGS and (Zn,Cd)S. Indeed, efficiencies of up to 19.5% have been achieved using (Zn,Cd)S buffer layer deposited by CBD [42]. The Cd:Zn ratio in that case was 80:20, corresponding to a band gap of 2.6 eV [41]. Even higher Zn content would be desirable to minimize the absorption in the buffer. However, this could possibly lead to the formation of a spike-like conduction band alignment, as in the case of pure ZnS, that would deteriorate devices.

Table 1. Efficiencies of devices and modules with alternative buffer layers and the current world record CdS device. The CIGS layers are made by different processes and the composition close to the interface to the buffer layer is not equal in all cases. [a] active area, [b] with antireflective coating, [c] MOCVD ZnO:B window

Cells	Buffer material	Deposition method	Efficiency (%)	Reference
	Zn(S,O,OH)	CBD	18.6[b]	[9]
	Zn(Se,OH)	CBD	13.7[b]	[10]
	In(OH,S)	CBD	15.7[a,b]	[11]
	CdZnS	CBD	19.5	[8]
	In_2S_3	ALD	16.4	[12]
	Zn(O,S)	ALD	18.5[b]	[13]
	ZnO	ALD	13.9	[14]
	(Zn,Mg)O	ALD	18.1[b]	[15]
	ZnS	Ilgar	14.2	[16]
	ZnS:In_2S_3	Ilgar	15.3	[17]
	ZnInSe	evaporation	15.1	[18]
	InS	evaporation	15.2[b]	[19]
	(Zn,Mg)O	sputtering	12.5	[20]
	InS	sputtering	13.3	[21]
	ZnO	sputtering	15.0	[22]
	CdS	CBD	19.9[b]	[23]
Modules Area (cm^2)				
855	Zn(S,O,OH)	CBD	15.2[c]	[24]
900	Zn(S,O,OH)	CBD	12.5	[25]
62.7	Zn(S,O,OH)/(Zn,Mg)O	CBD/sputtering	15.2	[26]
77	Zn(S,O,OH)	ALD	14.7	[13]
714	In_2S_3	ALD	12.9	[27]
100	In_2S_3	Ilgar	12.4	[28]

3.2. Overview of Buffer Layer Deposition Techniques

A large number of materials and deposition methods have been investigated as alternatives to the standard CBD-CdS. In addition to the ones shown in table 1, processes such as metal-organic chemical vapour deposition, MOCVD [61], ultrasonic spray pyrolysis [62], metal-organic vapour phase epitaxy, MOVPE [63], electro-deposition [64] and sputtering in combination with partial electrolyte treatments of the absorber [65] have been investigated. In addition to the materials listed in table 1 also In(OH)$_3$ [66], SnO$_2$, SnS$_2$ [67], TiO$_2$, Ta$_2$O$_5$ and Al$_2$O$_3$ [68] have been considered. From an industrial point of view, the ideal process would be fast with high materials yield and low cost of equipment and source material. The requirements for forming a good interface, leading to high efficiency, are sufficient coverage of the CIGS surface, minimization of recombination centres at the interface, proper band alignment assuring inversion of the CIGS surface without creating energetic barriers at the interface and finally high resistivity layer to minimize the influence of non-uniformities in the CIGS and pn-junction. All of these requirements are heavily dependant on the properties of the CIGS such as surface roughness, composition and uniformity. These properties will in turn depend on the process used for CIGS deposition, why buffer layer optimization is closely connected to absorber optimization. One exception is CBD-CdS for which the standard process works surprisingly well for a large variety of CIGS absorbers. Fine-tuning of the process for specific absorbers can improve efficiency, but the robustness and tolerance to absorber variations is clearly one of the benefits of the CBD-CdS buffer layer.

Two processes with advantages due to industrial maturity and fast processing are sputtering and evaporation. Both processes are vacuum-based techniques. In the case of evaporation, the material can be evaporated from compounds or co-evaporated from elemental sources. Sputtering can be performed using R.F or pulsed D.C sputtering from compound targets or reactive sputtering using metal targets. The main challenge when using sputtering for buffer layer deposition is damage of the CIGS by energetic ions. Conformal coverage of rough surfaces is difficult in particular for evaporation but also using sputtering. Despite these difficulties, relative high efficiencies have been obtained using sputtering of ZnO, (Zn,Mg)O and In_2S_3. In the case of evaporation, high efficiency devices have been obtained for ZnInSe and In_2S_3 (table 1).

Figure 7. Transmission electron micrograph of a cross section of a CIGS/ALD-ZnO/sputtered ZnO stack. The deep trench in the CIGS is perfectly covered by the ALD buffer layer while the sputtered ZnO does not penetrate the trench.

The most studied buffer layer processes that are able to provide conformal coverage and high quality interface formation at low deposition temperature are CBD, atomic layer deposition, ALD, and Ion-layer gas reaction, Ilgar. The CBD process is based on aqueous solutions of the precursor chemicals. These solutions are mixed and heated to the reaction temperature which is typically below 100 °C. When introducing the substrates into the solution, film deposition occurs on all surfaces. In some cases the reaction can be self-terminating, but in most cases the reaction is stopped by removing the substrate from the solution and rinsing in water or ammonia solution. In the case of CdS deposition, the solution is basic due to the use of ammonia as complexing agent [69]. When the substrate is immersed into the ammonia containing solution, surface oxides on the CIGS are etched.

ALD is a vacuum-based method similar to chemical vapour deposition, CVD [70]. Precursors in the gas-phase are introduced sequentially into a reactor that is constantly purged with inert gas. The reactions that occur on the surface of the substrate are ideally surface terminated, resulting in very good step coverage and thickness control. The desired thickness is obtained by repeating the process cycles. Buffer layer deposition is typically performed at 100-200 °C. In the case of the Ilgar technique, the most promising results have been obtained using spray-Ilgar [71]. In this process a solution containing one of the precursors is sprayed onto the substrate using an inert gas as carrier. The sprayed layer is then converted into a film by reaction with a precursor such as H_2S gas.

3.3. Zn(S,O,OH)

ZnS and ZnO are II-VI semiconductors with wide bandgap, 3.8 and 3.3 eV respectively. The most successful buffer layers from this group of materials contain oxygen and sulphur in combination with hydrogen in the form of hydroxides. In the case of ALD Zn(O,S), the composition can easily be varied from ZnO to ZnS by varying the pulsing sequence of the precursors. The structure of these films range from hexagonal ZnO structure with increasing unit cell for the O-rich side (S/Zn<0.5) followed by an amorphous or nano-crystalline region for S/Zn≈0.7 and a hexagonal (and cubic) ZnS structure for S/Zn>0.8 as shown in figure 8. In the case of CBD films, the composition is harder to control.

The band gap of $ZnO_{1-x}S_x$ exhibits a large bowing with a minimum band gap of 2.6 eV for x=0.5 as shown for films deposited by pulsed laser deposition and ALD [72,73]. In the case of multiphase systems including for example amorphous $Zn(OH)_2$, or segregated ZnS or ZnO, the buffer layer will be characterized by more than a single band gap.

Since ZnO is already a part of the CIGS device structure, the possibility of applying it directly onto the absorber by various techniques has been widely studied. However, due to the relative position of the conduction bands of CIGS and ZnO, increased interface recombination is expected even for deposition of ZnO using soft deposition techniques such as CBD and ALD.

The situation can be improved by using a Ga-free surface, since the conduction band of CIGS is lowered by decreasing the Ga content. The addition of sulphur at the interface is another possibility to reduce the risk for interface recombination. This is due to lowering of the valence band for $CuInS_2$ as compared to $CuInSe_2$, decreasing the concentration of holes at the interface. Passivation of trap states in the CIGS by sulphur has also been suggested [74] and could be beneficial at the interface. In table 2, relative positions of valence and conduction bands in the Cu(In,Ga)Se$_2$ and (Zn,Mg)(O,S) systems are given.

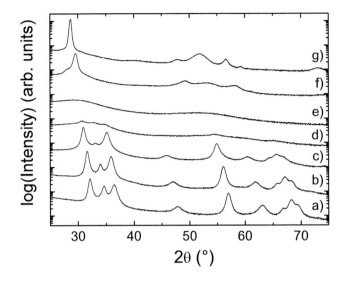

Figure 8. Grazing incidence diffractograms of ALD ZnO$_{1-x}$S$_x$ films: a) ZnO, b) x=0.10, c) x=0.28, d) x=0.48, e) x=0.71, f) x=0.84 and g) x=0.97.

Table 2. Measured VBO relative to CIGS (30% Ga), buffer layer band gap and calculated CBO for various buffer layers. ZnO, ZnS and Zn(O,S) values are based on measurements using ultraviolet and x-ray photoelectron spectroscopy, UPS/XPS, coupled to an ALD reactor. (Zn,Mg)O values are based on UPS/XPS measurements of the CdS/sputtered (Zn,Mg)O interface and CIGS/sputtered ZnMgO interface

buffer	S or Mg content	VBO (eV)	Eg (buffer) (eV)	CBO (eV)	Ref
ZnO		-2.3	3.3	-0.2	[43]
ZnO/CIS		-2.2	3.3	0.1	[43]
ZnO$_{1-x}$S$_x$	x=0.3	-1.8	2.7	-0.2	[44]
ZnO$_{1-x}$S$_x$	x=0.7	-1.7	3.0	0.2	[40]
ZnO$_{1-x}$S$_x$	x=0.8-0.9	-1.3	3.1	0.7	[40]
ZnS		-1.2	3.6	1.2	[40]
Zn$_{1-x}$Mg$_x$O	x=0.15, 0.17	-2.3, -2.2	3.6	0.2, 0.3	[45, 46]

The conduction band offset, CBO, is negative (i.e buffer layer conduction band below that of the CIGS) for ZnO and ZnO$_{0.7}$S$_{0.3}$. For higher sulfur contents the position of the buffer layer conduction band increases giving a small positive value of the CBO for ZnO$_{0.3}$S$_{0.7}$ and large CBO of 1.2 eV for ZnS. For Zn$_{0.8}$Mg$_{0.2}$O a small positive CBO is obtained. These values are consistent with device results with losses in open circuit voltage for negative CBO, high performance for small positive CBO and blocked photo-current for large positive CBO [72,79,80].

Devices with efficiencies above 18% have been obtained using CBD-Zn(S,O,OH) [43], CBD-Zn(S,O,OH) in combination with sputtered (Zn,Mg)O [58] and ALD-Zn(O,S) [12] (see table 3). In all these cases the devices were optimized using an antireflective coating. The reference devices with CdS buffer layer did not have this AR coating, why direct comparison of performance is difficult. In figure 9, IV and QE characteristics of devices with identical CIGS material but in one case CBD-CdS and the other ALD-Zn(O,S) buffer layer are compared. None of the devices contain AR coating. It can be observed that the gain in J_{sc} can be attributed to the blue wavelength region only, corresponding to reduced absorption in the buffer layer. There is a gain in V_{oc} and a loss in FF for the Zn(O,S) cell as compared to CdS. These results indicate that an overall gain in efficiency could be expected from replacement of CdS with Zn(O,S). However, attempts to surpass the world record efficiency by using CBD-Zn(S,O,OH) on CIGS by the three-stage process from NREL have so far failed.

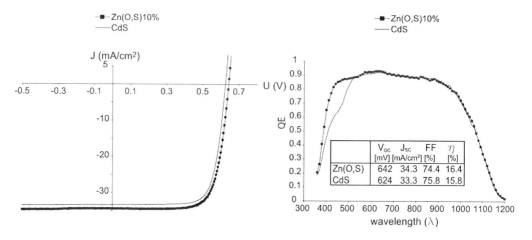

Figure 9. Current-voltage and quantum efficiency characteristics of device with ALD Zn(O,S) buffer layer compared to reference CdS device.

Table 3. Summary of best devices and modules obtained with Zn(S,O,OH) buffer layers by different deposition techniques for co-evaporated CIGSe and sputtered and post treated CIGSSe absorbers. * MOCVD window, ** sputtered (Zn,Mg)O window

Process	Module area (cm^2)	Efficiency (%)	Absorber	Ref
CBD		18.6	CIGSe	[43]
		14.2	CIGSSe	[57]
		18.0**	CIGSe	[58]
	855	15.2 module*	CIGSSe	[56]
	900	12.5 module*	CIGSSe	[82]
	63	15.8 module**	CIGSe	[83]
Ilgar		14.2	CIGSSe	[49]
evaporation		9.1	CIGSe	[84]
ALD		18.5	CIGSe	[12]

In industry, CBD-Zn(S,O,OH) has been used for large area modules by Showa Shell for many years. Their buffer layer is less than 10 nm thick [81] and the device is completed by deposition of i-ZnO and ZnO:B by MOCVD in order not to damage the buffer layer. Recently an efficiency of 15.2% was obtained for a large area module [56].

3.4. In$_2$S$_3$

β-In$_2$S$_3$ has a spinel-like crystalline structure, which can act as a host for impurity elements. Different values have been reported for the band gap such as direct band gap of 2.7-2.8 eV [85,86], direct band gap of 2.0 eV [87] and indirect band gap of 2.1 eV [88]. One reason for the large scatter in reported band gap could be the strong influence of impurities such as sodium diffusing from the substrate, or oxygen contamination. In the case of deposition on CIGS, diffusion of Cu into the buffer layer has been reported by several authors (see for example refs in [60]). For example it has been reported that Na$_x$Cu$_{1-x}$In$_5$S$_8$ is formed at the interface between CIGS and In$_2$S$_3$ [89]. This phase has a band gap of 1.5 eV for x=0 and 2.4 eV for x=1. Inclusion of oxygen causes an increase of the band gap to 2.9 eV for In$_2$S$_{3-3x}$O$_{3x}$ with x=0.14. In$_2$O$_3$ has a direct band gap of 3.75 eV [90], but a smaller indirect band gap of 2.1-2.7 eV has also been determined experimentally but questioned based on calculations [91]. For In(S,OH) buffer layers deposited by CBD, band gap values ranging from 2.0 to 3.7 eV have been reported depending on sulphur to oxygen composition [92].

Due to the observed interdiffusion at CIGS/In$_2$S$_3$ interfaces, the interface formation and device performance is very process dependant. The band alignment at such an interface is also complicated and could vary for different buffer deposition processes or absorber surface properties. Conclusions on the band alignment were reported based on the dominating recombination path in devices with varying Ga content in the absorber and evaporated In$_2$S$_3$. [93,94] In one case, interface recombination was found to dominate for Ga/In+Ga > 0.1, while in the other case devices were reported to be dominated by recombination in the space charge region. Pronounced interface recombination could possibly be due to a cliff-like band alignment at the CIGS/buffer interface. Direct measurements of the band alignment have been performed [88], however, since interdiffusion at the interface is expected to different extent depending on absorber and buffer deposition process, the band alignment is expected to vary in these different cases.

Table 4. Summary of best devices and modules obtained with indium sulfide buffer layers by different deposition techniques for co-evaporated CIGSe and sputtered and post treated CIGSSe absorbers. *with AR coating and active area

Process	Module area (cm^2)	Efficiency (%)	Absorber	Ref
ALD		16.4	CIGSe	[46]
	714	12.9 module	CIGSe	[59]
CBD	11208	11.2 module	CIGSe	[95]
		15.7 *	CIGSe	[45]
Ilgar		14.7	CIGSSe	[71]
	100	12.4 module	CIGSSe	[60]
Evaporation		15.2	CIGSe	[52]
		12.0 module	CIGSSe	[96]
Sputtering		13.3	CIGSe	[54]
	48	11.2	CIGSe	[54]

Indium sulfide buffer layers have been deposited by ALD, CBD, Ilgar, sputtering and evaporation as shown in table 4. A highest efficiency of 16.4% was obtained by ALD for a 30 nm thick buffer layer deposited at 220 °C. In many cases, devices with indium sulphide buffer layers show comparable or improved V_{oc} and FF as compared to CdS references. The J_{sc} is often reduced however due to losses both in the short and long wavelength regions. Industrially, CBD-InS(O,OH) is used by Honda Soltec in Japan with reported module efficiency of 11.2% [95].

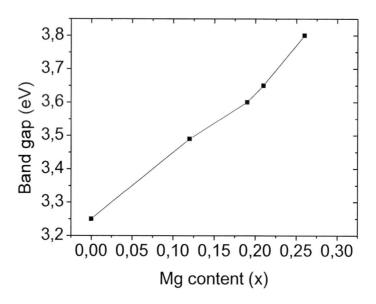

Figure 10. Band gap of ALD-$Zn_{1-x}Mg_x$ film as a function of magnesium content.

3.5. (Zn,Mg)O

The band gap of ZnO can be increased by addition of Mg. By deposition of $Zn_{1-x}Mg_xO$ by pulsed laser deposition, linear increase of the bandgap was shown from 3.3 eV for ZnO until phase segregation of MgO for x=0.33, where the band gap was 4.0 eV [97]. The use of $Zn_{1-x}Mg_xO$ as buffer layer in CIGS was proposed by Minemoto et al [98]. For $Zn_{1-x}Mg_xO$ deposited by ALD, single phase wurtzite (Zn,Mg)O was obtained for x<0.2 [79] for deposition at 150 °C. The variation in band gap for ALD $Zn_{1-x}Mg_xO$ is shown in figure 10.

The band alignment at the CIGS/ $Zn_{1-x}Mg_xO$ interface can be estimated from measurements of the valence band alignment at the CdS/ $Zn_{1-x}Mg_xO$ interface [77]. From those results, the position of the valence band is expected to be unchanged at least for x< 0.15. Measurements of the CIGS/sputtered $Zn_{1-x}Mg_xO$ interface [78] indicate an increase of 0.1 eV in the position of the valence band for x=0.17. This implies that the conduction band is raised relative to that of CIGS for increasing Mg content, resulting in optimum band alignment for x=0.15-0.2.

$Zn_{1-x}Mg_xO$ deposited by sputtering and ALD has been investigated as buffer layer in CIGS devices (table 5). In the case of sputtering, absorbers with sulphur-containing and Ga-free interface have been shown compatible with relatively high efficiency without use of intermediate layers. For co-evaporated CIGS, where the surface contains Ga but no S, the use of intermediate layers or interface treatments appears necessary in order not to compromise efficiency when using sputtering of $Zn_{1-x}Mg_xO$. Efficiencies of over 18% were obtained using ALD $Zn_{1-x}Mg_xO$ without the sputtered i-ZnO layer [48]. The devices show a very large gain in J_{sc} as compared to CdS references, but a slight loss in V_{oc} and FF. In the case of devices with CBD-Zn(S,O,OH)/sputtered $Zn_{1-x}Mg_xO$ buffer combination, V_{oc} and FF of the best devices was comparable to CdS references while the J_{sc} was improved [58].

Table 5. Summary of best devices and modules obtained with $Zn_{1-x}Mg_xO$ buffer layers by different deposition techniques for co-evaporated CIGSe and sputtered and post treated CIGSSe absorbers. *with AR coating and active area

Process	Efficiency (%)	absorber	Ref
Sputtering	13.2	CIGSe/Cd	[78]
	12.5	CIGSSe	[53]
	7.3	CIGSe	[99]
	18.0 *	CIGSe/Zn(O,S)	[58]
ALD	18.1 *	CIGSe	[48]

3.6. Metastabilities

CIGS devices can exhibit changes in performance as a function of annealing, biasing, illumination or exposure to humidity. Some of the effects are reversible, or metastable, while others are irreversible. Degradation of devices in humidity is irreversible why commercial CIGS modules require sufficient encapsulation including edge sealing. Another effect that is non-reversible, at least in the case of CdS buffers, is the improvement of devices after a few minutes of annealing in air at around 200 °C. Reversible effects can be related to illumination and biasing where improvement of devices is seen after illumination with white light and degradation after reverse biasing and illumination with red light. For state of the art devices with CdS buffer layer, the positive light-soaking effect is so fast that it is hardly noticeable in standard measurements. For many devices with alternative buffer layers however, the improvements are slow, on the order of minutes or even up to hours. This must be avoided in commercial modules since system dimensioning requires predictable output. Below, an overview of metastable effects in devices with Zn(S,O,OH), In_2S_3 and $Zn_{1-x}Mg_xO$ buffers is given, followed by a brief discussion of possible mechanisms for the effects.

3.6.1. Zn(S,O,OH)

Large transient effects have been reported for devices with Zn(S,O,OH) buffers deposited by CBD [100]. These devices show improvement in FF after light soaking and relaxation after storage in the dark. A repeated light-soaking treatment at elevated temperature of 100-140 °C was reported to stabilise the device performance at the optimized state [101]. Another way to improve the stability is to use a (Zn,Mg)O layer on top of the CBD-Zn(S,O,OH) buffer instead of i-ZnO [83,98]. The stability depends on the Mg content. Encapsulated modules with CBD-Zn(S,O,OH)/ (Zn,Mg)O layers showed degradation after 1000 h damp-heat testing and the initial performance was restored after less than 30 minutes of lightsoaking at 80 °C. Devices with ALD-Zn(S,O,OH) buffer layers normally do not exhibit transient behaviour as-deposited and encapsulated modules have shown stable performance [102]. However, after dark storage of about 2-6 months, FF degradation was observed on unencapsulated devices [103]. The performance could be restored to the initial level after light soaking at about 100 °C. Light soaking at room temperature or annealing at 100 °C in the dark did not improve the FF.

3.6.2. In$_2$S$_3$

Devices with indium sulfide buffer layers have been reported to be stable, without the transient effect observed with other alternative buffer layers [52]. Encapsulated modules with ALD-In$_2$S$_3$ [59] and 10x10 cm^2 modules with sputtered In$_2$S$_3$ [54] have passed damp heat test. One exception is reports on metastabilities using CBD-In(OH$_x$,S$_y$) buffer layers [104], where the observed hysteresis in current-voltage characteristics was related to defects at the buffer/ZnO interface [105].

3.6.3. Zn$_{1-x}$Mg$_x$O

As mentioned above for the CBD-Zn(S,O,OH)/ (Zn,Mg)O buffer combination [83], and equally observed for devices with ALD Zn$_{1-x}$Mg$_x$O [106] or sputtered Zn$_{1-x}$Mg$_x$O [107], the lightsoaking effect is strongly dependant on Mg content. This has been correlated with the band alignment at the buffer/absorber interface showing that a small spike is preferred for stable devices. However, even if devices with the optimum Mg content are much more stable than for other compositions, improvements are still observed after short lightsoaking treatments. It could be worth mentioning that devices with ZnO devices by ALD show a slightly different behaviour. In this case, annealing in air at 200 °C causes a slow and reversible improvement in V$_{oc}$. The same treatment can also result in a reversible removal of severe shunts. Reduction of the light-soaking metastability has been shown for increased resistivity of ALD-ZnO buffered devices [47].

As seen above, the main effect of light-soaking in devices with Zn(S,O,OH) or Zn$_{1-x}$Mg$_x$O buffer layers is in the FF. The observed low FF in relaxed devices and increasing FF with illumination indicates the existence of a barrier for carriers, and that this barrier can be reduced or eliminated by illumination. Two possible origins of this barrier have been suggested. The first is in the surface region of the CIGS [108,109] and the second due to conduction band offsets at the CIGS/buffer interface [110,111]. A barrier could also be formed by acceptor states at the buffer/i-ZnO interface as suggested for CBD-In(OH,S)/ i-ZnO [105]. In the first model, pinning of the Fermi level close to the conduction band at the CIGS surface is suggested to cause the (V$_{Cu}$ + V$_{Se}$) defect complex to be charged 2- due to trapping of electrons [112]. This causes a formation of a p+ layer close to the interface that can act as a barrier for electrons generated in the bulk. On illumination, the negative charges can be reduced by photo generated holes in the surface region or holes injected from the buffer layer.

In the second model, the barrier is caused by a spike-like conduction band alignment at the absorber/buffer interface. Illumination does not change the CBO itself, but photo-doping of the buffer layer, so called persistent photo conductivity (PPC), can lower the conduction band of the buffer layer relative to the position in the bulk of the absorber. This reduces the effective barrier seen by electrons in the absorber [111].

It is of course possible that both models are valid and that the overall effect is due to a combination of the two. Depending on materials and deposition processes one of the effects could be dominating. For example, in the case of $Zn_{1-x}Mg_xO$ buffer with large Mg content, it is probable that the huge LS improvement can be related to the CBO at the buffer/absorber interface and PPC in $Zn_{1-x}Mg_xO$ [113] while the small effect seen for optimum Mg content could be related to the CIGS surface layer.

The reason for the absence of large transient effects in devices with In_2S_3 buffer layers could be due to non-existing barriers at the two buffer interfaces or different defect distribution in the CIGS surface region. The latter could possibly be caused by the observed intermixing at $CIGS/In_2S_3$ interfaces with In diffusing into the absorber and Cu and Na diffusing into the buffer. One way of avoiding barriers at the buffer/ZnO interface is to apply the ZnO:Al layer directly onto the buffer. Since the ZnO:Al is degenerately doped, this can also pin the conduction band of the buffer layer closer to the Fermi level and thereby reduce the effective barrier for electrons. Since the role of the i-ZnO is to decrease the influence of shunts and inhomogenieties, the removal of the i-ZnO requires a sufficiently thick and resistive buffer layer.

REFERENCES

[1] Shafarman, W. and L. Stolt, *Cu(In,Ga)Se2 solar cells*, in *Handbook of photovoltaic science and engineering*, A. Luque and S. Hegedus, Editors. 2003, Wiley: Chichester.

[2] Klenk, R. and M. Lux-Steiner, *Chalcopyrite based solar cells*, in *Thin film solar cells: Fabrication, characterization and applications*, J. Poortmans and V. Arkhipov, Editors. 2006, Wiley: Chippenham.

[3] Shafarman, W. and J. Zhu, *Thin Solid Films* 2000; 361-2: 473-477.

[4] Mickelsen, R. and W. Chen. *Proceedings of the 16th IEEE Photovoltaic Specialist Conf.* 1982. p. 781-785

[5] Repins, I., M. Contreras, B. Egaas, C. DeHart, J. Scharf, C. Perkins, B. To, and R. Noufi, 19.9%-efficient ZnO/CdS/CuInGaSe2 solar cell with 81.2% fill factor *Prog. Photovoltaics* 2008; 16: 235.

[6] Lundberg, O., M. Bodegård, J. Malmström, and L. Stolt, Influence of the Cu(In,Ga)Se2 thickness and Ga grading on solar cell performance *Progress in Photovoltaics: Research and application* 2003; 11: 77-88.

[7] Harsha, K.S.S., *Chapter 05 Thermal evaporation sources*, in *Principles of Vapor Deposition of Thin Films*. 2006, Elsevier.

[8] Kawamura, M., T. Fujita, A. Yamada, and M. Konagai, Cu(In,Ga)Se2 thin film solar cells grown with cracked selenium *JOurnal of crystal growth* 2009; 311: 753-756.

[9] Ishizuka, S., H. Shibata, A. Yamada, P. Fons, K. Sakurai, K. Matsubara, and S. Niki, Growth of polycrystalline Cu(In,Ga)Se2 thin films using radio frequency-cracked Se-radical beam source and application for photovoltaic devices *Appl Phys Lett* 2007; 91: 041902.

[10] Stolt, L., J. Kessler, M. Ruckh, K. Velthaus, and H.W. Schock, ZnO/CdS/CuInSe2 thin film solar cells with improved performance *Appl Phys Lett* 1993; 62: 597-9.

[11] Kessler, J., M. Bodegård, J. Hedström, and L. Stolt, New World Record Cu(In,Ga)Se2 based mini-module: 16,6%. *Proceedings of the 16th Euro Conf Photovoltaic Solar Energy Conv.* 2000. Glasgow. p. 775-778

[12] Zimmermann, U., M. Ruth, and M. Edoff, Cadmium-free CIGS mini-modules with ALD-grown Zn(O,S)-based buffer layers. *Proceedings of the 21st European Photovoltaic Solar Energy Conference.* 2006. Dresden, Germany. p. 1831

[13] Hergert, F., S. Jost, R. Hock, and M. Purwins, A crystallographic description of experimentally identified formation reactions of Cu(In,Ga)Se2 *Journal of Solid State Chemistry* 2006; 179: 2394-2415.

[14] Contreras, M., B. Egaas, D. King, A. Swartzlander, and T. Dullweber, Texture manipulation of CuInSe2 thin films *Thin Solid Films* 2000; 361-362: 167-171.

[15] Noufi, R., Y. Yan, K.M. Jones, M. Al-Jassim, B. Keyes, J. Alleman, K. Ramanathan, and J. Abu-Shama, Investigation of the microstructure of Cu(In,Ga)se2 thin films used in high-efficiency devices. *Proceedings of the 29th IEEE PV Specialists Conf.* 2002. New Orleans, Lousiana

[16] Wada, T., Microstructural characterization of high-efficiency Cu(In,Ga)Se2 solar cells *Solar Energy Materials & Solar Cells* 1997; 49: 249-260.

[17] Jackson, P., R. Wurz, U. Rau, J. Mattheis, M. Kurth, T. Schlötzer, G. Bilger, and J. Werner, High quality baseline for high efficiency Cu(In,Ga)Se2 solar cells *Progress in Photovoltaics: Research and application* 2007; 15: 507-519.

[18] Powalla, M., G. Voorwinden, D. Hariskos, P. Jackson, and R. Kniese, Highly efficient CIS solar cells and modules made by yje co-evaporation process *Thin Solid Films* 2009; 517: 2111-2114.

[19] Stolt, L., J. Hedström, and D. Sigurd, Coevaporation with rate control system based on quadrupole mass spectrometer *Journal of Vacuum Science and Technology* 1985; 3: 403-407.

[20] Nishitani, M., T. Negami, and T. Wada, Composition monitoring method in CuInSe2 thin film preparation *Thin Solid Films* 1995; 258: 313.

[21] Repins, I., D. Fisher, W. Batchelor, L. Woods, and M. Beck, A non-contact low-cost sensor for improved repeatability in co-evaporated CIGS *Progress in Photovoltaics: Research and application* 2005; 13: 311-23.

[22] Godecke, T., T. Haalboom, and F. Ernst, *Z. Metallkd* 2000; 91: 622-634.

[23] Kessler, J., J. Schöldström, and L. Stolt, Analysis of CIGS films and devices resulting from different Cu-rich to Cu-poor transitions. *Proceedings of the 17th EPVSEC.* 2001. Munich

[24] Scheer, R., A. Neisser, K. Sakurai, P. Fons, and S. Niki, CuInGaSe2 growth studies by in situ spectroscopic light scattering *Appl Phys Lett* 2003; 82: 13.

[25] Sakurai, K., R. Scheer, S. Nakamura, Y. Kimura, T. Baba, C. Kaufmann, A. Neisser, S. Ishizuka, A. Yamada, K. Matsubara, K. Iwata, P. Fons, H. Nakanishi, and S. Niki, Structural changes of CIGS during deposition investigated by spectroscopic light scattering: A study on Ga concentration and Se pressure *Solar Energy Materials & Solar Cells* 2006; 90: 3377-3384.

[26] Bodegård, M., O. Lundberg, J. Lu, and L. Stolt, Re-crystallisation and interdiffusion in CGA/CIS bilayers *Thin Solid Films* 2003; 431-432: 46-52.

[27] Gloecker, M. and J.R. Sites, Band-gap grading in Cu(In,Ga)Se2 solar cells *Journal of Physics and chemistry of solids* 2005; 66: 1891-1894.

[28] Bodegård, M., O. Lundberg, J. Malmström, and L. Stolt, High voltage CIGS devices with Ga-profiling fabricated using co-evaporation. *Proceedings of the 28th IEEE Photovoltaic Specialists Conf.* 2000

[29] Malmström, J., O. Lundberg, and L. Stolt, Potential for light trapping in Cu(In,Ga)Se2 solar cells. *Proceedings of the 3rd World Conf on PV Energy Conversion.* 2003. Osaka, Japan

[30] Malmström, J., S. Schleussner, and L. Stolt, Enhanced back reflectance and quantum efficiency in Cu(In,Ga)Se2 thin film solar cells with a ZrN back reflector *Appl Phys Lett* 2004; 85: 2634-6.

[31] Dullweber, T., U. Rau, M. Contreras, R. Noufi, and H.W. Schock, Photogeneration and carrier recombination in graded gap Cu(In,Ga)Se2 solar cells *IEEE Transactions on electron devices* 2000; 47: 2249.

[32] Bodegård, M., L. Stolt, and J. Hedström, The influence of sodium on the grain structure of CuInSe2 films for photovoltaic applications. *Proceedings of the 12th European Photovoltaic Solar Energy Conf.* 1994. Amsterdam. p. 1743

[33] Rudmann, D., A. da Cunha, M. Kaelin, F.-J. Haug, H. Zogg, and A. Tiwari, Effects of Na on the growth of Cu(In,Ga)Se2 thin films and solar cells. *Proceedings of the Materials Research Soc* 2003. p. 53-64

[34] Rudmann, D., D. Bremaud, A. da Cunha, G. Bilger, A. Strohm, M. Kaelin, H. Zogg, and A. Tiwari, Sodium incorporation strategies for CIGS growth at different temperatures *Thin Solid Films* 2005; 480-481: 55-60.

[35] Niles, D., M. Al-Jassim, and K. Ramanathan, Direct observation of Na and O impurities at grain surfaces for CuInSe2 thin films *Journal of Vacuum Science and Technology* 1999; 17: 291-296.

[36] Lundberg, O., J. Lu, A. Rockett, M. Edoff, and L. Stolt, Diffusion of indium and gallium in Cu(In,Ga)Se2 thin film solar cells *Journal of Physics and chemistry of solids* 2003; 64: 1499-1501.

[37] Siebentritt, S., Alternative buffers for chalcopyrite solar cells *Solar Energy* 2004; 77: 767.

[38] Hariskos, D., S. Spiering, and M. Powalla, Buffer layers in Cu(In,Ga)Se2 solar cells and modules *Thin Solid Films* 2005; 480-481: 99-109.

[39] Naghavi, N., D. Abou-Ras, N. Allsop, N. Barreau, S. Buecheler, A. Ennaoui, C. Fischer, C. Guillen, D. Hariskos, J. Herrero, R. Klenk, K. Kushiya, D. Lincot, R. Menner, T. Nakada, C. Platzer-Björkman, S. Spiering, A. Tiwari, and T. Törndahl, Buffer layers and transparent conducting oxides for chalcopyrite Cu(In,Ga)(S,Se)2 based thin film photovoltaics: present status and current developments *Progress in Photovoltaics: Research and application* submitted 2009.

[40] Romeo, A., R. Gysel, S. Buzzi, D. Abou-Ras, D. Bätzner, D. Rudmann, F. Kurdesau, H. Zogg, and A. Tiwari. *Proceedings of the 14th International Photovoltaic Science and Engineering conference.* 2004. Bankok, Thailand. p. 705

[41] Kumar, V., V. Singh, S. Sharma, and T. Sharma, Structural and optical properties of sintered Cd1-xZnxS films *Otical Materials* 1998; 11: 29-34.

[42] Bhattacharya, R.N., M. Contreras, B. Egaas, and R. Noufi, High efficiency thin-film CuIn1-xGaxSe2 photovoltaic cells using a Cd1-xZnxS buffer layer *Applied physics letters* 2006; 89: 253503.

[43] Contreras, M.A., T. Nakada, M. Hongo, A.O. Pudov, and J.R. Sites, ZnO/ZnS(O,OH)/Cu(In,Ga)Se2/Mo solar cell with 18.6% efficiency. *Proceedings of the 3rd world conference on photovoltaic energy conversion.* 2003. Osaka, Japan. p. 570-573

[44] Ennaoui, A., U. Blieske, and M.C. Lux-Steiner, 13.7%-efficient Zn(Se,OH)x/Cu(In,Ga)(S,Se)2 thin film solar cell *Prog Photovolt Res Appl* 1998; 6: 447-451.

[45] Hariskos, D., M. Ruckh, U. Ruhle, T. Walter, H.W. Schock, J. Hedström, and L. Stolt, A novel cadmium free buffer layer for Cu(In,Ga)Se2 based solar cells *Solar Energy Materials and Solar Cells* 1996; 41/42: 345-353.

[46] Naghavi, N., S. Spiering, M. Powalla, B. Canava, and D. Lincot, High-efficiency copper indium gallium diselenide (CIGS) solar cells with indium sulfide buffer layer deposited by atomic layer chemical vapor deposition (ALCVD) *Progress in Photovoltaics: Research and application* 2003; 11: 437-443.

[47] Chaisitsak, S., A. Yamada, and M. Konagai, Comprehensive study of light-soaking effect in ZnO/Cu(In,Ga)Se2 solar cells with Zn-based buffer layers. *Proceedings of the Materials Research Society Symposium.* 2001. San Fransisco. p. H9.10.1-5

[48] Hultqvist, A., C. Platzer-Björkman, T. Törndahl, M. Ruth, and M. Edoff, Optimization of i-ZnO window layers for Cu(In,Ga)Se2 solar cells with ALD buffers. *Proceedings of the 22nd European Photovoltaic Solar Energy Conference.* 2007. Milano. p. 2381

[49] Muffler, H., M. Bär, C. Fisher, R. Gay, F. Karg, and M. Lux-Steiner, Ilgar technology, VIII: Sulfidic buffer layers for Cu(In,Ga)(Se,S)2 solar cells prepared by ion layer gas reaction (Ilgar). *Proceedings of the 29th IEEE Photovoltaics Specialists Conference.* 2000. p. 610-613

[50] Allsop, N., C. Camus, A. Hänsel, S. Gledhill, I. Lauermann, M. Lux-Steiner, and C. Fischer, Indium sulfide buffer/CIGSSe interface engineering: Improved cell performance by the addition of zinc sulfide *Thin Solid Films* 2007; 515: 6068-6072.

[51] Ohtake, Y., T. Okamoto, A. Yamada, M. Konagai, and K. Saito, Improved performance of Cu(In,Ga)Se2 thin-film solar cells using evaporated Cd-free buffer layer *Solar Energy Materials and Solar Cells* 1997; 49: 269-275.

[52] Pistor, P., R. Caballero, D. Hariskos, V. Izquierdo-Roca, R. Wächter, S. Schorr, and R. Klenk, Quality and stability of compound indium sulfide as source material for buffer layers in Cu(In,Ga)Se2 solar cells *Solar Energy Materials & Solar Cells* 2009; 93: 148-152.

[53] Glatzel, T., H. Steigert, R. Klenk, and M. Lux-Steiner, Zn1-xMgxO as a window layer in completely Cd-free Cu(In,Ga)(S,Se)2 based thin film solar cells. *Proceedings of the 14th International Photovoltaic Scince and Engineering Conference.* 2004. Thailand

[54] Hariskos, D., R. Menner, E. Lotter, S. Spiering, and M. Powalla, Magnetron sputtering of indium sulphide as the buffer layer in Cu(In,Ga)Se2-based solar cells. *Proceedings of the 20th European Photovoltaic Solar Energy Conference.* 2005. p. 1713-1716

[55] Contreras, M., B. Egaas, K. Ramanathan, J. Hiltner, A. Swartzlander, F.S. Hasoon, and R. Noufi, *Progress in Photovoltaics: Research and application* 1999; 7: 311.

[56] Kushiya, K., Y. Tanaka, H. Hakuma, Y. Goushi, S. Kijima, T. Aramoto, and Y. Fujiwara, Interface control to enhance the fill factor over 0.70 in a large-area CIS-based thin-film PV technology *Thin Solid Films* 2009; 517: 2108-2110.

[57] Saez-Araoz, R., A. Ennaoui, T. Kropp, E. Veryaeva, T. Niesen, and M. Lux-Steiner, Use of different Zn precursors for the deposition of Zn(S,O) buffer layers by chemical bath

for chalcopyrite based Cd-free thin-film solar cells *Physica Status Solidi (a)* 2008; 205: 2330-2334.

[58] Hariskos, D., B. Fuchs, R. Menner, M. Powalla, N. Naghavi, C. Hubert, and D. Lincot, The Zn(S,O,OH)/ZnMgO buffer in thin film Cu(In,Ga)(S,Se)2-based solar cells, Part II: Magnetron sputtering of the ZnMgO buffer layer for in-line co-evaporated Cu(In,Ga)Se2 solar cells *Progress in Photovoltaics: Research and application* 2009.

[59] Spiering, S., D. Hariskos, S. Schröder, and M. Powalla, Stability behaviour of Cd-free CIGS solar modules with In2S3 buffer layer prepared by atomic layer deposition *Thin Solid Films* 2005; 480-481: 195-198.

[60] Naghavi, N., Buffer review *Progress in Photovoltaics: Research and application* 2009.

[61] Olsen, L., W. Lei, F. Addis, W. Shafarman, M. Contreras, and K. Ramanathan, High efficiency CIGS and CIS cells with CVD ZnO buffer layers. *Proceedings of the 26th PVSC*. 1997. Anaheim, Canada. p. 363-366

[62] Buecheler, S., D. Corica, D. Guettler, A. Chirila, R. Verma, U. Muller, T. Niesen, and J. Palm, Ultrasonically sprayed indium sulfide buffer layers for Cu(In,Ga)(S,Se)2 thin-film solar cells *Thin Solid Films* 2009.

[63] Engelhardt, F., L. Bornemann, M. Köntges, T. Meyer, J. Parisi, E. Pschorr-Schoberer, B. Hahn, W. Gebhardt, W. Riedl, and U. Rau, Cu(In,Ga)Se2 solar cells with ZnSe buffer layer: Interface characterisation by quantum efficiency measurements *Progress in Photovoltaics: Research and application* 1999; 7: 423-436.

[64] Gal, D., G. Hodes, D. Lincot, and H.W. Schock, Electrochemical deposition of zinc oxide films from non-aqueous solution: a new buffer/window process for thin film solar cells *Thin Solid Films* 2000; 361-362: 79-83.

[65] Ramanathan, K., F.S. Hasoon, S. Smith, A. Mascarenhas, H.A. Al-Thani, J. Alleman, H. Ullal, J. Keane, P. Johnson, and J.R. Sites. *Proceedings of the 29th IEEE Photovoltaic Specialist Conference*. 2002. New Orleans

[66] Tokita, Y., S. Chaisitsak, H. Miyazaki, R. Mikami, A. Yamada, and M. Konagai, *Japanese Journal of Applied Physics* 2002; 41: 7407.

[67] Hariskos, D., R. Herberholz, M. Ruckh, U. Ruhle, R. Schäffler, and H.W. Schock. *Proceedings of the 13th European Photovoltaic Solar Energy Conference*. 1995. Nice, France. p. 1995

[68] Yousfi, E., T. Asikainen, V. Pietu, P. Cowache, M. Powalla, and D. Lincot, Cadmium-free buffer layers deposited by atomic layer epitaxy for copper indium diselenide solar cells *Thin Solid Films* 2000; 361-362: 183-186.

[69] Ortega-Borges, R. and D. Lincot, Mechanism of chemical bath deposition of cadmium sulfide thin films in the ammonia-thiourea system *J. Electrochem. Soc* 1993; 140: 3464.

[70] Suntola, T., Surface chemistry of materials deposition at atomic layer level *Appl Surf Sci* 1996; 100/101: 391-398.

[71] Allsop, N., A. Schönmann, H. Muffler, M. Bär, M. Lux-Steiner, and C. Fischer, Spray-Ilgar indium sulfide buffers for Cu(In,Ga)(S,Se)2 solar cells *Progress in Photovoltaics: Research and application* 2005; 13: 607.

[72] Platzer-Björkman, C., T. Törndahl, D. Abou-Ras, J. Malmström, J. Kessler, and L. Stolt, Zn(O,S) buffer layers by atomic layer deposition in Cu(In,Ga)Se2 based thin film solar cells: Band alignment and sulfur gradient *Journal of Applied Physics* 2006; 100: 044506.

[73] Meyer, B.K., A. Polity, B. Farangis, Y. He, D. Hasselkamp, T. Krämer, and C. Wang, Structural properties and bandgap bowing of ZnO1-xSx thin films deposited by reactive sputtering *Applied physics letters* 2004; 85: 4929-4931.

[74] Rau, U., M. Schmitt, D. Hilburger, F. Engelhardt, O. Seifert, and J. Parisi, Influence of Na and S incorporation on the electronic transport properties of Cu(In,Ga)Se2 solar cells. *Proceedings of the 25th IEEE Photovoltaics Specialists Conference*. 1996. Washington D. C. p. 1005-1008

[75] Platzer-Björkman, C., J. Lu, J. Kessler, and L. Stolt, Interface study of CuInSe2/ZnO and Cu(In,Ga)Se2/ZnO devices using ALD ZnO buffer layers. *Thin Solid Films* 2003; 431-432: 321-325.

[76] Persson, C., C. Platzer-Björkman, J. Malmström, T. Törndahl, and M. Edoff, Strong valence-band offset bowing of ZnO1-xSx enhances p-type nitrogen doping of ZnO-like alloys *Phys. Rev. Lett* 2006; 97: 146403.

[77] Venkata Rao, G., F. Säuberlich, and A. Klein, Influence of Mg content on the band alignment at CdS/(Zn,Mg)O interfaces *Applied physics letters* 2005; 87: 032101.

[78] Minemoto, T., Y. Hashimoto, T. Satoh, T. Negami, H. Takakura, and Y. Hamakawa, Cu(In,Ga)Se2 solar cells with controlled conduction band offset of window/Cu(In,Ga)Se2 layers *J Appl Phys* 2001; 89: 8327-8330.

[79] Törndahl, T., C. Platzer-Björkman, J. Kessler, and M. Edoff, Atomic Layer Deposition of Zn1-xMgxO buffer layers for Cu(In,Ga)Se2 solar cells *Progress in Photovoltaics: Research and application* 2006; 15: 225-235.

[80] Minemoto, T., Y. Hashimoto, W. Shams-Kolahi, T. Satoh, T. Negami, H. Takakura, and Y. Hamakawa, Control of conduction band offset in wide-gap Cu(In,Ga)Se2 solar cells *Solar Energy Materials and Solar Cells* 2003; 75: 121-126.

[81] Kushiya, K., Development of Cu(In,Ga)Se2 based thin-film PV modules with a Zn(O,S,OH)x buffer layer *Solar Energy* 2004; 77: 717-724.

[82] Saez-Araoz, R., D. Abou-Ras, T. Niesen, A. Neisser, K. Wilchelmi, M. Lux-Steiner, and A. Ennaoui, In situ monitoring the growth of thin-film ZnS/Zn(S,O) bilayer on Cu-chalcopyrite for high performance thin film solar cells *Thin Solid Films* 2009; 517: 2300-2304.

[83] Hariskos, D., B. Fuchs, R. Menner, M. Powalla, N. Naghavi, and D. Lincot, The ZnS/ZnMgO buffer combination in CIGS-based solar cells. *Proceedings of the 22nd European Photovoltaic Solar Energy Conference*. 2007. Milan. p. 1907-1910

[84] Romeo, A., R. Gysel, S. Buzzi, D. Abou-Ras, D. Bätzner, D. Rudmann, H. Zogg, and A. Tiwari. *Proceedings of the 14th Photovoltaic Science and Engineering Conference*. 2004. Bankok, Thailand. p. 705

[85] Donsanti, F., B. Weinberger, P. Cowache, M. Bernard, and D. Lincot, Atomic layer deposition of Indium Sulfide layers for Copper Indium Gallium Diselenide solar cells. *Proceedings of the Materials Research Society Symposium*. 2001. San Fransisco. p. H8.20.1-8

[86] Naghavi, N., R. Henriquez, V. Laptev, and D. Lincot, Growth studies and characterisation of In2S3 thin films deposited by atomic layer deposition (ALD) *Applied surface science* 2004; 222: 65-73.

[87] Rehwald, W. and G. Haarbeke, On the conduction mechanism in single crystal B-indium sulfide *Journal of Physics and chemistry of solids* 1965; 26: 1309-1324.

[88] Sterner, J., J. Malmström, and L. Stolt, Study on ALD In2S3/Cu(In,Ga)Se2 interface formation *Prog. Photovoltaics: Research and Application* 2005; **13**: 179-193.

[89] Barreau, N., C. Deudon, A. Lafond, S. Gall, and J. Kessler, A study of bulk NaxCu1-xIn5S8 and its impact on the Cu(In,Ga)Se2/In2S3 interface of solar cells *Solar Energy Materials & Solar Cells* 2006; 90: 1840-1848.

[90] Weiher, R., *Journal of Applied Physics* 1962; 33: 2834.

[91] Erhart, P., A. Klein, R. Egdell, and K. Albe, Band structure of indium oxide:indirect versus direct band gap *Phys Rev B* 2007; 75: 153205.

[92] Bayon, R., C. Guillen, M. Martinez, M. Gutierrez, and J. Herrero, Preparation of Indium hydroxy sulfide thin films by chemical bath deposition *Journal of the Electrochemical Society* 1998; 145: 2775.

[93] Strohm, A., T. Schlötzer, Q. Nguyen, K. Orgassa, H. Wiesner, and H.W. Schock, New approaches for the fabrication of Cd-free Cu(In,Ga)Se2 heterojunctions. *Proceedings of the 19th European Photovoltaic Solar Energy Conference.* 2004. Paris. p. 1741

[94] Jacob, F., N. Barreau, S. Gall, and J. Kessler, Performance of CuIn1-xGaxSe2/(PVD)In2S3 solar cells versus gallium content *Thin Solid Films* 2007; 515: 6028-6031.

[95] Matsunaga, K., T. Komaru, Y. Nakayama, T. Kume, and Y. Suzuki, Mass-production technology for CIGS modules. *Proceedings of the 17th Photovoltaic Solar Energy Conference.* 2007. Fukuoka, Japan

[96] Palm, J., M. Fuerfanger, P. Morgensen, W. Stetter, C. Fischer, and F. Karg, *A rigorous approach to scale up and optimization of a CIS manufacturing process: The Avancis route*, in *23rd European Photovoltaic Solar Energy Conference, oral presentation.* 2008: Valencia, Spain.

[97] Ohtomo, A., M. Kawasaki, T. Koida, K. Masubuchi, and H. Koinuma, MgxZn1-xO as a II-VI widegap semiconductor alloy *Applied physics letters* 1998; 72: 2466-2468.

[98] Minemoto, T., H. Takakura, Y. Hamakawa, Y. Hashimoto, S. Nishiwaki, and T. Negami, Highly efficient Cd-free Cu(In,Ga)Se2 solar cells using novel window layer of (Zn,Mg)O films. *Proceedings of the 16th EPVSEC.* 2000. Glasgow, UK. p. 686-689

[99] Naghavi, N., C. Hubert, A. Darga, G. Renou, C. Ruiz, A. Etcheberry, D. Hariskos, M. Powalla, J. Guillemoles, and D. Lincot, On a better understanding of the post-treatment effects on the CIGS/Zn(S,O,OH)/ZnMgO based solar cells. *Proceedings of the 23rd European Photovoltaic Solar Energy Conference.* 2008. Valencia, Spain. p. 2160-2164

[100] Kushiya, K., T. Nii, I. Sugiyama, Y. Sato, Y. Inamori, and H. Takeshita, Application of Zn-compound buffer layer for polycrystalline CuInSe2-based thin-film solar cells *Japanese Journal of Applied Physics* 1996; 35: 4383-4388.

[101] Kushiya, K. and O. Yamase, Stabilization of PN heterojunction between Cu(In,Ga)Se2 thin-film absorber and ZnO window layer with Zn(O,S,OH)x buffer *Jpn. J. Appl. Phys.* 2000; 39: 2577-2582.

[102] Powalla, M., B. Dimmler, and K.-H. Gros, CIS thin-film solar modules - an example of remarkable progress in PV. *Proceedings of the 20th European Photovoltaic Solar Energy Conference.* 2005. Barcelona. p. 1689-1694

[103] Platzer-Björkman, C., J. Kessler, and L. Stolt, Atomic layer deposition of Zn(O,S) buffer layers for high efficiency Cu(In,Ga)Se2 solar cells. *Proceedings of the 3rd world conference on photovoltaic energy conversion.* 2003. Osaka, Japan: World Conference on Photovoltaic Energy Conference (WCPEC), Japan. p. 461-464

[104] Nguyen, Q., U. Rau, M. Mamor, K. Orgassa, H.W. Schock, and J.H. Werner, Electrical metastabilities in Cu(In,Ga)Se2-based solar cells with Inx(OH,S)y, CdS and combined buffer layers. *Proceedings of the 17th EPVSEC*. 2001. Munich

[105] Nguyen, Q., K. Orgassa, I. Koetschau, U. Rau, and H.W. Schock, Influence of heterointerfaces on the performance of Cu(In,Ga)Se2 solar cells with CdS and In(OHx,Sy) buffer layers *Thin Solid Films* 2003; 431-432: 330-334.

[106] Pettersson, J., C. Platzer-Björkman, and M. Edoff, JVT and lightsoaking measurements on Cu(In,Ga)Se2 solar cells with ALD-Zn(1-x)MgxO buffer layers *Progress in Photovoltaics: Research and application, accepted* 2009.

[107] Minemoto, T., Y. Hashimoto, T. Satoh, T. Negami, and H. Takakura, Variable light soaking effect of Cu(In,Ga)Se2 solar cells with conduction band offset control of window/Cu(In,Ga)Se2 layers. *Proceedings of the Materials Research Society Symposium*. 2007. San Fransisco. p. 271-6

[108] Niemegeers, A., M. Burgelman, R. Herberholz, U. Rau, D. Hariskos, and H.W. Schock, Model for electronic transport in Cu(In,Ga)Se2 solar cells *Progress in Photovoltaics: Research and application* 1998; 6: 407-421.

[109] Igalson, M., M. Bodegård, L. Stolt, and A. Jasenek, The defected layer and the mechanism of the interface-related metastable behavior in the ZnO/CdS/Cu(In,Ga)Se2 devices *Thin Solid Films* 2003; 431-432: 153-157.

[110] Eisgruber, I., J. Granata, J.R. Sites, J. Hou, and J. Kessler, Blue-photon modification of nonstandard diode barrier in CuInSe2 solar cells *Solar Energy Materials and Solar Cells* 1998; 53: 367-377.

[111] Pudov, A.O., A. Kanevce, H.A. Al-Thani, J.R. Sites, and F.S. Hasoon, Secondary barriers in Cd-CuIn1-xGaxSe2 solar cells *Journal of Applied Physics* 2005; 97: 064901-6.

[112] Lany, S. and A. Zunger, Anion vacancies as a source of persistent photoconductivity in II-VI and chalcopyrite semiconductors *Physical Review B* 2005; 72: 035215.

[113] Zabierowski, P. and C. Platzer-Björkman, Influence of metastabilities on the efficiency of CIGSe-based solar cells with CdS, Zn(O,S) and (Zn,Mg)O buffers. *Proceedings of the 22nd European Photovoltaic Solar Energy Conference*. 2007. Milan. p. 2395-2400

In: Thin Film Solar Cells: Current Status and Future Trends　　ISBN 978-1-61668-326-9
Editors: Alessio Bosio and Alessandro Romeo　　© 2010 Nova Science Publishers, Inc.

Chapter 4

CIGS ABSORBER LAYERS PREPARED BY SPUTTERING BASED METHODS

António F. da Cunha and Pedro M. P. Salomé
Department of Physics, University of Aveiro, 3810-193 Aveiro, Portugal.

ABSTRACT

In this chapter we attempt to give a coherent overview of the various approaches to the deposition of CIGS absorber using solely the sputtering technique or in combination with other processing methods. The advantage of a sputtering related approach is the use of a technology already known to industries which produce large area thin film coatings for various functionalities. This latter fact makes, in principle, up scaling simpler.

After a brief introduction we discuss in section 2 the hybrid sputtering/evaporation method giving also some examples of the results in terms of material properties and solar cells. In section 3 we discuss the selenization/sulfurization of sputtered precursors method and again give some examples of the most important results. Finally we conclude by making a reference to industrial initiatives to produce CIGS solar panels using some of the methods discussed.

1. INTRODUCTION

Worldwide various methods have been explored with the aim of preparing good quality CIGS. Without intending to be exhaustive we can mention the ones that, in our view, are the most successful, namely: coevaporation, selenization/sulfurization of sputtered precursors, hybrid sputtering/evaporation, nano-powder based printing, electrodeposition, spray pyrolysis, etc. If we take the highest efficiency achieved both in laboratory and at an industrial scale as the main criterion to order the several methods, we can safely say that coevaporation has been the most successful of all given that 19.9% [1] and 12.5% [2] have been reported in laboratory and at the industrial scale, respectively. The selenization/sulfurization of sputtered precursors does not lag too far behind given that at the industrial scale 12.8% [2] module efficiency has been reported. Although some of the other methods present the very attractive feature of being non-vacuum, at least partially, they still have to produce the kind of results that make them attractive for large scale application.

In this chapter we will discuss, in some depth, CIGS deposition methods, in some way, based on the sputtering technique again without attempting to be exhaustive because that is beyond the scope the chapter. Rather, we will try to present a coherent picture of the area. Therefore below we will discuss the hybrid sputtering/evaporation and selenization/sulfurization of sputtered precursors methods.

2. Hybrid Sputtering/Evaporation

2.1. The Method

In the literature two approaches are found that are defined as hybrid sputtering/evaporation methods for CIGS deposition. There are, however, considerable differences between the two even though in both cases evaporation and sputtering are combined.

In the approach followed at the present author's laboratory, henceforth named approach 1, all the metal elements are sputtered while the chalcogen, Se, is simultaneously evaporated [3]. In the second approach, henceforth named approach 2, only Cu is sputtered while the other elements are evaporated [4].

In a laboratory scale approach 2 has the advantage of being more versatile than approach 1, especially regarding the incorporation of Ga into the CIGS films which is important in terms of device performance. However from the up scaling point of view approach 1 has advantages over approach 2 given the reduced use of the evaporation technique.

2.1.1. Hybrid CIGS - Approach 1

In approach 1 CIGS layers are deposited in a modified sputtering system as the one schematically shown in figure 1. It contains two magnetrons for 3-inches diameter targets and a substrate holder/heater set, which can be rotated and positioned over any of the magnetrons. It contains also a Se evaporation source that is mechanically coupled to the substrate holder. The mechanical coupling of the Se source with the substrate holder ensures a Se vapour overpressure in the area where the CIGS layer is forming at approximately constant conditions throughout the growth run. The spacing between substrate and the heater is 3-4 mm. A thermocouple used to monitor the substrate temperature and to perform end point detection (EPD) is placed in between the substrate back side and the heater. A Cu target and an In/GaSe composite target consisting of GaSe pellets distributed on the surface of a metallic In target are used for the sputtering.

The CIGS deposition is made on Mo coated soda lime glass (SLG). The 0.5 μm Mo layers were deposited by DC magnetron sputtering in an Ar plasma (p=8.5×10^{-3} mbar) with discharge power density of 10 W/cm^2.

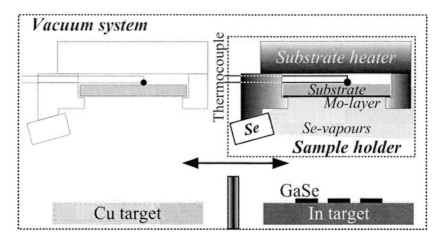

Figure 1. Schematic representation of the Hybrid Sputtering/Evaporation System set-up (after [3]).

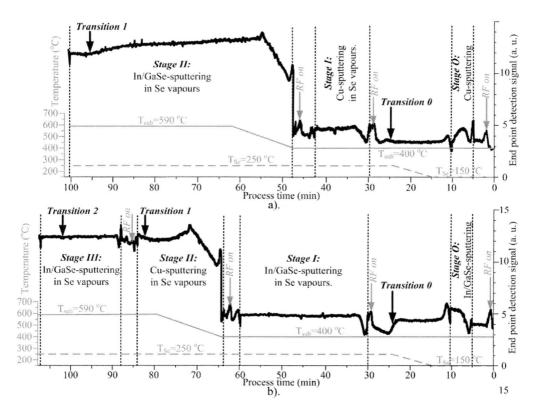

Figure 2. End point detection signal measured for two-stage (a) and three-stage (b) CIGS deposition processes (after [3]).

Two CIGS deposition sequences named two and three-stage were tested and the end point detection technique has been applied. Typical EPD signals are presented in figure 2 for two and three-stage sequences.

In both cases a thin metallic layer (Cu for two-stage and In/GaSe for three-stage sequences) is deposited on the substrate maintained at 400 °C when the Se source temperature was at 150 °C. This is done in order to prevent the pre-selenization of the Mo back contact which could happen for the applied substrate temperatures [5] and since no shutters are available in this system. The deposition of these thin layers is indicated as stage 0 in figure 2. It can be seen in figure 2, for both sequences, that the EPD signal varies as the Se source is ramped up to 250 °C (transitions 0). A slight substrate heating power increase is observed for the two-stage (thin Cu layer) and significant power decrease is observed for three-stage (thin In/GaSe layer) processes. This EPD signal behaviour can be explained by the selenization of the thin metallic layers with formation of Cu_2Se and $(In,Ga)_2Se_3$ selenide layers with different formation energies and emissivities [6]. Short pulses are detected in the EPD signal when the RF power is switched on or off. This is the result of a slight RF interference with the thermocouple circuit.

For the two-stage deposition sequence (figure 2-a) the first stage is performed by sputtering from the Cu target while evaporating Se during 10 – 20 minutes and maintaining the substrate temperature at 400 °C. The RF power density is around 10-11 W/cm^2 while the Se source is kept at 250 °C. In the beginning of the second stage the substrate temperature is ramped up to 590 °C within 15 minutes and In/GaSe is sputtered onto the SLG/Mo/$Cu_{2-x}Se$ formed in the first stage. The RF power density is 8-10 W/cm^2. The second stage duration is controlled by the end point detection technique and typically lasts around 40 minutes until transition 1 is detected (see figure 2-a). At the end of the deposition the samples are cooled down maintaining the Se source at 250 °C until the substrate temperature decreases down to 270 °C at which point the Se source is switched off.

For the three-stage deposition sequence (figure 2-b) the first stage corresponds to sputtering from an In/GaSe target onto the substrate maintained at 400 °C with RF power density around 8-10 W/cm^2 during 20-30 minutes in Se vapours (Se source at 250 °C). In the second stage Cu is sputtered onto the formed SLG/Mo/$(In,Ga)_2Se_3$ structure at discharge power density around 10-11 W/cm^2 while the substrate temperature is ramped up to 590 °C within the first 15 minutes and kept at this level afterwards. The second stage duration is controlled by the EPD technique through the detection of transition 1. Sputtering continues beyond that point until the CIGS layer became Cu-rich with about 10 % Cu excess (figure 2-b). In the third and final stage In/GaSe is sputtered using similar conditions as for the first stage but with substrate temperature at 590 °C. In this stage the CIGS film is returned to a slightly Cu-poor state. The transition could also be detected by EPD (transition 2) but the heating power difference between Cu-rich and Cu-poor CIGS structures is difficult to detect in this stage (see figure 2-b). The Se source is maintained at 250 °C during all three stages and after the end of deposition the sample is cooled down maintaining Se vapour overpressure until the substrate temperature decreases down to 270 °C.

Hybrid CIGS Properties

In approach 1 the morphology of the two and three-stage CIGS films has been studied with electron microscopy (SEM) and typical results are shown in figure 3.

Clearly with this method it is possible to obtain compact films with crystallites about 0.5μm across. The surface of these films is, however, rougher than a typical coevaporated CIGS film. This feature has a negative impact on the performance of the cells prepared with these films since the effective area of the interface with CdS is bigger leading to increased recombination through the interface states.

The crystalline structure of the approach 1 CIGS layers analysed by X-ray diffraction reveals that they are single phase chalcopyrite oriented preferentially along the (112) direction as shown in figure 4.

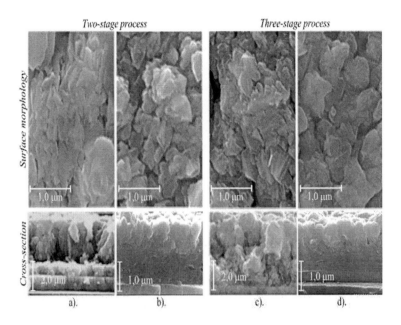

Figure 3. SEM micrographs of approach 1 hybrid CIGS deposited with different Ar pressure profiles: a) low pressure, b) high pressure, c) low pressure, d) high pressure.

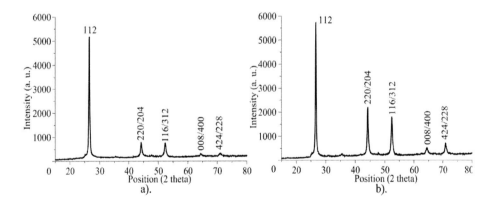

Figure 4. Typical X-ray diffraction results for approach1 hybrid CIGS a) two-stage and b) three-stage.

The tailoring of the Ga profile which is a major advantage of coevaporation becomes difficult to accomplish in this method. Depth composition analysis through Auger electron spectroscopy shows that Ga tends to have a rather flat profile or accumulate at the back as shown in figure 5. This limits the solar cell performance especially in terms of a comparatively low Voc.

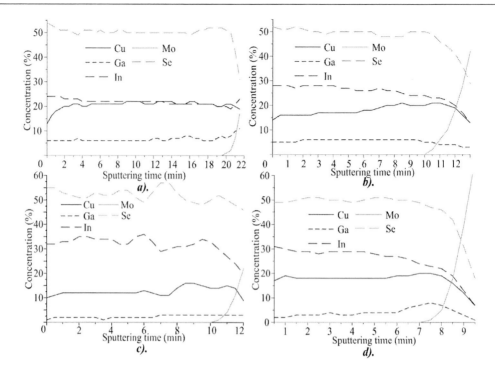

Figure 5. Depth composition profile obtained by Auger electron spectroscopy for approach 1 hybrid CIGS (after [7]).

Solar Cell Results

For the solar cell processing a CdS buffer layer was deposited by chemical bath deposition (CBD) performed at 70 °C bath temperature from a working solution containing 1.6 mM of cadmium acetate (Cd(CH$_3$COO)$_2$), 36.1 mM of thiourea (CS(NH$_2$)$_2$) and 571.4 mM of ammonium hydroxide (NH$_4$OH). The deposition time for 50 nm thick CdS layers was about 16-18 min. The ZnO front contact consisted of a bi-layer of i-ZnO (50nm) on top of which a thicker layer of ZnO:Al (300nm). Both were deposited by RF magnetron sputtering under conditions similar to those in [8].

The solar cell parameters have been estimated from I-V curves measured under simulated standard conditions: AM1.5, 100 mW/cm^2 light intensity and at room temperature. Typical results are shown in figure 6.

Solar cell results based on CIGS films shown in figure 3 are summarized in table I. The best efficiency was obtained for two-stage CIGS deposited at constant low Ar pressure of 4x10^{-3} mbar.

2.1.2. Hybrid CIGS – Approach 2

In approach 2 CIGS layers are prepared by selenization, in Se vapour, of a precursor with the following structure SLG/Mo/(In$_{1-x}$Ga$_x$)$_2$Se$_3$/Cu. The (In$_{1-x}$Ga$_x$)$_2$Se$_3$ layer is obtained by simultaneous evaporation of In, Ga and Se with substrate temperature at around 400 °C. Then a Cu layer is sputtered at room temperature in another system. The precursor structure is selenized at high temperature (between 500 °C and 600 °C) and the CIGS film is terminated by again evaporating In, Ga and Se to return the film to a Cu poor composition [9]. The authors found that Cu-rich (Cu/In+Ga~1.27) precursors selenized at high temperature, around 600 °C result in CIGS films with bigger columnar grains and ultimately in better cell performances.

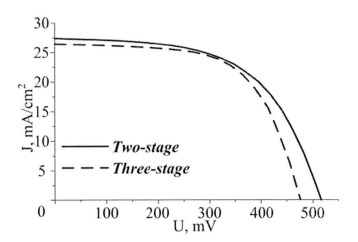

Figure 6. Typical J-V curves for approach 1 hybrid CIGS solar cells (after [3]).

Table I. Solar cell parameters

Absorbers (Figure 3)	Cu, %	In, %	Ga, %	Se, %	U_{oc}, mV	J_{sc}, mA/cm^2	FF, %	η, %
a	23.1	16.7	8.7	51.5	513.6	27.4	57.1	8.0
b	19.4	19.6	12.9	48.1	374.4	30.5	47.4	5.4
c	17.3	24.0	5.4	53.3	474.7	25.5	61.9	7.8
d	21.8	13.9	15.7	48.6	456.4	36.6	40.6	6.8

Hybrid CIGS Properties

Compact CIGS layers are obtained which demonstrates that through this method it is possible to obtain high quality absorber material in terms of morphology and eventually crystallinity.

Again in this approach the difficulty in controlling the Ga depth profile seems to be present. However given the growth sequence described above where Ga is evaporated it is expected that it should be easier to overcome than in the case of approach 1. Ga depth profiles, by Auger electron spectroscopy, were produced for CIGS films selenized at 580 °C and 600 °C. Those profiles revealed a relatively flat depth distribution of Ga for the samples selenized at 580°C while those selenized at 600 °C showed a considerable accumulation of Ga at the back contact [9]. The latter profile is favourable for an enhanced cell performance due to reduced recombination at the back contact as a result of the quasi-field arising from that profile which favours diffusion of photogenerated electrons towards the junction. In the case of the film selenized at 600 °C the increased concentration of Ga at the back was achieved by increasing the Ga evaporation rate during the deposition of the first precursor layer, $(In_{1-x}Ga_x)_2Se_3$. These films also show a lower concentration of Ga at the front than the films obtained at 580°C. This feature generally leads to a decrease in Voc and that is indeed the case for the cells made with these two absorbers.

Solar Cell Results

In table II the parameters of solar cells based on approach 2 hybrid CIGS are presented. The results clearly show that the overall performance of the cells improves as the selenization temperature is increased from 525 °C up to 600 °C.

Generally, as the Ga content of the layers goes up the Voc of the cells tends to increase as well. However, the trend is somewhat erratic, most likely, because of the variation in Ga concentration near the top surface of CIGS. The data shows a clearer trend for the current density whereby it increases significantly as the selenization temperature and the Ga content goes up. It is, however, also clear that for the Ga average values presented the Voc is about 100 mV lower than the values achieved with coevaporated CIGS.

Clearly the reported cell efficiency results are higher than in the approach 1.

Table II. J-V parameters for solar cells prepared with Cu-rich precursors, at different selenization temperatures and different Ga distributions (after [9])

Ts (°C)	Ga/(Ga+In)	Voc (mV)	Jsc (mA/cm^2)	FF (%)	Eff (%)
525	0.26	529	23.1	64.7	7.9
550	0.28	558	26.9	72.7	10.9
580	0.30	554	30.7	70.6	12.0
600	0.30	560	32.3	68.6	12.3
600	0.32	543	33.4	70.1	12.7

3. SELENIZATION/SULFURIZATION OF SPUTTERED PRECURSORS

3.1. The Method

The selenization of sequentially sputtered precursors is an attractive method for preparing CIGS on large area substrates due the industrial ability to make large sputtering machines and to its intrinsic ability to deposit large area layers with good uniformity and reproducibility. However, a few challenges had to be overcome before CIGS based on this method became an industrial reality. It happens that incorporation of Ga is more complicated in this method than in coevaporation and the tailoring of its depth profile is very difficult. On the other hand, as we will show below, sputtering of In also poses some challenges due to its tendency to form droplets on the surface of the substrate which is then a mechanism responsible for the low compactness of the CIGS layers.

The simplest approach to the preparation of CuInGa precursors on a Mo coated glass substrate is by sputtering alternatingly from CuGa and In targets. A CuGa target is used because elemental Ga is difficult to be deposited by sputtering due to its low melting point at

30 °C. Hence by alloying Cu and Ga in the right proportion (typically 75%Cu, 25%Ga) it is possible to prepare a solid target from which to deposit a precursor with Cu and Ga.

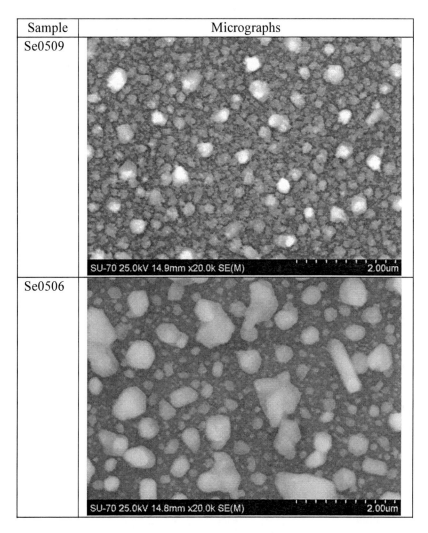

Figure 7. SEM micrographs of SLG/Mo/CuGa/In precursors (after [10]).

Figure 7 shows an example of two precursors with the following structure SLG/Mo/CuGa/In. The first, labelled Se0509, was Cu-poor that is Cu/(In+Ga)<1 and the second, labelled Se0506, was Cu-rich that is the Cu/(In+Ga)>1. Their compositions are shown in table III. It is a common feature of these precursors that the In top layer is very inhomogeneous forming scattered In grains on the surface which are particularly big in the Cu-rich precursor Se0506.

Table III. EDS results of the precursor layers (after [10])

Sample	Mo %	Cu %	In %	Ga %	Cu/(In+Ga)	Ga/(In+Ga)
Se0509	55.93	19.55	19.95	4.57	0.80	0.18
Se0506	37.06	35.52	18.55	8.87	1.30	0.32

When the precursors above are selenized in a vacuum chamber where Se vapour is supplied to the surface the resulting CIGS films have a morphology shown in figure 10 after treatment in a KCN solution to remove excess Cu or spurious $Cu_{2-x}Se$ phases. The CIGS films are not compact as we can see in figure 10. The films Se0509 and Se0506 are however distinct in the sense that the first has smaller grains and bigger voids than the second. This correlates well with the structure of the In layer in the precursors where the Se0509 precursor shows smaller In grains than the Se0506. The Se0506 film which results from a Cu-rich precursor is clearly more compact and shows bigger grains, which is in agreement with what is found in coevaporated CIGS.

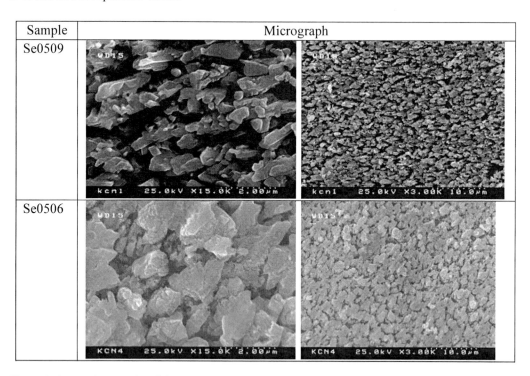

Figure 8. SEM micrographs of the samples after the KCN treatment (after [10]).

When the crystalline structure of the CIGS films above is analysed through x-ray diffraction we see that they are apparently single phase chalcopyrite CIGS, as shown in figure 11. Those results also show that these CIGS films grow preferentially in the (112) direction. The chalcopyrite structure is further confirmed by the Raman scattering results shown in figure 12. Clearly only chalcopyrite characteristic vibrational modes appear in those results with particular emphasis for the A1 mode.

In order to circumvent the problems identified above two modifications to the CIGS preparation process described above have been adopted. The first modification concerns the preparation of the precursor layers and the second concerns the selenization step. The results discussed above refer to precursors with only one CuGa layer and one In layer. For the results that follow the precursors were prepared with several alternating CuGa and In layers in order to minimize the effect, observed and discussed above, of inhomogeneous In distribution. Examples of precursors prepared in such a way are shown in figure 13, below. Figure 13-a) show a precursor film formed with 200 alternating layers of CuGa and In. Figure 13-b) show a precursor film formed with less than 10 alternating layers of CuGa and In. Although the SEM pictures have different scales it appears that the surface morphology of the two is very much identical with a grainy morphology that must be determined by the In layers.

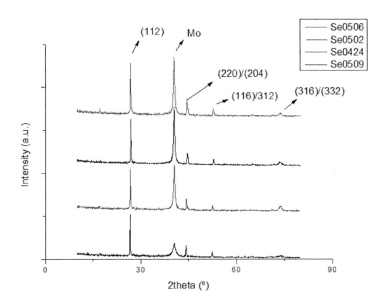

Figure 9. XRD spectra after the selenization of the precursor layers (after [10]).

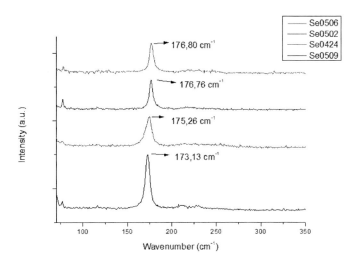

Figure 10. Raman scattering spectra after the KCN treatment (after [10]).

When compared with the morphology of the single layer precursors shown in figure 7 these ones appear to have much smaller grains and of course they have a better mixing of CuGa and In. Above we made reference to a modification in the selenization process. Indeed, now, the selenization is now performed in a rapid thermal processing furnace (RTP) in Se vapour. It must also be said that the actual precursors are finished by evaporating a layer of Se on top of the CuGa and In multi-layer structure before the RTP selenization. This means heating rates around 10 °C/s. One obvious advantage of these high heating rates is the reduced effect of the possible de-wetting of the precursors which may be contributing to the lack of compactness seen in the films obtained by slow heating selenization discussed above. Another important aspect is that with fast heating it is possible to minimize the formation highly conducting binary phases of the type of $Cu_{2-x}Se$ which occur at temperatures around 350 °C.

The CIGS films obtained by the modified process are much more compact than before with very big grains about 2µm across. In both pictures a) and b) we can see that at the interface between CIGS and Mo the CIGS layer shows smaller grains, perhaps due to the higher Ga concentration at the back. This has also been seen in coevaporated CIGS films [12]. The top layer is very compact in both cases. The CIGS surface is still more irregular than in coevaporated CIGS. The results also show that the multi-layer structure of the precursors combined with RTP selenization/sulfurization play an important role in producing high quality CIGS. Still, those results also show that the 200 alternating CuGa and In layers' precursors do not produce better CIGS than the precursors with less than ten alternating layers which is good news.

In this method the difficulty of tailoring the Ga concentration depth profile remains. However, since what is important is not so much the Ga profiling but the conduction band edge profiling in this case a different approach is followed and that is the partial substitution of S for Se at the CIGS top surface in order to increase the band gap. This is known to increase the cell open circuit voltage, Voc, without degrading the short circuit current density, Jsc.

3.2. Solar Cell or Module Results

The solar cells prepared with the CIGS obtained with the modified method showed enhanced performance. In figure 15, below, we show the performance results of a mini-module formed by integrating in series 12 cells on a 10 cm x 10 cm substrate [13]. This mini-module reached an efficiency of 14.7% which is a very good result given the values reported for other technologies.

4. CONCLUSION

As clearly stated at the beginning the aim of this chapter was not to present an exhaustive review of all the work published in the area of sputtering related CIGS deposition but rather present a coherent picture of the subject. In attempting that we left out important contributions by many authors as a quick literature survey would show. This was a personal choice questionable as it may be.

At this point it comes to mind for instance that the results and discussion presented above make no reference to selenization and/or sulfurization of precursors using the toxic gasses H$_2$Se and H$_2$S. Indeed much of the early work on selenized/sulfurized CIS/CIGS had these gasses has the supplier of Se and S. However, given their toxicity presently the groups moving to an industrial scale of CIGS production are adopting solutions where these toxic gasses are avoided as much as possible.

The most well known industrial initiatives to bring sputtering based CIGS solar panels to the market are the AVANCIS, GmbH, project (undertaken by a consortium formed by Shell solar and Saint Gobain), Sulfurcell, Johanna Solar Technology, GmbH, all in Germany and Showa Shell in Japan.

5. REFERENCES:

[1] I. Repins, M. A. Contreras et al, *Prog. Photovolt: Res. Appl.* 2008, 16: 235-239.
[2] K. Kushiya, *Solar Energy 77* (2004), 717-724.
[3] A. F. da Cunha, D. Rudmann, et al, *Proceedings of the 20th European Photovoltaics Solar Energy Conference and Exhibition*, Barcelona, 6-10 June, 2005, 1819-1822.
[4] A. E. Delahoy, L. Chen, et al, *Solar Energy* (2004), 785-793.
[5] D. Abou-Ras, G. Kostorz, D. Bremaud, M. Kälin, F.V. Kurdesau, A.N. Tiwari, M. Döbeli, *Thin Solid Films*, Vol. 480-481 (2005), 433.
[6] D. Cahen, R. Noufi, *J. Phys. Chem. Solids, Vol. 52* (1991), № 8, 947.
[7] P.M.P. Salomé and A.F. da Cunha, *Materials Science Forum*, Vols. 587-588 (2008) pp 323-327.
[8] D.Rudmann, A.F. da Cunha, M. Kaelin, F. Kurdesau, H. Zogg, A.N. Tiwari, *Applied Physics Letters, Vol. 84* (2004), № 7, 1129.
[9] A. E. Delahoy, L. Chen, et al, *Proceedings of the 20th European Photovoltaics Solar Energy Conference and Exhibition*, Barcelona, 6-10 June, 2005, 1843-1846.
[10] "*Growth and Characterization of CuIn$_{1-x}$Ga$_x$Se$_2$ Thin Films as Absorber Layer for Solar Cells*", J. C. B. Malaquias, 1st Degree final year project report, Department of Physics, University of Aveiro, Portugal, 2008.
[11] J. Palm et al, *Thin Solid Films*, vol. 431-432 (2003), 514-522.
[12] "Effects of sodium on growth and properties of Cu(In,Ga)Se2 thin films and solar cells"; D. Rudmann, *PhD DISS. ETH Nr. 15576*, 2004.
[13] V. Probst et al, *Thin Solid Films, 387* (2001), 262-267.

Chapter 5

ELECTRICAL CHARACTERIZATION OF CU(IN,GA)SE2 - BASED THIN FILM PHOTOVOLTAIC DEVICES

Paweł Zabierowski
Warsaw University of Technology, 00 662 Warszawa, Poland.

ABSTRACT

The goal of this chapter is to present an overview on metastable changes in basic electrical characteristics of CIGSe–based photovoltaic devices: capacitance – voltage profiling, defect spectroscopy (admittance and DLTS) and most important from the point of view of photovoltaic applications, light current – voltage curves. Present models combine the metastable behaviour with two intrinsic defects in CIGSe material, In_{Cu} and V_{Se}. The properties of these defects predicted by theory are juxtaposed with experimental results. It is demonstrated that metastable defects must be taken into account while interpreting electrical measurements since they determine the shape of space charge profiles, significantly influence the defect spectra and are the key element for understanding fill factor losses.

INTRODUCTION

Thin film photovoltaic devices based on $Cu(In,Ga)Se_2$ compound (CIGSe) have reached almost 20% efficiency on the lab scale [1] and full size CIGSe PV–modules are being produced on an industrial scale by several companies [2]. In many respects, however, the excellent performance is much ahead of the understanding of electro–optical properties of CIGSe – based solar cells. It appears that still using try–and–error approach may not suffice and future progress will require a deeper knowledge of electronic processes limiting the conversion efficiency. This pertains especially to deep defects which were recently shown to be responsible for a metastable behaviour of CIGSe – based devices [3-5], commonly understood as persistent changes in electrical characteristics induced by application of voltage bias and/or illumination. Metastabilities in CIGSe – based thin film solar cells as well as CIGSe single crystals are detected by many electro–optical experiments: capacitance – voltage profiling and drive level capacitance profiling [6-12], dark and light current – voltage characteristics [7], [13-[21], AC and DC conductivity measurements [22-26], admittance spectroscopy and deep level transient spectroscopy [8,13,27-30], electroluminescence [17,31], and EBIC [32]. Some of the experiments indicate clearly that the concentration of defects in the CIGSe absorber layer which are able to adjust their charge state to the actual electron and hole concentrations within seconds in the wide temperature range may exceed the net shallow acceptor concentration even by orders of magnitude [12,17]. Hence, sooner or later, any interpretation of electrical characteristics in CIGSe – based devices will be faced with metastable phenomena. This is usually quite difficult because there are many factors which can influence the state of the sample, and thus the measurements results, such as the electrical bias (reverse, forward), the energy of incident photons (red, blue), illumination conditions (short, open), as well as temperature and duration of sample treatment. Moreover, the two intrinsic defects which are supposed to be the origin of metastabilities, V_{Se} and In_{Cu} and their complexes with V_{Cu}, behave in a very similar way with respect to illumination and bias treatment [4,5,17], which significantly increases the complexity of the problem. Nevertheless, as shown in this chapter, a careful analysis taking into consideration the properties of metastable defects may bring a valuable insight into the processes determining electrical characteristics which in turn can be helpful in unveiling the factors limiting the conversion efficiency of the investigated cells. As the examples of such studies the following electrical characteristics are discussed: capacitance – voltage profiling, capacitance space charge spectroscopy and light current – voltage characteristics. Below, before dealing with more sophisticated issues concerning metastabilities, an introductory overview of these techniques in application to CIGSe – based heterojunctions is presented.

Capacitance – Voltage Profiling

Interpretation of most electrical measurements of semiconductor devices requires the knowledge of shallow doping levels. A standard technique that is widely used for determination of space charge distributions in semiconductor junctions is capacitance–voltage (C–V) profiling [33]. However the presence of deep levels makes the interpretation of C–V curves less straightforward because such defects may cause large distortions in measured charge density profiles. If not analyzed with care, these non–uniformities can be easily misinterpreted as variations of defect distributions [34]. This is especially true for CIGSe–based devices, where C–V characteristics very often dramatically depend on the direction of the bias scan (Figure 1) and deep defects of densities exceeding free carrier concentrations can persistently change their charge states in response to voltage bias or illumination [12,17]. All this makes the interpretation of capacitance space charge profiles a tough task and causes that the conclusions which can be drown from such analysis are in most cases only qualitative. Hence, even net shallow acceptor concentrations are usually known only roughly and many aspects of capacitance profiling in CIGSe devices are still under debate [10,12,35]. Therefore, before studying metastable charge distributions, the general interpretation of low temperature capacitance profiles in CIGSe devices will be discussed. Low temperature range means here T < 240 K: as shown in Figure 1, below this temperature the response from deep defects giving rise to large hysteresis is frozen out, and one can believe that experimental profiles correspond to true net acceptor concentrations. High temperature C–V curves will be analyzed in the context of metastable defects properties.

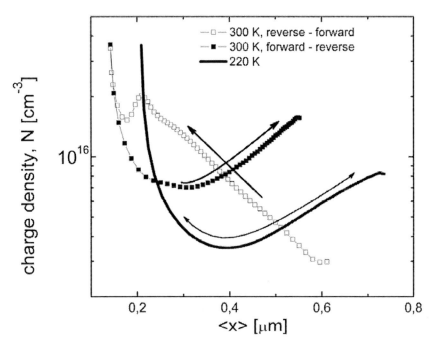

Figure 1. Capacitance space charge profiles of a standard CIGSe – based solar cell. Above ~250 K the shape of the profile depends on the direction of the bias scan (squares). At lower temperatures the hysteresis vanishes (solid line).

Space Charge Spectroscopic Techniques

Deep level transient spectroscopy (DLTS) and admittance spectroscopy (AS) – are very effective and sensitive tools for detection of deep levels in semiconductor junctions [33]. In their elementary applications they allow for determination of deep trap parameters such as the type of a defect (donor or acceptor) [6], its concentration, activation energy, and capture cross sections. However, as discussed further in this chapter, the interpretation of DLTS and AS measurements in CIGSe–based devices is definitely far from standard and even the basic analysis must be viewed with extreme caution. In modern high efficiency devices the defect spectra are surprisingly poor if one considers the possible number of only intrinsic defects in a quaternary compound. For higher temperatures DLTS spectra are usually characterised by a broad, featureless signal which can be assigned to charging/discharging of metastable defects followed by lattice relaxations [36]. In the low temperature range (below 250 K) in most samples only a signal called in the literature N1 is observed [37,38]. It was attributed to the minority carrier traps basing on the following observations: (i) the signs of the DLTS and

---

[6] only DLTS

RDLTS N1 peaks indicate the emission/capture of electrons by minority carrier traps crossed by the Fermi–level at or close to the buffer/CIGSe interface [37-39], (ii) properties of the DLTS N1 signal are characteristic for a large density of interface states in MIS–like structures [39,40], (iii) activation energy of N1 states changes quasi–continuously after annealing the samples in oxygen atmosphere [41], and (iv) the height of the N1 admittance step correlates with the width of the CdS layer [12,38,42]. Since the property (iv) is the key element in the analysis of non–uniformities in low temperature C–V profiles (see below) some details of this interpretation are recalled here after [38]. Consider the schematic band diagram of a ZnO/buffer/CIGSe structure from Figure 2, where the origin of the N1 signal is marked by a circle. The admittance spectrum C(f) corresponding to N1 traps is shown in Figure 3. The capacitance depends on the frequency f of the ac voltage V_{ac} in the following way. For $f >> e_n$ (high frequency range, HF) only shallow levels respond to the test signal and the measured capacitance amounts $C_{HF} = A\varepsilon_s/(W_n+W_p)$ (e_n denotes the emission rate from the N1 states, A is the junction area, ε_s – the dielectric permitivity, assumed to be approximately the same for all layers, and W_n, W_p stand for space charge layer widths on the n– and p–side of the junction, respectively). If the frequency is low enough (low frequency range, LF) then also the N1 traps can follow V_{ac} contributing to the capacitance and $C_{LF} = A\varepsilon_s/W_p$. Combining the two expressions gives $W_n = A\varepsilon_s/(1/C_{HF}-C_{LF})$. Thus, in this configuration, the capacitance step height corresponds to the width of the space charge layer on the n – side of the junction, independently on the N1 trap density [37][38]. It will be demonstrated later in this chapter that a careful analysis of the metastable behaviour of the N1 signal can deliver some useful information on the defect distributions in the close–to–interface CIGSe region as well as electronic processes taking place therein under illumination and electrical bias.

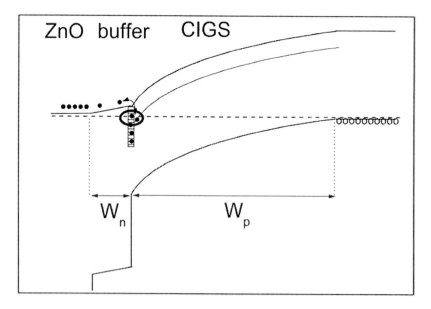

Figure 2. Schematic band diagram of a standard ZnO/buffer/CIGSe solar cell. The states giving rise to the N1 signal are marked by a circle. The space charge layer widths on the n– and p–site of the junction amount W_n and W_p, respectively.

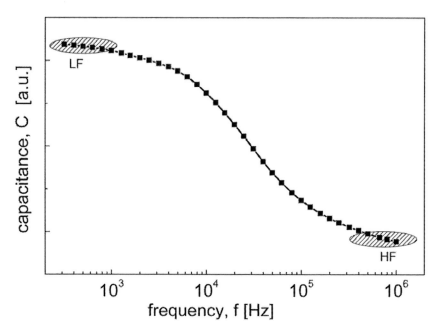

Figure 3. Admittance spectrum of N1 states. CF and LF denote the low and high frequency regimes, respectively.

Light Current – Voltage Characteristics

In modern CIGSe – based solar cells the parameter which still offers a room for significant improvement of conversion efficiency is the fill factor (FF) [43,44]. FF exhibits also the most pronounced metastable behaviour induced by light– or reverse bias soaking, while the two others, open circuit voltage (Voc) and short circuit current (Isc), are much more stable [14,16]. In this chapter the focus of attention is the FF. The description of the dark transport analysis as well as Voc–related issues can be found in [45-53].

It is well known that in the absence of photons absorbed in the buffer, CIGSe – based devices suffer from a lower FF due to a voltage–dependent carrier collection [14,19,20,54,55]. In extreme cases light current-voltage (I-V) characteristics display severe distortions, known as the "red kink" or "double diode" (DD) effects [14,55]. In this connetion "red light" denotes low energy photons absorbed exclusively in a CIGSe layer (hv < 1.6 eV) and "blue light" is understood as high energy photons absorbed also in a CdS buffer (hv > 2.4 eV). Although it is commonly accepted that DD is caused by a secondary barrier for photoelectrons which constitutes in the buffer/CIGSe interface region, the origin of this barrier as well as its position within the junction is still discussed. Basically two models coexist: (A) the barrier is attributed to the buffer – CIGSe conduction band offset (CBO) and as such is located exactly at the interface [55] and (B) it is situated a few tenths of nanometres further away at the virtual interface between the p^+–CIGSe layer and the p–CIGSe bulk region [54]. In the former approach (model A) the healing action of the high energy photons (blue light) is explained by the photo–doping of the buffer which shifts the spike below the conduction band minimum in CIGSe flat band region thereby removing the barrier for photo–electrons. In the model B the barrier originates from the inverse band bending in the bulk absorber. It arises under red illumination at forward bias and is caused by an accumulation of positive charge of free holes which are not able to penetrate the interface region from the absorber site due to a very strong band banding in the p+ layer [38,54]. Under blue illumination the negative charge in this layer is neutralized by photo–holes injected from the buffer into the interface region and the barrier disappears. The existence of such a close–to–interface thin layer with a surplus negative charge was initially postulated in [38,54] for the explanation of a non–ideal behaviour of light and dark I–V curves. Recent theoretical calculations of Lany and Zunger justify the p+ layer model [3-5]. They predict a non–uniform distribution of intrinsic V_{Se} and In_{Cu} defects in the absorber with the negative charge accumulated in the vicinity of the buffer/CIGSe interface [4]. A short review on the influence of the V_{Se} defect charge states distribution on electrical characteristics not analyzed here can be found in [56].

In this chapter the p+ layer model is employed for the analysis of room temperature light I–V curves for the following reasons: (i) it was shown in [17,55] that above 250 K the barrier at the buffer/CIGSe interface affects the photo–carrier transport for values of CBO>0.3 and in all investigated cells the CBO was significantly smaller, (ii) model A is in contradiction with the experimental result that the blue illumination always causes significant decrease of a junction capacitance [13,57], and (iii) the FF metastabilities are observed in devices with various buffer layers and closely correlate with persistent changes in negative charge distributions in the absorber induced by light absorbed exclusively in CIGSe [16]. It should be noted that the model A might be applicable for some cells in the low temperature range (below ~200 K).

In order to concentrate on effects which are likely to be observed in any standard high efficiency CIGSe – based heterojunction solar cell, the experimental results presented in this chapter come from the measurements performed on baseline photovoltaic devices of the structure ZnO:Al/i–ZnO/buffer/Cu(In,Ga)Se$_2$/Mo/SLG with the absorber composition with a Ga/In+Ga ratio close to 0.3. The devices were fabricated at Uppsala University and Nantes University. The CIGSe absorbers were deposited by co–evaporation from elemental sources [58,59] on a Mo–coated soda–lime glass substrate. The ZnO:Al/i–ZnO bi–layer, Mo and CBD–CdS layers were all deposited according to a baseline procedure as described in [58]. The need to replace CdS with another material is driven, among other things, by a possible gain in the short circuit current density (J_{SC}). This is achieved by an employment of a buffer with a larger bandgap, which reduces the current losses in the short wavelength range [60,61]. The (Zn,Mg)O and Zn(O,S) buffer layers in the investigated cells were fabricated by atomic layer deposition (ALD) as described in [62,63].The (Zn,Mg)O buffer layers have thicknesses of around 150 nm and a bandgap of 3.6 eV. The Zn(O,S) buffer layers have thicknesses of around 30 nm. The bandgap of the Zn(O,S) layer is a function of the sulfur content and varies from 2.7 eV to 3.0 eV due to increasing sulfur towards the CIGS surface [62]. The CBO at the CIGS/buffer interface has been determined by photoelectron spectroscopy (PES) and optical measurements to 0.2 eV for Zn(O,S) films [62]. For (Zn,Mg)O, a CBO of 0.2 eV is estimated from optical measurements and published PES data [63].

Space charge profiles were calculated in a standard way from C–V curves measured by use of HP4284A LCR meter. Capacitance transients were measured by the Boonton 72B capacitance bridge and digitized by the 12 bit AD converter (Advantech, PCL 818HD). The Lapalce–DLTS emission rate spectra were calculated by use of the commercial software CONTIN [65]. The transients for Laplace–DLTS analysis were induced by RDLTS voltage pulses (steady state bias $U_R=0$ V and amplitude $\Delta U=-0.3$ V). The pulse width was on the order of 1 s, which ensured the saturation of capacitance transient amplitude. Details of the Laplace–DLTS and RDLTS analysis can be found in [66,67,39]. White light IV characteristics were measured by using a halogen lamp of the intensity of about 1000 Wm^{-2} (AM1.5). For red illumination the same lamp with a 550 nm edge filter was used (the intensity of light was adjusted to the same value of I_{SC} current as under white illumination). As a simulation tool the SCAPS software was used [68]. The abbreviations for denoting metastable states of the samples used throughout this chapter are as follows:

RELAXED (REL) – sample annealed in dark at 0 V and 330 K for at least 1 hour,
LSO – light soaking at RT under open circuit conditions, no matters red or white,
LSS – light soaking at RT under short circuit conditions, no matters red or white,
RLSO – red light soaking at RT under open circuit conditions,
RLSS – red light soaking at RT under short circuit conditions,
WLS – white light soaking at RT,
REV – reverse bias soaking with –2V at RT,
ROB – "red on bias", sample illuminated with red light while reverse biased with –2V.

SPACE-CHARGE CAPACITANCE PROFILES

Typically, low temperature C–V measurements on CIGSe–based devices reveal characteristic U–shaped space charge profiles, as illustrated in Figure 4. The magnitude of the distortion may vary from sample to sample but this non–uniformity is present in all kind of CIGSe–based photovoltaic devices, regardless of the absorber composition and buffer layers employed [12,53]. Due to recent theoretical models, involving multi–charge defects with negative correlation energy, the acceptor concentration is expected to be much higher in the close–to–the–interface region than in the remaining part of the absorber [4,5]. Thus the increased acceptor density towards the interface may find its explanation in the properties of metastable defects, as described later in the text. However, contrary to theoretical predictions, in the backward part of the profile the negative charge density rises towards the back contact often even by an order of magnitude on a distance of only 100 nm – 200 nm. Now we will focus on this puzzling feature, partly following the thorough discussion conducted in [12]. According to Kimerling [34] such non–uniformities can be artefacts caused by deep states which cannot follow the test signal but accumulate a static charge during the voltage sweep. For low frequencies, as the traps can follow Vac and behave more like shallow dopants, the distortion should vanish. The defects commonly observed in admittance and DLTS spectra of CIGSe–based heterojunctions which potentially could cause the distortion in C–V profiles are traps called in the literature N1. Figure 4 shows exemplary C–V profiles with a very strong distortion measured at two frequencies, denoted as low (LF) and high (HF) with reference to N1 signal. What is important, the LF profile is translated horizontally towards the interface with respect to HF exactly by W_n and both are of the same shape. This implies that the width of the space charge region in CdS and ZnO layers does not depend on the applied voltage. Hence the ZnO/buffer/CIGSe devices behave in capacitance measurements analogically to MIS structures with the buffer playing the role of an insulator. Indeed, if we "shift" the HF profile by W_n, subtracting the inverse capacitance corresponding to W_n from each point of the

C–V curve measured at HF and then recalculate a "shifted" doping profile HF* (from $C^*_{HF}{}^{-1}$ = $C_{HF}{}^{-1}$ − $W_n/A\varepsilon_s$), we see that it perfectly coincides with the profile measured at LF. This rules out the possibility that the backward distortion of C–V profiles is due to N1 traps (in the Kimmerling's way) and gives very strong argument against the interpretation of N1 traps as bulk acceptor states as suggested in [10]. This procedure of profile translation by W_n works equally well for profiles measured on samples in metastable states[7] [12] as well as for cells with alternative buffers of thicknesses ranging from 30 nm (ZnOS) to 200 nm (ZnMgO) [53].

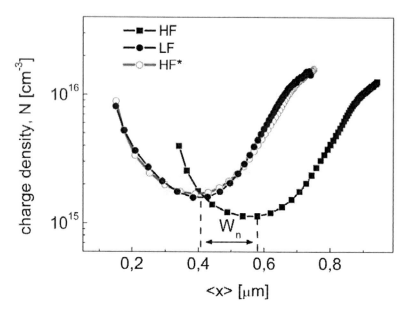

Figure 4. Typical low temperature space charge profiles measured at low (LF) and high (HF) frequencies with respect to N1 admittance step. The profile denoted as HF* was recalculated from HF capacitance–voltage curve as described in the text.

Although the contribution of N1 defects to non–uniformities of C–V profiles was excluded, some other traps could be involved. Detailed simulations have shown [12] that the best agreement between experimental C–V profiles and that modelled by the influence of deep traps using the Kimerling model could be achieved if the deep states were located close to 0.6 eV above the VB. Then the net acceptor concentration closest to the assumed one was found at the minimum of the profile (corresponding to small positive biases). However this attractive model imposes upon the traps contradictory constraints: on one hand they should be deep enough so as not to respond to low frequencies up to RT (traps other than N1 are not detected below 300 K) and on the other hand shallow enough to accumulate the charge during the dc voltage sweep at 100 K. We propose that this puzzle can be explained by the bi–stable behaviour of In_{Cu} DX centres. Thus, in order to continue the discussion, we should acquaint first with properties of metastable defects present in CIGSe.

---

[7] Only in the case of the metastability induced by reverse bias, where the apparent width W_n exceeds 300 nm, the LF and HF* profiles do not merge, which was attributed to the additional contribution of holes to the N1 signal [57]

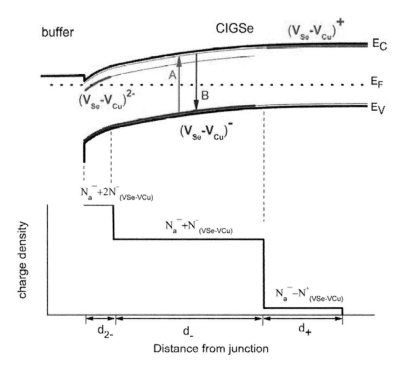

Figure 5. Schematic band diagram of the buffer/CIGSe region with the distribution of (V_{Se}–V_{Cu}) charge states. Detailed description in the text.

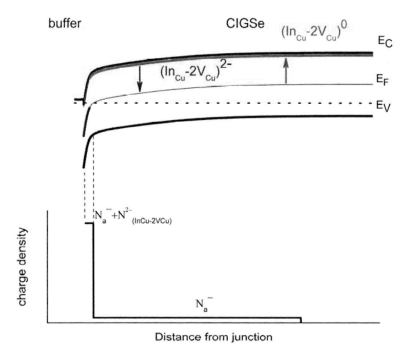

Figure 6. Schematic band diagram of the buffer/CIGSe region with the distribution of (In_{Cu}–$2V_{Cu}$) charge states. Detailed description in the text.

Recent first principles calculations revealed that two intrinsic defects, In_{Cu} and V_{Se}, might be involved in a metastable behaviour of CIGSe – based devices [3-5]. Both defects feature a negative – U energy property, i.e. the change of their charge state is accompanied by a large lattice relaxation which results in an inverted ordering of their energy levels. As a consequence both defects are assumed to give rise to large non–uniformities of charge distributions in the CIGSe absorber. Distributions of $(V_{Se} -V_{Cu})$ and $(In_{Cu} -2V_{Cu})$ charge states predicted by this model (called in the literature a Lany – Zunger (L–Z) model) together with schematic band diagrams of the space charge region of the Mo/CIGSe/buffer/ZnO/Al solar cell are shown in Figure 5 and Figure 6, respectively. For the sake of simplicity the defects are considered separately. Since most of both In_{Cu} and V_{Se} will form complexes with copper vacancies [3,5] and other configurations exhibit similar features, in the following we will concentrate on defect complexes.

As regards the divacancy complexes $(V_{Se}-V_{Cu})$, they act as positively charged compensating donors in the absorber region where the Fermi level position E_f is close to the valence band, i.e. farther away from the junction (region d_+ in Figure 5). Nearly in the space charge layer, as E_f rises about 0.2 eV above E_V (region d_-), the complexes form shallow acceptor states $(V_{Se}-V_{Cu})^-$. If only there is a type inversion at the absorber surface ($E_f > E_V +1$ eV), e.g. due to Fermi – level pinning at the buffer/CIGSe interface, there appears in the vicinity of the interface a thin layer in which very deep $(V_{Se}-V_{Cu})^{2-}$ or even $(V_{Se}-V_{Cu})^{3-}$ states prevail (region d_{2-}). As a consequence the charge profile is non–uniform as schematically depicted in Figure 5. A disturbance of the equilibrium by illumination or reverse bias results in redistribution of these charge states according to following reactions[8]:

A: $(V_{Se} - V_{Cu})^- + 2h \rightarrow (V_{Se} - V_{Cu})^+$: $\Delta E_A \approx 0.3$ eV
B1: $(V_{Se} - V_{Cu})^+ + e \rightarrow (V_{Se} - V_{Cu})^- +h$: $\Delta E_{B1} \approx 0.1$ eV
B2: $(V_{Se} - V_{Cu})^+ - 2h \rightarrow (V_{Se} - V_{Cu})^-$: $\Delta E_{B2} \approx 0.8$ eV

The predicted values of energy barriers associated with transitions A, B1 and B2 agree quite well with the results of transient capacitance [36], transient conductance and TSCAP experiments performed on CIGSe thin films as well as complete cell structures [69].

The metastable changes in charge distributions can easily be detected by capacitance profiling. Although the magnitude of these effects varies from sample to sample they are characteristic for all CIGSe–based devices, independent on Ga content (i.e. CISe, CIGSe or CGSe) [12] and the structure of the cells (buffer and window materials) [17,16,53]. As an example two sets of C–V profiles, measured for CdS– and (Zn,Mg)O–buffered devices are shown in Figure 7a and Figure 7b, respectively.

Light soaking increases the overall shallow acceptor concentration and the profiles are shifted to the left due to the shrinkage of the space charge layer width (LS, blue circles) as compared to the relaxed state (REL, black squares). Under illumination the donor → acceptor transition B1 prevails deeper in the bulk (roughly in the region d_+ and a part of d_- next to d_+) where electrons were in deficiency. On the other hand, close to the interface where almost all $(V_{Se}-V_{Cu})$ are in the acceptor configuration, process A of hole capture will dominate, reducing the negative charge in the region d_{2-}. As a result the concentration of the negative charge associated with $(V_{Se}-V_{Cu})^-$ becomes more uniform in the entire CIGSe layer. Since the structural energy barrier for capturing of electrons is very small (< 0.1 eV) process B1 occurs

---

[8] the values of energy barriers are calculated in [3] for CISe compound

also at low temperatures (blue open circles in Figure 7b). Similar effect can be obtained by forward biasing the junction [69].

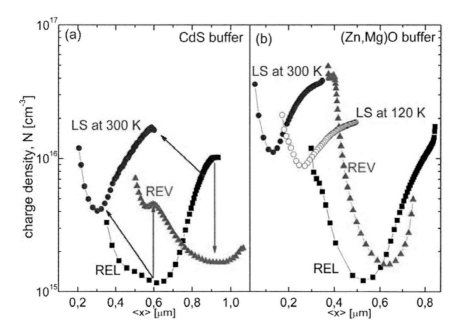

Figure 7. Low temperature (120 K) space charge profiles measured at HF in REL state and LS and REV metastable states for CdS– (a) and (Zn,Mg)O–buffered cells (b).

The profiles after reverse bias treatment (REV state, red triangles) exhibit larger non–uniformities. Apparently the negative charge concentration is increased merely in a limited region adjacent to the junction whereas for larger distances it is significantly reduced (as indicated by arrows in Figure 7a). The width of the region with increased charge density is closely related to the value of the reverse bias maintained during cooling [14]. Application of reverse bias removes free holes from a part of region d_+ next to d_-. If the temperature is high enough transition B2 will dominate and the defects $(V_{Se} - V_{Cu})^+$ will relax into the shallow acceptor configuration $(V_{Se}-V_{Cu})^-$ after emission of two holes. Due to the large value of the barrier ΔE_{B2} the REV metastability can be efficiently induced only at temperatures above ~250 K. Since this defect reconfiguration occurs exclusively in the transition region d_-/d_+ the increase of negative charge is more local as compared to changes induced by LS.

Both metastabilities LS as well as REV are stable below 200 K for a very long time and anneal out gradually above 250 K [69]. This is because in either case the return to the relaxed state requires recapture of holes and thermal emission over a structural barrier ΔE_A. Hence only heating up the sample above ~250 K brings the profiles back to the relaxed state. This feature makes them distinct from a metastability associated with In_{Cu} – related DX centres.

The defect complex $(In_{Cu} - 2V_{Cu})$ may exist in two charge states depending on the Fermi level position[9] [5]. For $E_F<1$ eV it forms a shallow state $(In_{Cu} - 2V_{Cu})^0$ just below the conduction band minimum producing free electrons in the conduction band. As E_F rises above 1 eV it becomes filled with two electrons and due to negative–U Hubbard correlation energy

---

[9] isolated In_{Cu} and $(In_{Cu} - V_{Cu})$ exhibit similar properties [5]

it resides in $(In_{Cu} -2V_{Cu})^{2-}$ deep DX state around midgap [5]. The resulting charge distribution is highly non–uniform as schematically shown in Figure 6. In almost entire CIGSe layer the complexes act as shallow compensating donors $(In_{Cu} -2V_{Cu})^0$ whereas in the vicinity of the buffer/CIGSe interface they form a very thin layer with excess negative charge due to $(In_{Cu} -2V_{Cu})^{2-}$. Under illumination the occupation of the shallow donors located in the SCL can change in a metastable manner by a capture of two electrons into the DX state. However, distinct from V_{Se} – related LS metastability, there is no barrier for hole capture by the DX states [5]. Thus, in order to observe any persistent changes in charge distribution due to accumulation of electrons on DX states one has to avoid recombination of trapped electrons with free holes, e.g. by reverse biasing the sample during illumination with photons absorbed exclusively in CIGSe, which for CdS–buffered cells means red light. This explains the reason why this effect was given the name *red–on–bias* (ROB) [71]. The ROB metastability is observed in devices with absorbers of different compositions (CISe, CIGSe and CGSe) and with various buffer layers (CdS, (Zn,Mg)O, Zn(O,S)) [14,72,73]. Examples are given for (Zn,Mg)O – and CdS–buffered devices with CIGSe absorber in Figure 8 (C–V profiles) and Figure 9 (Mott–Schottky plots), respectively. Clearly, if the samples are kept biased after switching the light off the electrons remain trapped at deep DX states giving rise to an increased negative charge (red lines) as compared to the initial states[10] (blue lines). As shown in the inset of Figure 8 the amount of charge trapped by DX centres increases with time of the ROB pulse: it rises by a factor of 2 after 2 seconds and saturates after about 200 s. In the case when neither forward bias nor blue illumination are applied after ROB, electrons remain trapped at the deep DX state until being eventually thermally released around 300 K due to the structural energy barrier of 0.3 eV [56,72]. However, during C–V profiling free holes start to appear in the space charge region as reverse bias is gradually decreased, and the $(In_{Cu} -2V_{Cu})^{2-}$ states are consecutively emptied by recombination even at low temperatures. In consequence, the negative charge is reduced in regions where free holes were introduced by the voltage sweep, as illustrated in Figure 8. Application of +1 V restores the initial charge distribution in the entire SCR (blue – initial, dark red – ROB after C–V up to +1 V). If C–V scan after ROB is stopped at 0 V (open circles), the negative charge in the subsequently measured profile (gray) is reduced only partly (for $<x> > 0.3$ µm) and for shorter distances it coincides with the profile taken directly after ROB (red). In some samples forward biasing the junction does not suffice (black curve in Figure 9) and the negative charge can be fully reduced at low temperatures only by injection of photo–holes from the buffer, i.e. after illumination with high energy "blue" photons (open squares).

The presence of the ROB effect does not depend on the state of the sample as regards V_{Se}–related metastabilities. It is illustrated in Figures 10 a,b,c which graph the ROB metastability superimposed on REL*, LS and REV states, respectively. As indicated by arrows, in either case DX centres can be filled with electrons under ROB conditions and emptied at low temperatures by application of a forward bias only. After the ROB/FWD cycle the C–V profiles return into the corresponding initial state, which reflects the fact that V_{Se}–related metastabilities are impervious to free holes below 250 K.

---

[10] here the initial state means that before ROB the sample was light soaked for 5 minutes at low temperature in order to saturate the V_{Se} – related LS metastability

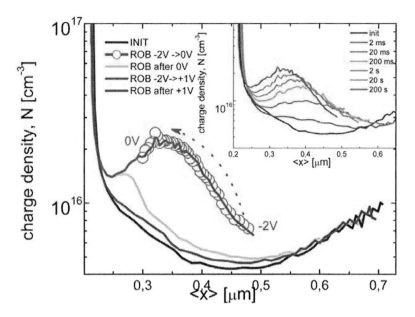

Figure 8. Influence of the red–on–bias effect on HF space charge profiles for (Zn,Mg)O – buffered cell measured at low temperature (120 K). In the inset a series of intermediate profiles corresponding to increasing duration of ROB treatment is shown. Detailed description in the text.

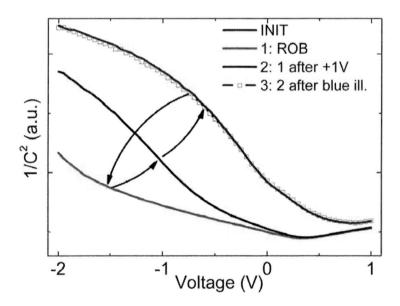

Figure 9. Influence of the red–on–bias effect on HF Mott–Schottky plot for CdS – buffered cell measured at low temperature (120 K).

Due to the low formation energy of In_{Cu} compensating donors their concentration is expected to be on the order of 10^{18} cm^{-3} or even higher [74]. Clearly, for distances accessible in C–V measurement the magnitude of the negative charge trapped by In_{Cu} DX centres is far below this value, which means that only a small fraction of In_{Cu} defects existing in the SCR participates in the ROB metastability. It follows also from Figure 10 that the occupancy of

DX states after ROB depends on the location in the SCR. Taking as an example the ROB profile in Figure 10a (red line), the charge density decreases from its maximal value of $\sim 5\times 10^{16}$ cm^{-3} at distances just below 0.3 µm to a plateau around 0.4 µm at $\sim 1\times 10^{16}$ cm^{-3} and further down to $\sim 5\times 10^{15}$ cm^{-3} at 0.55 µm. This behaviour can be understood within the framework of the L–Z model, since the number of defects that remain converted to DX states after switching the light off depends on the local relative electron and hole concentrations within the CIGSe layer, and the p/n ratio steeply increases by orders of magnitude from buffer/CIGSe interface towards CIGSe bulk. Similarly, the differences in the magnitude of the negative charge accumulated during ROB experiments on the sample in REV, LS and REL* states can be ascribed to variations between the corresponding metastable charge distributions. Different band bandings influence the p/n ratio and this in turn determines the fraction of existing In$_{Cu}$ defects participating in the ROB metastability.

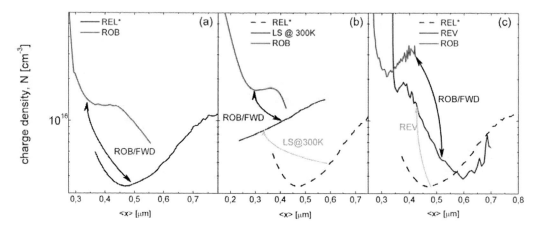

Figure 10. ROB effect induced in a CdS–buffered sample in relaxed (a), LS (b) and REV states (c). HF C–V measurements were performed at 120 K. REL* state: in order to saturate the V$_{Se}$–related LS metastability the sample was light soaked for 5 minutes at low temperature. The black arrows indicate the opposite transitions induced at 120 K by ROB treatment and forward biasing the junction.

At this moment, after having presented the characteristics of metastable defects, we are ready to get back to the discussion of the distortion in the rear part of low temperature C–V profiles. As we have seen, its explanation by regular defects is hardly likely since they would have to be simultaneously shallow and deep [12]. However, such properties are featured by In$_{Cu}$ antisite defects: (i) according to ROB experiments the negative charge can be accumulated at their deep DX states (E$_V$+0.6 eV), (ii) they can be filled and emptied sufficiently fast even at low temperatures due to electron capture by shallow compensating donors and hole capture by deep DX's, and (iii) since the DX states immediately disappear in presence of holes they will not be detected as deep levels in standard DLTS and AS experiments. Hence, we propose that the backward slope of the U–shaped C–V profiles is due to accumulation of negative charge at the deep DX levels of the In$_{Cu}$ defects. The number of unoccupied (shallow compensating donors) in relation to occupied (deep DX) In$_{Cu}$ defects depends on the ratio of the rates of their emptying (hole capture) and filling (electron capture) and thus on the ratio of hole to electron concentration p/n. Since under reverse bias the value of p is suppressed much stronger than n (p/n may decrease by several orders of magnitude in a part of the space charge layer) the filling process (requiring electrons) will predominate the emptying (requiring holes), and the number of negatively charged deep DX states may considerably increase [75]. As this charge is accumulated at some distance from the border of the space charge region the measured profiles do not reflect exactly the net acceptor distributions in the absorber layer towards the back contact [34].

So far only low temperature range was considered. Above about 250 K the C–V characteristics may strongly depend on the direction of the bias scan. Keeping in mind that one should be extremely cautious in the interpretation of C–V profiles in the case when deep levels are involved [34], we propose that the high temperature C–V hysteresis might be also due to charging and discharging of In_{Cu} defects. Figure 11 shows exemplary space charge profiles obtained from C–V curves measured at 300 K from forward to reverse (F–R, full symbols) and from reverse to forward biases (R–F, open symbols). Our suggestion is based on a close resemblance between the metastable behaviour of the low and high temperature C–V characteristics: (i) the F–R and low temperature profiles feature the same U–shape, whereas the R–F scans quite closely resemble the changes induced upon ROB conditions at low temperature, (ii) the defects giving rise to the hysteresis are sensitive to injection of holes in the same way as DX levels in ROB experiment, and (iii) the presence of the hysteresis, similarly to the ROB effect, does not depend on the state of the sample, REL, LS or REV (rectangles, circles and triangles in Figure 11a and 11b, respectively).

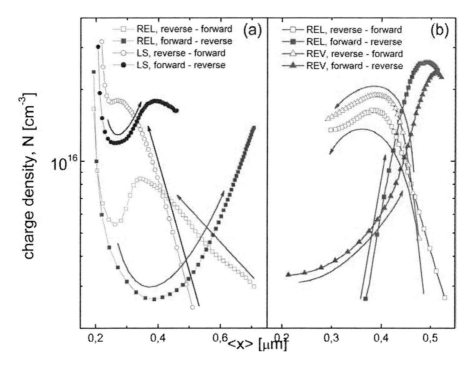

Figure 11. Dependence of the high temperature (300 K) capacitance space charge profiles on the direction bias scan in the REL state (green squares) compared to analogical measurements after 1 hour long light soaking (a) and reverse bias treatment (b).

Summing up, both defects V_{Se} as well as In_{Cu} give rise to the surplus negative charge accumulated in the close–to–interface CIGSe region – so called p+ layer. Both exhibit also the metastable behaviour which can markedly influence the space charge distributions. V_{Se}

give more rigid changes in the sense that after capturing of electrons the additional negative charge remains frozen up to room temperature. On the contrary, In$_{Cu}$ defects are able to adjust their charge state to the actual electron and hole concentrations even at low temperatures. The metastable defect distributions, especially the properties of the p+ layer, play an important role in the interpretation of other electrical characteristics as discussed further in this chapter.

DEFECT SPECTROSCOPY

One can distinguish two types of major difficulties hindering the defect characterization in CIGSe–based junctions: non–exponentiality of capacitance transients [36] and non–linearity of Arrhenius plots [30], each related to a specific temperature range. At temperatures above 250 K, as already mentioned while discussing capacitance–voltage profiles, any disturbance of the equilibrium leads to metastable change of defects charge states which is accompanied by lattice relaxations. Voltage or light pulses result thus in logarithmic in time, huge capacitance transients ($\Delta C/C \sim 1$) as exemplified in Figure 12. In most cases these signals cover any possible response from other deep defects. A thorough analysis of these processes conducted in [36,76,77] revealed values of activation energy in good agreement with theoretical predictions of structural barriers for hole capture by $(V_{Se} - V_{Cu})^-$ and hole emission from $(V_{Se} - V_{Cu})^+$.

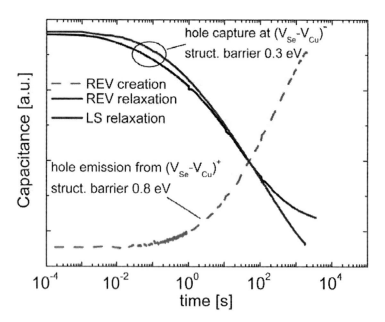

Figure 12. Capacitance transients due to hole capture recorded at 300 K after reverse bias soaking (black) and light soaking (blue). The transient caused by hole emission was monitored during reverse bias treatment (red).

In the low temperature range (T < 250 K) AS and DLTS spectra feature the aforementioned signal N1 (Figure 13). The sign of corresponding capacitance transients (positive and negative in DLTS and RDLTS modes, respectively) as well as the correlation of the admittance capacitance step with W_n, the width of the space charge region on the n–side of the junction (i.e. the buffer plus i–ZnO layers), strongly vote for the interpretation of N1 defects as minority carrier traps. Initially N1 states were considered only as a continuous distribution of donor–like interface states pinning the Fermi–level close to the conduction band at the buffer/CIGSe interface [37,38,42]. Further investigations revealed that the N1 signal has also other components, most probably due to CIGSe bulk compensating donors charged and discharged by electrons in the vicinity of the interface [30]. The problems that arise while analyzing N1 AS and DLTS data come from (i) non–linearity of corresponding Arrhenius plots [30], (ii) a variable position of N1 peaks on emission rate scale (a metastable change of emission rates induced by LS or REV) [22,30], and (iii) a steady increase of RDLTS signal height with rising voltage pulse amplitude [40]. The consequence of features (i) and (ii) is the apparent lack of common activation energies characterizing N1 traps whereas (iii) causes that the defect density can be estimated only roughly. In the following we will discuss these issues in more detail.

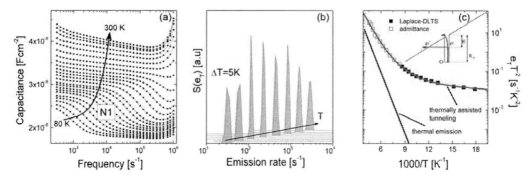

Figure 13. The sets of N1 admittance (a) and Laplace – RDLTS spectra (b) as a function of temperature. Corresponding Arrhenius plots are shown in Figure (c).

Amplitude of N1 RDLTS Signal

The formula for evaluation of N1 traps concentration should be derived from a general expression describing the relative capacitance change of the two sided p–n junction caused by trapping of carriers by deep defects of concentration $N_T(x)$ [35]:

$$\frac{\Delta C}{C} = -\int_{-W_n}^{W_p} A \frac{N_T(x)}{W^2}\left(\frac{x+W_n}{N^-} - \frac{W_p - x}{N^+}\right)dx \qquad (1)$$

x is the spatial coordinate ranging from $-W_n$ to W_p (the edges of the space charge region on both sides of the junction), N^- and N^+ stand for the net doping concentrations of the p and n type semiconductors, respectively, and A is the amplitude factor of the RDLTS mode. As indicated by simulations [35], the N1 signal is strongly localized in a region close to the buffer/CIGSe interface ($x_T \sim 20$ nm ÷ 80 nm and $\Delta x \sim 5$ nm ÷ 10 nm, for voltage pulse amplitudes between 0.1 V and 2 V). This explains high exponentiality of the RDLTS N1 signal[11] (delta like peaks in the Laplace RDLTS spectra in Figure 13b) and allows for the estimation of the concentration of defect states N_T giving rise to the N1 capacitance transient by the expression [35]

$$\frac{\Delta C}{C} = -\frac{N_T \Delta x}{2N^-} \frac{(x_T + W_n)}{W^2}, \qquad (2)$$

where $W = W_n + W_p$ is the total space charge layer width and Δx the width of the region located at the distance x_T from the CdS/CIGS interface where the charging/discharging of the defect takes place. The form of Equation 2 confirms that, as regards capacitance measurements, ZnO/buffer/CIGSe devices should be considered as MIS–like structures (recall the discussion above concerning the influence of N1 traps on capacitance – voltage profiles) [78]. Since all of the quantities on the right–hand side of Equation 2 are known at least roughly, one can make an attempt to estimate the $N_T / \overline{N^-}$ ratio: the widths of the space charge layer W and the depletion region on the n side of the junction W_n correspond to the junction capacitance at 0 V and to the height of the capacitance step N1, and according to simulations, Δx and x_T are on the order of a few nm and tenths of nm, respectively. Taking as an example the values for a sample in the relaxed state ($\Delta x = 10$ nm, $x_T = 20$ nm, $W = 700$ nm, $W_n = 120$ nm, and $\Delta C/C = 0.14$) we arrive at $N_T/\overline{N^-} \approx 100$. This means that the concentration of N1 states at the point where they are crossed by the Fermi level exceeds the net doping acceptor concentration in the bulk of the absorber by two orders of magnitude. Since N1 traps are donors located close to CBM, it implies that CIGSe is highly compensated. Analogical calculations for the LS state indicate that light soaking results in a decrease of the degree of compensation (in this example from 0.99 to 0.96), which is in accordance with the behaviour of metastable defects and their influence on space charge profiles described above. As several approximations were required, the accuracy of these computations is limited. Nevertheless they provide lower bounds for the values of shallow acceptor and compensating donor concentrations.

Nonlinearity of Arrhenius Plots

As shown in Figure 13c the time constants τ_n obtained by L–RDLTS (full symbols) and AS (open symbols) coincide perfectly. This is because the trap response is localized around the point where the Fermi–level and the trap level intersect, i.e. in the region where capture equals emission (see Figure 2). Hence, also for RDLTS we have $\tau_n = (e_n + nc_n)^{-1} = (2e_n)^{-1}$, where n is the free electron concentration and c_n is the capture coefficient. However, Arrhenius plots for N1 signal are very often curved in the whole temperature range. It was shown that these deviations from linearity can be understood within the framework of the thermally assisted tunnelling model (TAT) when the carriers may

---

[11] In the standard situation (considered in [67]) capacitance transients in RDLTS mode originate from a capture of carriers by bulk traps. In general such transients are strongly non-exponential due to the dependence of the capture rate on the spatial coordinate. Here the N1 compensating donors are charged/discharged in a very narrow region around x_T (an order of magnitude thinner than for bulk acceptors) and the time constant is dominated by $\tau(n(x_T))$ [35].

be emitted across the potential barrier in the way depicted in the inset of Figure 13c [30]. Then the modified emission rate e_{TAT} is given by [79]

$$e_{TAT} = e_n \left(1 + \int_0^{E_T/kT} \exp\left[z - z^{3/2} \left(\frac{4}{3} \frac{(2m^*)^{1/2}(kT)^{3/2}}{q\hbar F} \right) \right] dz \right), \quad (3)$$

$e_n = \sigma_n \upsilon_{th}^n N_C \exp\left(\frac{-E_T}{kT} \right)$ being the emission rate in the zero field limit. E_T is the trap depth, k is the Boltzman constant, T is temperature and $\sigma_n, \upsilon_{th}^n$, and N_C are electron capture cross section, thermal velocity of electrons and the effective density of states in the conduction band, respectively, m^* is the the electron effective mass, q – the elementary charge, h –the Planck constant and F – the electric field at the point where the tunnelling takes place. The result of a fit to Equation 3 is represented in Figure 13c by a solid line. The parameters found by the fitting procedure [30] are: $E_T = (210 \pm 10)$ meV, $\sigma_n = (7\pm4)\times10^{-16}$ cm^2, and $F = (1.14\pm0.06)\times10^7$ Vm^{-1}. The magnitude of the electrical field on the order of 10^7 Vm^{-1} sounds reasonable since the N1 signal is localized in the vicinity of the CdS/CIGSe interface. It should be pointed out that using the standard linear regression method for AS data alone, i.e. for the apparently most linear part of the Arrhenius plot, gives strongly underestimated values of $E_T = (116 \pm 6)$ meV and $\sigma_n = 1\times10^{-18}$ cm^2 as compared to the results of the fit in the TAT model. In order to visualize this discrepancy the emission rates not modified by tunnelling, i.e. e_n calculated for the trap parameters obtained from the fit to Equation 3 with F=0, are marked in Figure 13c by a blue line. Clearly, even at higher temperatures the impact of tunnelling on the emission rates cannot be neglected and limiting the analysis only to admittance data may lead to significant errors in trap parameters.

Metastable Behaviour of N1 Minority Carrier Traps

Illumination and reverse bias affect the N1 response by changing persistently its time constant, as illustrated in Figure 14. The shifts of RDLTS and admittance spectra exemplify typical behaviour which consists in an increase after LS (circles) or decrease after REV (triangles) of the N1 peak emission rates as compared to REL state (squares). The height of the admittance step in metastable states is conserved or slightly diminished (Figure 14a) which is in accordance with the interpretation of N1 defects as traps charged in the buffer/absorber interface region[12]. The return to REL state requires annealing above 250 K. The corresponding Arrhenius plots in metastable states are usually also non–linear (Figure 14c). Using fitting in the TAT model it was found that the shift of the N1 peak is due to the change of *both* parameters: activation energy and electrical field [30]. The illumination induced metastable increase of the emission rate can be explained in terms of TAT effect only by the simultaneous lowering of the barrier height and the decrease of the electrical field. Reverse bias causes always an increase of F, and depending on the sample, no change[13] or an

---

[12] The increase of net acceptor concentration in the absorber due to LS and REV (note reduced W after illumination and reverse bias in Figure 14a) should affect the depletion width W$_n$ only slightly.
[13] occasionally the emission rate of the N1 transition in the REV metastable state can be higher than that of REL

increase of E_T. The observed changes of the electrical field can be understood in the framework of the L–Z model as resulting from accumulation (after REV) and reduction (after LS) of the negative charge in the close-to-interface p^+ layer. Since the large charge redistribution in the interface region is able to move the Fermi level at the buffer/CIGSe interface, the change of activation energy may be ascribed to the upward (LS) or downward (REV) shift of E_F. This assignment is straightforward only if N1 signal is interpreted as the response from a continuous distribution of interface states. Then $E_T = E_C - E_F$, i.e. the barrier height is given as a distance of the Fermi level from the conduction band minimum at the interface [38]. Quite often, however, the N1 signal consists of two or three other components and then this interpretation is not sufficient [80]. It was proposed, basing on the analysis in the TAT model, that in addition to interface states three bulk defects of activation energies 340 meV, 260 meV, and 190 meV are present in the interface region and that their contribution to N1 signal may metastably change after illumination or reverse bias treatment [30]. Arrhenius plots for these levels together with best fit lines to Equation 3 are shown in Figure 15. It should be noted that the characteristics of N1 peak after REV treatment deviate in some samples from the typical described above: (i) the sign of the RDLTS signal does not invert to a positive one after switching to DLTS mode, and (ii) the AS step height exceeds considerably the width of the buffer and i–ZnO layers. As discussed in [57] such behaviour can be explained by the influence of holes discharging the deepest level (340 meV) from the absorber side. This transition was tentatively attributed to the $(V_{Se} - V_{Cu})^{2-}$ acceptor level located 0.8 eV above the valence band [57].

Summing up DLTS and AS can provide valuable information about processes taking place in the buffer/CIGSe interface region. The mechanism of metastable changes in defect spectra that emerges from the analysis presented above is consistent with the L–Z model used for explanation of metastabilities in capacitance space charge profiles. Here the redistribution of the negative charge in the p+ layer influences minority carrier N1 signal and is reflected in the value of the electrical field (increased after REV and reduced after LS) and the position of the Fermi–level at the buffer/CIGSe interface (shifted towards VB and CB after REV and LS, respectively).

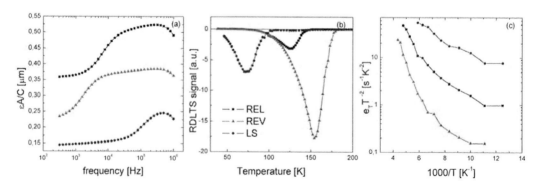

Figure 14. Metastable changes induced by LS and REV treatments in N1 admittance (a) and RDLTS (b) spectra. Figure (c) shows corresponding Arrhenius plots.

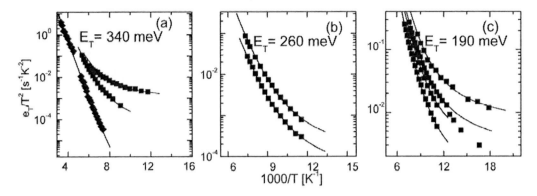

Figure 15. Arrhenius plots for close–to–interface donor states of activation energy 340 meV (a) 260 meV (b) and 190 meV (c). Different emission rates for each trap correspond to differences in electrical field as revealed by fitting in the TAT model (best fit indicated by lines).

LIGHT CURRENT-VOLTAGE CHARACTERISTICS

Most importantly from the point of view of applications, light current–voltage characteristics of CIGSe–based solar cells also exhibit metastable behaviour. Typically, if light absorbed exclusively in the CIGSe layer is used, LS under open and short circuit conditions result in a metastable increase and decrease of FF, respectively [16,81]. These changes are stable at low temperatures and anneal out gradually above ~250 K [81]. As usual in CIGSe devices the magnitude of these effects may vary from sample to sample (in rear cases FF does not change under illumination) but they are observed in structures with absorbers of different compositions and buffer layers [16,81]. In order to avoid FF deterioration induced by charging the In_{Cu} defects with electrons [14] all light I–V curves were measured for positive voltages only (from 0 V).

The difference between the action of light soaking[14] under short and open circuit conditions for CIGSe cells with ALD–Zn(O,S) buffer is illustrated in Figure 16. The light I–V curves were taken at 300 K under white light illumination. The transition between the two metastable states is reversible and continuous, i.e. during illumination under open/short c.c. the fill factor increases/decreases gradually as indicated by arrows, and finally saturates either at the FF_{LSO} (open circles) or FF_{LSS} (full circles) value. It should be noted that the only parameter which is influenced is the FF whereas V_{oc} and I_{sc} remain unaltered. The metastable changes in light I–V curves are correlated with capacitance space charge profiles. As depicted in Figure 17, FF changes induced by LS are accompanied by the redistribution of the negative charge in the absorber. LSS results in an increase (full circles) of the net acceptor concentration by a factor of ~2 as compared to LSO (open circles) whereas the surplus negative charge is accumulated in a 50 nm thin close–to–interface layer (p+ layer). Interestingly, reverse bias soaking influences space charge profiles in the interface region in a similar way as LSS does but the width of the p+ layer extends up to ~100 nm (REV, full triangles).

---

[14] Since there is no or negligible absorption of white light in ZnOS [62], for such cells white light plays the role of red light.

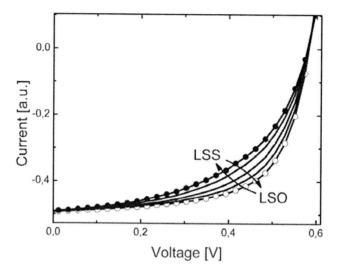

Figure 16. White light current–voltage curves of the Zn(O,S)–buffered device recorded at 300 K during 15 min. of white light soaking under short (transition LSO–LSS) and open (transition LSS–LSO) circuit conditions.

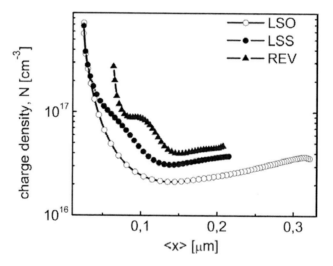

Figure 17. Low temperature (120 K) space charge profiles for a Zn(O,S)–buffered cell in LSO, LSS and REV states.

The FF degradation caused by LS can be observed in standard cells with CdS buffer provided that the red light, i.e. not absorbed in CdS, is used for inducing the metastabilities as well as for measurements. Also here light I–V characteristics (Figure 18b) correlate with space charge distributions (Figure 18a). The photo–current begins to deteriorate exactly at the same voltage for which the increased concentration of shallow acceptors in the vicinity of the CIGSe/CdS interface shows up in the C–V profiles (as indicated by arrows: +0.1 V and +0.3V for RLSS and RLSO, respectively). The strongest FF improvement is observed after white light soaking (WLS, minor difference between open and short) and it is accompanied by the largest reduction of the negative charge in the close–to–interface absorber region (solid lines in Figure 18).

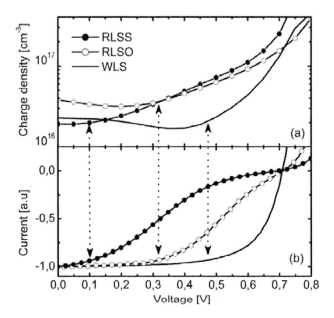

Figure 18. Correlation between metastable space charge distributions (a) and red light I–V curves (b) in a CdS–buffered cell. Measurements were done at 250 K.

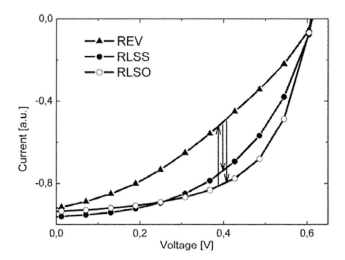

Figure 19. Red light I–V curves for a CdS–buffered cell in REV, RLSS and RLSO metastable states. Measurements done at 300 K. Current transients corresponding to the transitions between metastable states indicated by arrows are shown in Figure 20.

Application of reverse bias at temperatures above about 250 K is detrimental for the cell performance and leads to a gradual deterioration of FF, as illustrated in Figure 19 and Figure 20 (triangles). This process is also reversible and during light soaking FF improves (circles). However, the maximal value at which FF saturates depends on the conditions of the illumination: in most devices red light soaking under open circuit conditions (RLSO, open circles) results in higher FF values as compared to short circuit conditions (RLSS, full circles). This shows that the action of LSS depends on the initial state of the sample and can lead to either an improvement (after REV) or deterioration (after LSO) of the FF. The recovery of the FF under RLSO is faster than under RLSS, which is shown in Figure 20. The three current transients (measured at U=+0.4V) represent transitions marked in Figure 19 by arrows: RLSO–>REV (triangles), REV–> RLSO (open circles) and REV–> RLSS (full circles).

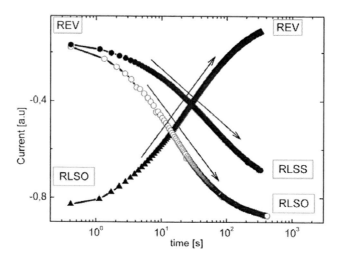

Figure 20. Current transients corresponding to transitions between the metastable states from Figure 19.

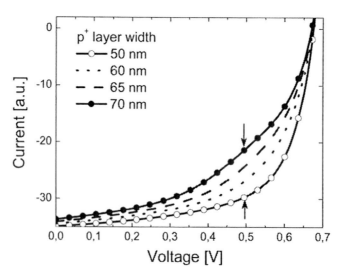

Figure 21. Simulated light I–V curves for different p+ layer widths. Band diagrams shown in Figure 22 and 23 correspond to 50 nm and 70 nm widths at 0 V and 0.5 V (indicated by arrows).

Thus, in samples with different buffers the LSS/LSO/REV treatments affect FF and space charge profiles in a similar way, and the only requirement for inducing LSS/LSO transition is that the light must be absorbed exclusively in CIGSe. Hence, the metastable behaviour of light current – voltage curves originates from the absorber layer itself rather than from other parts of the junction (a buffer, a window or some interface). Moreover, FF is closely correlated to the width of the close–to–interface absorber layer with the concentration of a surplus negative charge on the order of 10^{17} cm^{-3}. The LSS/LSO–induced changes are stable at low temperatures and anneal out above ~250 K. This leads to the conclusion that also FF–related metastabilities can be best explained in the p+ layer model with the excess negative charge stemming from (–) and (2–) (V_{Se}-V_{Cu}) charge states (In_{Cu} – related defects can be charged/discharged within seconds even at low temperatures).

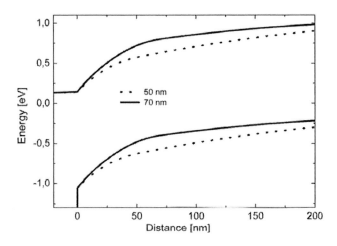

Figure 22. Simulated band diagrams at 0 V corresponding to 50 nm (dotted line) and 70 nm (solid line) widths of the p+ layer from Figure 21.

The insight into the mechanisms influencing FF was provided by numerical modelling of light current–voltage characteristics [82]. The metastable behaviour of light I–V curves can be reproduced by changing the thickness of the p+ layer (W_{p+}). The exemplary set of simulated light I–V characteristics for four values of W_{p+} between 50 nm and 70 nm is shown in Figure 21. The origin of FF losses can be understood from Figure 22 and Figure 23 which graph simulated band diagrams of the junction region for W_{p+} equal to 50 nm and 70 nm, at 0 V (Figure 21) and +0.5 V (Figure 22). These thicknesses of the p+ layer correspond to the experimental charge distributions after LSO and LSS, respectively. It is plain to see that the extension of the p+ layer results in a significant decrease of the electrical field in a part of the absorber ranging from 70 nm to 200 nm. Hence, in this model, the photo–current loss at forward biases can be ascribed to the reduced collection of photo–generated carriers in the absorber region adjacent to the p+ layer. As can be seen from Figure 21 in the simulated light I–V characteristics the open circuit voltage does not change, which is also in accordance with experimental results. Considering the employed model this might be surprising since one would expect that Voc will drop after increasing the number of deep $(V_{Se}$-$V_{Cu})^{2-}$ states in the space charge region. However, as shown in Figure 24, the recombination through these defects effectively takes place only in a restricted region very close to the interface, whereas the changes generated by LSS/LSO/REV treatments occur at larger distances. The reason why the action of light soaking depends so strongly on illumination conditions can be inferred from Figure 25. It displays simulated free electron to hole (n/p) concentration ratios in the interface region upon illumination under short (black) and open (gray) circuit conditions for two widths of the p+ layer, 50 nm and 70 nm, respectively. The distance 0 nm stands for the buffer/CIGSe interface. Under illumination electrons and holes are generated in the space charge region and the distribution of (V_{Se}-V_{Cu}) states depends on the local n/p ratio. Due to quite high activation energy for hole capture (~0.3 eV) the reaction $(V_{Se} - V_{Cu})^- + 2h \rightarrow (V_{Se} - V_{Cu})^+$ will prevail only in regions where p>>n, whereas the inverse transformation $(V_{Se} - V_{Cu})^+ + e \rightarrow (V_{Se} - V_{Cu})^- + h$ requires much lower energy (~0.1 eV) and will dominate in regions for which n>p or n≤p. Hence, the changes of (V_{Se}-V_{Cu}) occupation induced by illumination under open and short circuit conditions, which result in the shrinkage or widening of the p+ layer, respectively, require that in the narrow close–to–interface region the hole concentration changes by orders of magnitude. Indeed, as expected from experimental results in Figure 17, such large changes in the n/p ratio are restricted to the absorber region between ~40 nm and ~100 nm (indicated by arrows). Application of reverse bias (without illumination) depletes the interface region from both holes and electrons. The holes will be emitted then from positively charged (V_{Se}-V_{Cu}) according to the reaction $(V_{Se} - V_{Cu})^+ - 2h \rightarrow (V_{Se} - V_{Cu})^-$ and as a consequence the p+ layer will become much thicker (c.f. Figure 17). Subsequent light soaking will lead to the reduction of this surplus negative charge. However, due to large differences in the n/p ratio in the interface region during LSO and LSS, the final width of the p+ layer (determining the value at which FF saturates) as well as the time needed for reaching the quasi–equilibrium will depend on illumination conditions. This is in accordance with experimental results displayed in Figure 19 and Figure 20. The largest reduction of the p+ layer thickness takes place after white light soaking of CdS–buffered cells (see Figure 18). This is because WLS results in a strong injection of photo–generated holes from the buffer into the interface region. The stability of the p+ layer model against variations of important material parameters was demonstrated [82] and very good qualitative agreement with experimental results was obtained for wide ranges of absorber and buffer net shallow doping concentrations, carrier mobilities and the position of the Fermi level at the buffer/absorber interface.

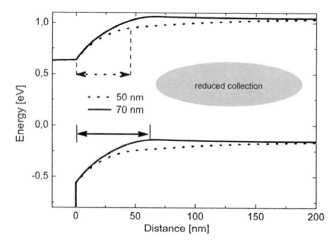

Figure 23. Simulated band diagrams at 0.5 V corresponding to 50 nm (dotted line) and 70 nm (solid line) widths of the p+ layer from Figure 21.

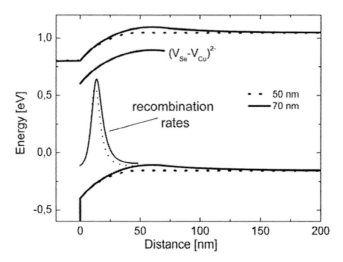

Figure 24. Simulated band diagrams at V_{oc} corresponding to 50 nm (dotted line) and 70 nm (solid line) widths of the p+ layer from Figure 21. The recombination rates for the deep acceptor states associated with $(V_{Se}-V_{Cu})^{2-}$ are indicated.

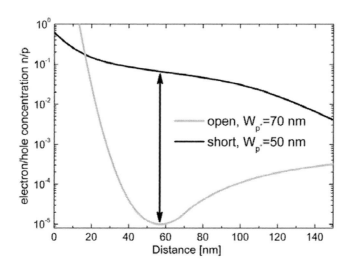

Figure 25. Simulated electron to hole concentrations in the interface region upon illumination under open and short circuit conditions. The large difference in the n/p ratio is indicated.

CONCLUSION

The impact of two intrinsic defects, V_{Se} and In_{Cu}, on basic electrical characteristics of CIGSe – based solar cells was reviewed. The analysis of capacitance space charge profiles confirms the theoretical predictions concerning the non–uniform charge distribution within the absorber layer and reveals that their concentrations may considerably exceed the net acceptor concentration. The negative charge accumulated at these defects in their metastable configurations forms the p+ layer. It appears that understanding the properties of this thin close–to–interface CIGSe region is a key for correct interpretation of other electrical measurements. The experimental results presented in this chapter can be summarized as follows:

(i) There exists a relatively simple criterion for a distinction between the influence of In_{Cu} and V_{Se} defects on metastable changes in electrical characteristics. If the metastable state disappears at low temperature (below ~250 K) due to hole injection (after blue illumination or application of forward bias) it can be assigned to In_{Cu}. On the contrary, due to a relatively large barrier for hole capture, negatively charged V_{Se} are not sensitive to free holes below ~250 K and the metastabilities associated with this defect remain frozen practically up to room temperature. This is best illustrated by the comparison of ROB and REV metastable states: despite the apparent similarity of the corresponding space charge distributions the ROB state vanishes almost immediately after hole injection at low temperature whereas REV slowly anneals out only above 250K.

(ii) There is a close correlation between the metastable behaviour of capacitance space charge profiles, light I–V characteristics and the N1 response of minority carrier traps. The largest amount of negative charge in the p+ layer appears in the REV state. It is accompanied by the lowest FF, the highest electrical field in the interface region and the highest activation energy of N1 traps (reflecting the shift of the Fermi level towards the VB at the buffer/CIGSe interface). The opposite changes are induced by light soaking. Apart the overall increase of doping level in the bulk absorber, it leads to the metastable shrinkage of the p+ layer, FF improvement, decrease of electrical field in the vicinity of the interface and the lowering of N1 states activation energy. Since these phenomena are stable below ~250 K, according to point (i) they are attributed to the redistribution of V_{Se} charge states.

(iii) The metastable FF changes induced at RT by LSS and LSO are ascribed to the redistribution of V_{Se} charge states in the p+ layer due to large differences in the electron to hole concentration ratio upon illumination under short and open circuit conditions in the interface region.

(iv) It was proposed that the distortion in the rear part of the U–shaped low temperature C–V profiles is due to accumulation of negative charge at the deep DX levels of the In_{Cu} defects.

ACKNOWLEDGMENTS

This work was partially supported by the UPB projects at WUT. The author wishes to thank Dr. C. Platzer–Björkman and Dr. N. Barreau for providing the samples.

REFERENCES

[1] Repins, I.; Contreras, M.A.; Egaas, B.; DeHart, C.; Scharf, J.; Perkins, Noufi, R. *Prog. Photovolt. Res. Appl.* 2008, 16 (3), 235–239.

[2] Ullal, H.S.; von Roedern, B. *Proceedings of 22nd European Photovoltaic Solar Energy Conference* 2007, 1926–1932.

[3] Lany, S.; Zunger, A. *Phys. Rev. B* 2005, 72, 035215.
[4] Lany, S.; Zunger, A. *J. Appl. Phys.* 2006,100, 113725.
[5] Lany, S.; Zunger, A. *Phys. Rev. Lett.* 2008, 100, 016401.
[6] Herberholz, R.; Rau, U.; Schock, H.W.; Haalboom, T.; Godecke, T.; Ernst, F.; Beilharz, C.; Benz, K.W.; Cahen, D. *Eur. Phys. J. AP* 1999, 6, 131–137.
[7] Rau, U.; Jasenek, A.; Herberholz, R.; Schock, H.W.; Guillemoles, J.F.; Kronik, L. *Proc. 2nd World Conf. Solar Energy Conv., Ed. by J. Schmid, H.A. Ossenbrink, P. Helm, H.Ehmann, E.D. Dunlop* 1998, 428–436.
[8] Herberholz,R. *Inst. Phys. Conf. Ser. Ed.by R.D. Thomlinson, A. E. Hill, R. D. Piklington* 1998,152, 733–739.
[9] Igalson, M.; Zabierowski, P. *Thin Solid Films* 2000, 361, 371–377.
[10] Heath, J. T.; Cohen, J. D.; Shafarman, W. N. *J. Appl. Phys.* 2004, 95, 1000.
[11] Igalson, M.; Ćwil, M.; Edoff, M. *Thin Solid Films* 2007, 515, 6142–6146.
[12] Ćwil, M.; Igalson, M.; Zabierowski, P.; Siebentritt, S. *J.Appl. Phys.* 2008, 103, 063701.
[13] Zabierowski, P.; Rau, U.; Igalson, M. *Thin Solid Films* 2001, 387, 147–200.
[14] Igalson, M.; Bodegård, M.; Stolt, L. *Sol. En. Mat. Sol. Cells 2003*, 80, 195–207.
[15] Igalson, M. *Mater. Res. Soc. Symp. Proc.* vol. 1012, 1012–Y04–01.
[16][16] Zabierowski, P.; Platzer–Björkman, C. *Proc. 22nd EC Photovoltaic Solar Energy Conf.* 2007, 2395–2399.
[17] Igalson, M.; Zabierowski, P.; Przado, D.; Urbaniak, A.; Edoff, M.; Shafarman, W.N. 2009 *Sol. En. Mat. Sol. Cells* 93 (8), 1290–1295.
[18] Meyer, T.; Engelhardt, F.; Parisi, J.; Schmidt, M.; Rau, U. Proc. *2nd World Conf. Solar Energy Conv.;Ed. by J. Schmid, H.A. Ossenbrink, P. Helm, H. Ehmann, E.D. Dunlop* 1998, 1157–1160.
[19] Ruberto, M.N.; Rothwarf, A. *J. Appl. Phys.* 1987, 61, 4662–4665.
[20] Sasala, R.A.; Sites, J.R. *Proc 23rd IEEE Photovolt. Spec. Conf. (IEEE, New York 1993)*,1993, 543–5436.
[21] Igalson, M. *phys. stat. sol. (a)* 1993, 139, 481–494.
[22] Engelhardt, F.; Schmidt, M.; Meyer, Th.; Seifert, O.; Parisi, O.; Rau, U. *Phys. Lett. A* 1998, 245(5), 489–493.
[23] Meyer, Th.; Engelhardt, F.; Parisi, J.; Riedl, W.; Karg, F.; Schmidt, M.; Rau, *2nd World Conf. Solar Energy Conv.; Ed. by J. Schmid, H.A. Ossenbrink, P. Helm, H. Ehmann, E.D. Dunlop* 1998, 435–438.
[24] Seifert, O.; Engelhardt, F.; Meyer, T.; Hirsch, M.T.; Parisi, J.; Beilharz, C.; Schmidt, M.; Rau, U. *Proc. 11th Int. Conf. on Ternary and Multinary Compounds, Ed. by R. Tomlinson, E.A. Hill, R.D. Piklington* 1998, 253–6.
[25] Schmitt, M.; Rau, U.; Parisi, J. *Phys. Rev. B* 2000, 61, 16052–16058.
[26] Rau, U.; Schmitt, M.; Parisi, J.; Riedl, W.; Karg, F. *Appl. Phys. Lett.* 1998, 73 (2), 223–225.
[27] Igalson, M.; Schock, H. W. *J. Appl. Phys.* 1996, 80, 5765–5772.
[28] Zabierowski, P.; Igalson, M.; Schock, H.W. *Solid State Phenomena* 1999, 67, 403–408.
[29] Igalson, M.; Zabierowski, P. *Opto–electronics Review* 2003, 11 (4), 261–267.
[30] Zabierowski, P.; Edoff, M. *Thin Solid Films* 2005, 480–481, 301–306.
[31] Przado, D.; Igalson, M.; Bacewicz, R.; Edoff, M. *Acta Physica Polonica A* 2007, 112 (2), 183–189.
[32] Kniese, R.; Powalla, M.; Rau, U. *Thin Solid Films* 2007, 515, 6163–6167.

[33] Schroder, D. K., *Semiconductor Material and Device Characterization* 3.Edition – February 2006, John Wiley & Sons, Inc. Publ.
[34] Kimerling, L.C. *J. Appl. Phys* 1974, 45 (4), 1839–1845.
[35] Zabierowski, P.; Ćwil, M. Edoff,M. *MRS Symposia Proceedings,* 2005, 865, 1231–1235.
[36] Urbaniak, A.; Igalson, M. *J. Appl. Phys.* 2009, 106 (6), 063720.
[37] Herberholz, R.; Igalson, M.; Schock, H.W. *J. Appl. Phys.* 1998, 83 (1), 318–325.
[38] Niemegeers, A.; Burgelman, M.; Herberholz, R.; Rau, U.; Hariskos, D.; Schock, H.–W. *Prog. Photovolt. Res. Appl.* 1998, 6 (6), 407–421.
[39] Igalson, M.; Zabierowski, P.; Romeo, A.; Stolt, L. *Opto–electronics Review* 2000, 8 (4), 346–349.
[40] Igalson, M.; Stolt,L. *Proc. 14th Europ. Photovolt. Solar Energy Conf., Ed. by H.A. Ossenbrink,P. Helm, H. Ehmann* 1997, 2153–2156.
[41] Rau, U.; Braunger, D.; Herberholz, R.; Schock, H.W.; Guillemoles, J.F.; Kronik, L.; Cahen, D. *J. Appl. Phys* 1999, 86 (1), 497–505.
[42] A. Niemegers, S. Gillis, M. Burgelman, *Proc. 2nd World Conf. Solar Energy Conv. (Vienna), Ed. by J. Schmid, H.A. Ossenbrink, P. Helm, H. Ehmann, E.D. Dunlop* 1998, 1071–1075.
[43] Kushiya, K.; Tanaka, Y.; Hakuma, H.; Goushi, Y.; Kijima, S.; Aramoto, T.; Fujiwara, Y. *Thin Solid Films* 2009, 517 (7), 2108–2110.
[44] Kushiya, K.; Tanaka, Y.; Hakuma, H.; Goushi, Y.; Kijima, S.; Aramoto, T.; Fujiwara, Y. *MRS Symposia Proceedings,* 2009, in print
[45] Scheer, R. *J. Appl. Phys* 2009,105 (10), 104505.
[46] Nadenau, V.; Rau, U.; Jasenek, A.; Schock, H.W. *J. Appl. Phys* 2000, 87 (1), 584–593.
[47] Malm, U.; Malmström, J.; Platzer–Björkman, C.; Stolt, L. *Thin Solid Films* 2005, 480–481, 208–212.
[48] Rau, U.; Jasenek, A.; Schock, H.W.; Engelhardt, F.; Meyer, T. *Thin Solid Films* 2000, 361, 298–302.
[49] Rau, U. *Appl. Phys. Lett.* 1999, 74 (1), 111–113.
[50] Rau, U.; Schmidt, M.; Jasenek, A.; Hanna, G.; Schock, H.W. *Sol. En. Mat. Sol. Cells* 2001, 67, 137–143.
[51] Hanna, G.; Jasenek, A.; Rau, U.; Schock, H.W. *Physica Status Solidi A Appl. Res.* 2000, 179 (1), R7–R8.
[52] Hanna, G.; Jasenek, A.; Rau, U.; Schock, H.W. *Thin Solid Films* 2001, 387, 71–73.
[53] Zabierowski, P.; Platzer–Björkman, C.; Ćwil, M. *Proceedings of 21st European Photovoltaic Solar Energy Conference* 2006, 2006 – 2009.
[54] Topič, M.; Smole, F.; Furlan, J. *Solar Energy Materials and Solar Cells* 1997, 49 311–317.
[55] Pudov, A.O.; Kanevce, A.; Al–Thani, H.A.; Sites, J.R.; Hasoon, F.S. *J. Appl. Phys* 2005, 97 (6), 064901–064906.
[56] Igalson, M. *Materials Research Society Symposium Proceedings* 2007, 1012, 211–222.
[57] Igalson, M.; Urbaniak, A.; Edoff, M. 2009 *Thin Solid Films* 517 (7), 2153–2157.
[58] Kessler, J.; Bodegård, M.; Hedström, J.; Stolt, L. (2001) *Solar Energy Materials and Solar Cells,* 67 (1–4), 67–76.
[59] Barreau, N.; Lähnemann, J.; Couzinié–Devy, F.; Assmann, L.; Bertoncini, P.; Kessler, J. *Sol. En. Mat. Sol. Cells* 2009, 93 (11), 2013–2019.
[60] Siebentritt, S. *Solar Energy* 2004, 77 (6), 767–775.

[61] Hariskos, D.; Spiering, S.; Powalla, M. *Thin Solid Films* 2005, 480–481, 99–109.
[62] Platzer–Björkman, C.; Törndahl, T.; Abou–Ras, D.; Malmström, J.; Kessler, J.; Stolt, L. *J. Appl. Phys* 2006, 100 (4), 044506.
[63] Törndahl, T.; Platzer–Björkman, C.; Kessler, J.; Edoff, M. *Prog. Photovolt. Res. Appl.* 2007, 15 (3), 225–235.
[64] Rao, G.V., Säuberlich, F., Klein, A. *Appl. Phys. Lett.* 2005, 87 (3), 1–3.
[65] Provencher, S.W. *Computer Physics Communications* 1982, 27 (3), 213–227.
[66] Dobaczewski, L., Peaker, A.R., Bonde Nielsen, K. *J. Appl. Phys* 2004, 96 (9), 4689–4728.
[67] Li, G.P.; Wang, K.L. *J. Appl. Phys* 1985, 57 (4), 1016–1021.
[68] Burgelman, M.; Nollet, P.; Degrave, S. *Thin Solid Films* 2000, 361, 527–532.
[69] Ćwil, M.; Igalson, M.; Zabierowski, P.; Kaufmann, C.A.; Neisser, A. *Thin Solid Films* 2007, 515, 6229–6232.
[70] Igalson,M.; Stolt,L. *Jpn. J. Appl. Phys.* 2000, 39, 426–428.
[71] Zabierowski, P.; Krysztopa, A. *unpublished.*
[72] Igalson, M.; Ćwil, M.; Edoff, M. *Thin Solid Films* 2007, 515, 6142–6146.
[73] Krysztopa, A.; Igalson, M.; Papathanasiou, N. *Physica Status Solidi C* 2009, 6, 1291–1294.
[74] Lany, S.; Zhao, Y.–J.; Persson, C.; Zunger, A. *Appl. Phys. Lett.* 2005, 86 (4), 042109.
[75] Lany, S. *private communication.*
[76] Urbaniak, A., Igalson, M., Siebentritt, S. *Materials Research Society Symposium Proceedings* 2007, 1012, 385–392.
[77] Urbaniak, A.; Igalson, M. *Thin Solid Films* 2009, 517 (7), 2231–2234.
[78] Murray, F.; Carin, R.; Bogdanski, P. *J. Appl. Phys* 1986, 60 (10), 3592–3598.
[79] Vincent, G.; Chantre, A.; Bois, D. *J. Appl. Phys* 1979, 50 (8), 5484–5487.
[80] Zabierowski, P. *PhD Thesis, 2002, Warsaw* (in polish)
[81] Zabierowski, P.; Platzer–Björkman, C.; *unpublished*
[82] Zabierowski, P.; *unpublished*

Chapter 6

CdTe SOLAR CELLS BY LOW TEMPERATURE PROCESSES

Alessandro Romeo[1] and Alessio Bosio[2]
[1]University of Verona Faculty of Science, 37134 Verona, Italy;
[2]University of Parma, 7A-43100 Parma, Italy.

ABSTRACT

CdTe polycrystalline thin film solar cells have a strong potential in scalability. They have shown long-term stable performance and high efficiency up to 16.5% under AM1.5 illumination. Amongst several attractive features, high chemical stability of CdTe and a simple compound formation are the most important ones for large area production of solar modules.

A further simplification has been done by substituting the $CdCl_2$ step by treating CdTe films in an atmosphere containing a non toxic gas that is inert at room temperature, like HCF_2Cl, that belongs to the Freon© family. The treatment temperature is typically 400°C, for a few minutes and in an atmosphere containing Cl, typically 100 mbar of Ar containing 15% of HCF_2Cl. The change in the morphology of CdTe films after treatment is very similar to that obtained with $CdCl_2$ treatment and an increase in the size of small grains is always observed. This process has been applied by N. Romeo et al. on CdTe deposited by close-spaced sublimation (CSS) with very interesting results (15.8% efficiency). The application of the regular $CdCl_2$ treatment and of this novel "activation process" on low and high temperature processed solar cells will be described

Moreover, there are new promising device configurations like bifacial solar cells, ultra-thin solar cells and flexible devices. The highest efficiencies in CdTe solar cells have been obtained using CSS deposition method, which requires a high substrate temperature (500÷550 °C). Instead, conventional physical vapor deposition (PVD) process where CdTe is evaporated in a high vacuum evaporation (HVE) system at lower substrate temperatures (typically 300°C) has provided solar cells with efficiencies of more than 12%. For these reasons HVE process is attractive for a very simple in-line deposition of large area CdTe solar modules on soda-lime glass substrates, as well as on polymer foils thereby facilitating the roll-to-roll manufacturing of flexible solar modules.

Flexible CdTe/CdS solar cells of 11% efficiency in superstrate and 7% efficiency in substrate configurations have been developed with a "lift-off" approach. However, roll-to-roll manufacturing is desired in future. Therefore, flexible superstrate solar cells have also been directly grown on commercially available polyimide foils. Solar cells with AM1.5 efficiency of 12.4% on Upilex™ foils (highest efficiency recorded for flexible CdTe cell) have been developed.

A different possibility in making flexible CdTe solar cells is to use a metal foil as substrate, provided that the stacks would be deposited in the opposite order to have light coming from the top and not through the substrate (the so-called "substrate configuration" process). However making a CdTe photovoltaic device in inverted structure implies a variety of different scientific issues, that will be addressed.

The latest development is the application of a transparent conducting oxide (TCO) ITO as a back electrical contact on CdTe leading to first bifacial CdTe solar cells, which can be illuminated from either or both sides. Accelerated long term stability tests show that light soaking improves the efficiency of CdTe solar cells with ITO back contacts and performance does not degrade. Such solar cells are attractive for tandem solar cells though larger band gap absorbers based on CdTe would be desired. Application of light trapping concepts can be employed to reduce the CdTe layer thickness to below 1 micron.

INTRODUCTION

For many years CdTe thin film solar cells have been one of the most promising technologies for low cost, high efficiency, large scale production photovoltaics. As a matter of fact CdTe thin film can be prepared by a variety of different deposition techniques. CdTe evaporates congruently and it grows stoichiometrically even at low substrate temperatures.

These features that have been stressed for many years by the researchers, have been now demonstrated by the enormous commercial success of an American CdTe thin film PV company and nowadays CdTe is not only a promising technology but a cheap and reliable product.

In particular what has allowed this technology to move from the laboratories to the fabrication plants has been the preparation of a stable and efficient back contact, that for a long time has been a difficult issue and it is still an interesting topic for researchers, and the high temperature deposition process which assures high efficiency.

Despite this, there is still a lot to be investigated and developed in order to still reduce production costs, improve efficiency and introduce new concepts like flexible or even bifacial cells.

The complete understanding and upgrade of the so called "activation process" could improve the performance of the cell. As a matter of fact CdTe solar cells could have efficiencies exceeding 20% if an appropriate doping of the CdTe would be applied.

CdTe high efficiency cells could be produced in flexible configuration, in order to have very light modules that would also be very easy to integrate in buildings. Cheap and real flexible modules would allow to a new concept in exploiting solar energy in buildings.

Moreover the thickness of CdTe can be reduced by optimizing the fabrication process and applying a reflecting back contact. This would result in a considerable reduction in the material costs.

Low temperature processes are typically less performing staying around 14% [1] in efficiency against the 16.5% [2] record efficiency for high temperature deposition but, on the other hand, they do give a series of advantages among which there are the lower energy demand in processing (resulting in lower fabrication costs and lower emissions) and the possibility of using innovative substrates (polymides for flexible cells) that do not withstand high temperatures.

In this chapter the state of the art of low temperature deposition processes for CdTe solar cell preparation will be presented, together with some new concepts like flexible and bifacial cells or new fabrication processes like an innovative recrystallization treatment.

THE FRONT CONTACT (TRANSPARENT CONDUCTIVE OXIDE, TCO)

The first layer to be deposited for the fabrication of a CdTe solar cell is the front contact.

The front contact has to be highly conductive and transparent, typically a transparent conductive oxide (TCO) with an electron affinity below 4.5 eV and with a good band alignment with the CdS. If the electron affinity of the TCO is higher than that of CdS, a blocking Schottky contact is formed.

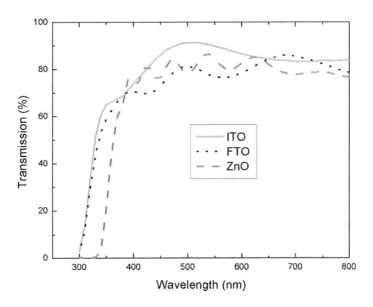

Figure 1. Transmission spectra of ITO, FTO, ZnO:Al layer deposited on glass [4].

The most commonly used TCOs for CdTe solar cells are F-doped SnOx [SnOx:F (FTO)] or ITO that have similar transparency, strongly depending on their thickness (see figure 1). They are often used in combination with a thin intrinsic SnOx layer between the TCO and the CdS window layer which works as a barrier for possible shunts caused by pinholes maintaining a high voltage [3]. Moreover the intrinsic (high-resistivity) transparent oxide facilitates the use of a thinner CdS layer for reducing photon absorption losses for wavelengths below 500 nm.

The Al doped ZnO instead, commonly used in CIGS cells, yields generally a high series resistance in CdTe devices, reducing the conversion efficiency [4]. However, even 14% efficiency cells have been developed on ZnO:Al with a sputtering method [5].

The properties of the TCO layer do not depend only from the deposition methods (usually RF sputtering method is used and the percentage of oxygen in the gas together with the sputtering power affect transparency and conductivity) but also from the CdS deposition and post-deposition annealing of the cell that also change the CdS/TCO characteristics

Moreover ITO front contacts are generally sensitive to annealing treatments and an increase of the electron affinity from around 4 to 5 eV, caused by oxidation or a post-deposition treatment brings to a blocking contact [6,7].

Finally the TCOs physical properties are also depending on the structural properties of the CdTe and CdS [4]. As a matter of fact, a more compact CdTe layer would limit the diffusion of $CdCl_2$ during the "recrystallization treatment" (see next section) or the diffusion of metals into the TCO after the deposition of the back contact. On the other hand high temperature deposition processes would enhance intermixing of the layers, inducing indium atoms to diffuse into the CdTe. For this reason high stability TCO is mandatory for high temperature processes but it is still very important in case of low temperature deposition.

A comparison between different commercial TCO layers such as ITO, FTO and ZnO Aluminum doped shows that they behave differently concerning the properties of the layers deposited successively, giving different performances on the solar cells.

As shown in figure 2, the as-deposited CdTe layers on ZnO have a uniform morphology and larger grains with a size of 1 to 1.5 μm are observed. On the other hand, the CdTe layers on ITO have smaller grains with a size in the range of 0.5 to 1 μm. On FTO the layers have some very small grains with a size of less than 0.5 μm and also a few big grains with a size of ≈2 μm. After the $CdCl_2$ treatment a large increase in the grain size is observed. In this case a 600 nm thick $CdCl_2$ layer was used for all the CdTe layers. $CdCl_2$ acts like a sintering flux in CdTe, small grains grow and coalesce together leading to narrow and large grain boundaries. The so called "$CdCl_2$ treatment" and its effects on the CdTe/CdS stacks will be discussed in the next sections.

Figure 2. SEM pictures of CdTe layers after $CdCl_2$ treatment deposited on CdS and on FTO (right), ITO (center) and Zno:Al (left) [4].

The shapes of grains on FTO and ZnO are almost rectangular with larger grain boundaries and a size of 3-7 μm, while grains of different dimensions are inhomogeneously distributed in CdTe layers on ITO [4].

THE CDS (WINDOW LAYER)

The CdTe high efficiency solar cells use generally CdS for window n-type layer forming a heterojunction.

Indeed homojunction CdTe devices do not provide high efficiency devices, because of the high absorption coefficient of CdTe and its limited dopability. The CdS thin film layers grow easily, with the majority of the deposition methods, in a stoichiometric phase and a hexagonal wurtzite structure.

CdS has a band gap of 2.4 eV. This makes it not completely transparent in the range between 400 and 550 nm, and its transparency is then strongly depending on the layer thickness. A very thin buffer layer, on the other hand, would enhance the formation of pin-holes, especially if the homogeneity of the deposition is critical. For this reason the thickness for the CdS deposition is always a tradeoff between transparency and conformal coverage.

Layers of n-conducting CdS are easily grown by various deposition methods including chemical bath deposition (CBD) as well as physical vapor deposition (PVD). Chemical bath deposition is made at 60°-80°C in a solution of Cd salt, ammonia and thiourea, but the physical properties strongly depend on the reaction that takes place in the bath. CBD yields the highest efficiency devices, since it provides very thin (from 5 up to 50 nm), but continuous layers, giving a high transmission through the window layer also for low-wavelength photons.

By PVD-method CdS generally evaporates congruently (like CdTe) and recomposes as 1:1 cadmium and sulfur even with low substrate temperatures. If high-pressure growth conditions are used (like in the close space sublimation case, see next section), CdS is typically in the cubic, metastable zincblende structure [8].

With high-vacuum evaporation (HVE) CdS films show small columnar grains below one micron [9], growing with a preferred [2110] orientation parallel to the substrate. CdS polycristals grown with HVE method are deposited with a substrate temperature of maximum 150 °C, above which the layer does not stick to the substrate (atoms re-evaporate). Typically the thickness of the HVE-CdS can be varied between 50 and 500 nm, however the optimum thickness is about 100 nm [10].

For the chemical bath deposition case CdS can be much thinner since the process assures a higher conformal coverage, typically from 50 to 100 nm with an optimum thickness of 70 nm [10].

However, high substrate temperatures improve the crystalline quality of the layer and its stability to the following processes. The morphology of the as-deposited CBD-CdS and HVE-CdS layers are quite different (Figure 3). The grain size of HVE-CdS is in the range of 0.1-0.3 μm and the layers are rough. The CBD-CdS consists of clusters of up to 0.5 μm but clearly these clusters are formed due to a coalescence of small grains of about 0.1 μm. The CBD-CdS layers are rather smooth.

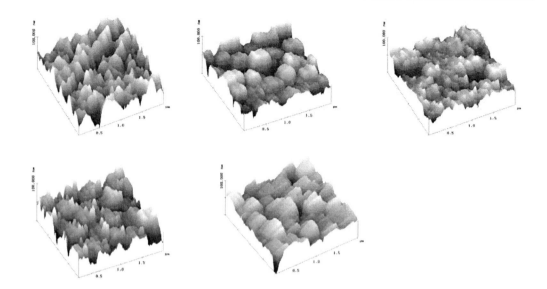

Figure 3. Atomic force microscopy images of as-deposited HVE-CdS (top left), of as-deposited CBD-CdS (top center), vacuum annealed HVE-CdS (top right), vacuum annealed CBD-CdS (left bottom) and CdCl2 treated HVE-CdS (right bottom) [9].

To improve the crystalline quality the CdS layer can be treated in vacuum at higher temperature or can be recrystallized with deposition of $CdCl_2$ and subsequent annealing in air.

As shown on Figure 3, the grain size of vacuum annealed at 450°C HVE-CdS (recrystallized) is in the range of 0.3-0.5 μm and the layer is smoother than in the as-deposited case resulting in a more compact morphology. The effect of annealing at high temperature in vacuum is different for the CBD-CdS. The grains, that on the as-deposited sample have a compact and smooth structure, are recrystallized in such a way that a flourished and rougher structure with still high compactness is observed.

However the morphology of CdS strongly depends also on the morphology of the underlying TCO layer. The CdS grains on ITO are smaller and more homogenous than those on FTO.

If a $CdCl_2$ treatment, as described above, is applied to the HVE-CdS layers, the shape and size of the CdS grains change considerably (see again figure 3). In this case some of the small grains coalesce together to form bigger grains of 0.5 μm width. Moreover the surface of the treated CdS is smoother and the shape of the grains is mostly rectangular [9].

THE CdTe (ABSORBER LAYER)

The main advantage in case of CdTe thin film devices is that CdTe can be successfully grown by a variety of vacuum and non-vacuum deposition methods. Generally they are divided in high-temperature processes, such as close space sublimation (CSS) or close-spaced vapor transport with deposition temperature above 500 °C, and low-temperature processes, such as electrodeposition, HVE and sputtering with deposition temperature below 400 °C.

The main difference between the two processes is that layers deposited at high-temperature exhibit large grain size of up to 10 μm owing to the enhanced mobility of atoms at the growth surface, while, in low-temperature-deposited CdTe, the grain size is about 0.1–0.5 μm [9,10].

In the last case, grain growth is enhanced by a post-deposition annealing (the "CdCl$_2$ treatment"), while, in case of high substrate temperature, this process does not increase the grain size, but reduces the structural defects and affects the grain boundaries, resulting in a higher acceptor concentration [10].

As a matter of fact, with any growth process as-deposited CdTe cells exhibit poor electrical performance and it is common routine to subject the CdTe/CdS stacks to heat treatments under Cl-containing ambient at temperatures between 350 and 600 °C. The presence of oxygen during annealing helps to passivate the grain boundaries.

The heat treatment in chlorine atmosphere generates a flux that reduces atomic diffusion barriers at the grain boundaries improving the crystalline quality and increases the intermixing between CdS and CdTe films removing the lattice defects at the junction region [11,12,13].

It has been observed an increase in grain size by a factor from 5 to 20 in low-temperature grown CdTe. The overall density of stacking faults and misfit dislocations is reduced by recovery. The CdTe-CdCl$_2$ phase diagram predicts only sub-at.% solubility of Cl in CdTe at temperatures below 525 °C so that Cl is assumed to diffuse into the CdTe layer preferentially along grain boundaries [12,13].

The reaction that takes place with chlorine flux in presence of O$_2$, CdTe and CdS is a formation of CdO and TeCl$_2$ at the grain boundaries as it will be shown in the next paragraphs [14].

Indeed, due to the increase of Cd and Te atoms mobility recrystallization starts from the CdTe surface, leading to a loss of texture in low-temperature CdTe deposited layers. On the other hand, due to their higher crystalline quality, high-temperature deposited CdTe layers show few crystallographic changes after CdCl$_2$ treatment [10,15]. A similar reaction leads to the recrystallization of CdS and promotes the diffusion of S into CdTe [15].

The intermixing of CdTe and CdS brings to the formation of a CdS$_x$Te$_{1-x}$ layer at the interface reducing the lattice mismatch between CdS and CdTe, typically x does not exceed 0.06 according to the CdTe/CdS phase diagrams though layers with higher sulfur content may also grow, under non-equilibrium conditions [16,17].

Therefore, a thermal treatment of the CdS layer prior to CdTe deposition is frequently employed (as shown before). CdCl$_2$ treatment also provides electrical changes to the CdTe induced by Cl, O and S: an overall increase of the shallow-acceptor concentration leading to enhanced p-doping after annealing with Cl and O is observed [18,19]. Especially the grain boundary regions become more p-doped, due to the preferred diffusion and segregation of Cl and O at the grain boundaries. For this reason an increased photo-carrier collection efficiency is also measured [20-22].

The most common procedure for the activation treatment is the deposition of a CdCl$_2$ layer on top of the CdTe either by thermal evaporation or chemical bath deposition and subsequent air annealing at 400–440 °C for 15–30 min or the annealing of the CdTe/CdS stack directly in a CdCl$_2$ or Cl$_2$ vapor [23]. CdCl$_2$ is a very volatile and carcinogen material and it requires expensive safety procedures in a line production. A cheaper and safer method for the activation step has been introduced by N. Romeo et al. [24] on close space sublimation (CSS)- CdTe deposited samples where CdTe is treated in the presence of a gas of the Freon© family containing Cl.

Freon© and CdCl$_2$ activation treatments were applied to the CSS-CdTe/CdS, as well as to the high vacuum evaporation (HVE)-CdTe/CdS deposited stacks on transparent conducting oxide (TCO) coated glass.

As described by McCandless [14] the reaction that takes place during the CdCl$_2$ treatment between the two layers, at temperatures above 400 °C, is the following:

$$CdCl2\ (s) + O2\ (g) + CdTe\ (s) <=> TeCl2\ (g) + 2CdO\ (s)$$

For the Freon© treatment the CdTe/CdS/TCO system is placed in a quartz ampoule. The ampoule is evacuated to 10^{-6} mbar and then a non-toxic gas containing chlorine (both chloro-fluorocarbons and hydro-chloro-fluorocarbons can be used) and argon is inserted. Best results for CSS-CdTe samples were obtained with pressures in the range between 100 and 500 mbar of Argon containing 15% of HCF$_2$Cl (difluorochloromethane) at 400 °C for 5 min.

Since HCF$_2$Cl is dissociated at 400 °C and CdTe starts to dissociate at around 400 °C, the following process is assumed to take place:

$$CdTe\ (s) + 2Cl_2\ (g) <=> CdCl_2\ (g) + TeCl_2\ (g)\ [25]$$

After the treatment, the sample is kept in high vacuum at 400 °C for 10 min in order to re-evaporate the CdCl$_2$ which could be deposited on the CdTe surface.

CdTe films deposited by HVE and by CSS show a very distinct morphology as shown by SEM pictures (see figure 4). As already mentioned the grain size of HVE-CdTe (see figure 4) is in the range of 0.1–1 µm and the structure is very compact, while the CdTe deposited by CSS shows a bigger grain size (from 1 to 4 µm).

But also the shape of the grains is different between the two layers: the CSS grains are hexagonally-shaped while the HVE ones exhibit a more random structure. The difference in grain size and the randomized grain structure of the HVE-CdTe give rise to a very rough surface, which gives place to a different reactivity under the recrystallization treatment.

As already mentioned the CdCl$_2$ treatment modifies the size and the shape of the CdTe grains. In particular, the HVE- grains coalesce together forming enlarged grains (in the range of 2 to 5 µm). In contrast, no enhancement of grain size is observed for the CSS-CdTe but a strong reduction of the hexagonal shape is found (see figure 4).

Observing the Freon© treated HVE-CdTe layers in figure 4, the enlargement of the grain size is similar to that obtained by CdCl$_2$ treatment: the grains coalesce together to form bigger grains.

On the other hand there is a different effect on the grain boundaries: the Freon© treated layers have a compact structure and the grain boundaries are reduced, the CdCl$_2$ treated layers show much more pronounced boundaries.

For the case of Freon© recrystallized CSS-CdTe layers, we have round grains which are very similar to the ones of the as deposited case only a little bit enlarged.

If we analyze the evolution of the recrystallization treatment (in particular the Freon© one) in time as shown in figure 5 (15 min, 20 min and 30 min Freon© treated HVE-CdTe) we can observe that the small grains are coalescing forming very large flat grains. Together with this the edges of the grains are being smoothed resulting in a compact morphology that will have very reduced grain boundaries for the longer process. AFM measurements not shown here demonstrate that the Freon© treatment is more effective in reducing the roughness, providing a smoother surface and more compacted grains [26].

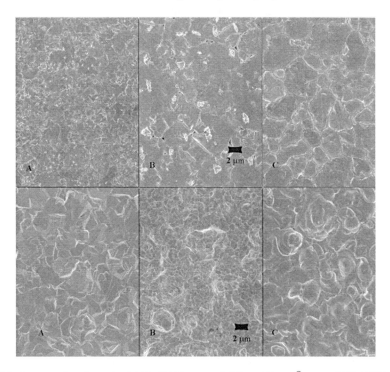

Figure 4. SEM pictures of as deposited (A), CdCl$_2$ treated (B) and Freon© treated (C) HVE-CdTe layers (top) and CSS-CdTe layers (bottom) [25].

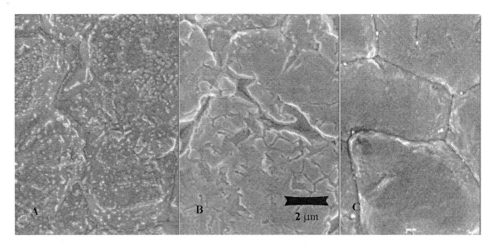

Figure 5. SEM pictures of 15 minutes- (A), 20 minutes- (B) and 30 minutes-(C) Freon© treated HVE-CdTe layer [25].

Finally we can also observe that for the CdCl$_2$ treated samples there are remaining particles that are not present on the Freon© treated samples suggesting that in the last case the etching treatment is not necessary.

Since the crystalline structure of the CdTe layer changes with the post-deposition treatment, also the crystallographic orientations of different CdTe/CdS stacks are changing.

Generally the as-deposited HVE-CdTe exhibits a strong (111) preferred orientation which is gradually lost after the application of the CdCl$_2$ treatment.

But the decrease of preferential orientation appears also in case of Freon© treated HVE-CdTe samples: after a 10–15 minutes activation process the (311) and (220) peaks show much higher intensities compared to the as-deposited case. When a longer Freon© treatment is applied a more random orientation is recorded.

For the CSS-CdTe, no re-orientation of the layers was observed due to the Freon© treatment. However, after the treatment, the loss in the preferential orientation takes place as well, but with a lower extent compared to HVE-CdTe layers. This is probably due to the high substrate temperature growth of CSS deposited layers that grow with a large grain size.

XRD measurements can be used also to study the intermixing and the stress of the CdTe/CdS stacks. The in-plane lattice parameters can be accurately determined using the Nelson–Taylor plot method [10,27].

Table 1. Lattice parameters of CdTe with different activation treatments [25]

CdTe layer	Treatment	CdTe lattice parameter
HVE-CdTe	As deposited	6.499
HVE-CdTe	CdCl$_2$ Treated	6.446
HVE-CdTe	Freon© treated (5 minutes)	6.488
HVE-CdTe	Freon© treated (15 minutes)	6.487
HVE-CdTe	Freon© treated (30 minutes)	6.489
CSS-CdTe	As deposited	6.491
CSS-CdTe	CdCl$_2$ treated	6.488
CSS-CdTe	Freon© treated	6.489

As given in Table 1, the as-deposited CdTe layers have a high in-plane lattice constant compared to the recrystallized layers. For example, the lattice parameter of the as-deposited CdTe on HVE-CdS decreases from 6.499 to 6.446 Å after the CdCl$_2$ treatment, due to the relaxation of the compressive stress generated by the lattice and thermal mismatch between the CdTe and the underlying substrate. The lattice parameter of the CdTe layer deposited at 150°C is 6.484 Å which is very near to the value obtained from the powder diffraction data (6.481 Å). This indicates that there is a very low stress in the low-temperature grown layer. The lattice parameter may also decrease because of the intermixing of CdS into the CdTe layer. These effects are also evident for CdTe on CBD-CdS. In the case of CdTe on the CdCl$_2$ treated CdS, the lattice parameter of the as-deposited CdTe is not as high as for the one deposited on the vacuum annealed CdS, and after the CdCl$_2$ treatment the lattice parameter does not decrease so much. This is due to a change in the structure of the CdS, probably the lattice constants of vacuum annealed and CdCl$_2$ treated CdS are different, because of crystallographic structural transformation of phases. It appears that the influence of the CdCl$_2$ treatment on the recrystallization of CdTe and CdS-CdTe intermixing is less pronounced for CdCl$_2$ treated CdS window layers.

This effect is also less pronounced in case of Freon© treated HVE CdTe layers (lattice parameters are between 6.487 and 6.489), this is because of the difference in the reaction process between the activating material and the CdTe/CdS stacks. Indeed, as the $CdCl_2$ treatment is performed, a material diffusion occurs into the CdTe during the annealing step down to the junction. Conversely, for the Freon© treatment, the gas that is used would not diffuse as much through the grain boundaries as for the $CdCl_2$ treatment.

For CSS-CdTe, the relaxation of the compressive stress is much lower since the higher temperature growth generates a CdS and CdTe intermixing already during CdTe deposition, consequently reducing the change of the solid solution in the treatment process. Indeed for the HVE process case, the recrystallization treatment has also the task of rearranging the grains and regulating the CdTe/CdS intermixing, while this is not obtained during the low substrate temperature deposition process.

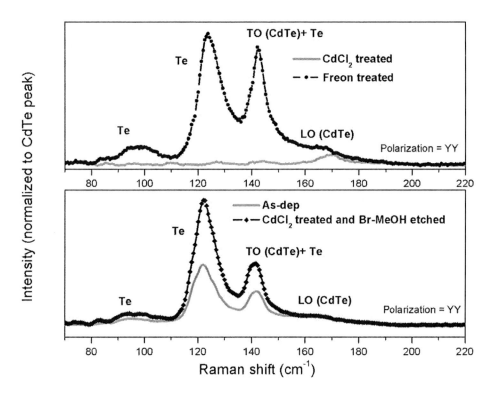

Figure 6. Low frequency micro-Raman spectra of as-dep, $CdCl_2$ and Freon© treated CdTe samples [25].

Polarized Raman spectra[15] in the low-frequency region (between 70 and 250 cm−1) of as-deposited, $CdCl_2$ (with and without etching) and Freon© treated HVE-CdTe are shown in Figure 6. The peak at 123 cm^{-1} is attributed to Te, the one at 140 cm^{-1} is a mix of both Te and CdTe, while the 165 cm^{-1} peak is due to CdTe [25,28]. It can be observed that in $CdCl_2$ treated samples only CdTe signal is achieved while Freon© treated samples show stronger Te

---

[15] Room-temperature micro-Raman spectra were recorded in backscattering geometry by exciting the samples with 514.5 nm line of a mixed Ar–Kr ion gas laser, using a triple-monochromator coupled to a nitrogen cooled CCD detector (1024×256 pixels). The laser power at the sample surface was kept below 1 mW and focused on a spot of 2 μm.

signal compared to the same CdTe one. The CdCl₂ treated CdTe layer, after bromine–methanol etching (which is generally known to make a Te rich layer), shows again a very strong Te signal (the Te peak is very high compared to the mixed Te+CdTe peak).

Instead, the as-deposited CdTe exhibits a weaker Te signal.

Therefore, the Freon© treatment, as compared to the CdCl₂ one, not only allows to get cleaner surfaces, but also acts as an etcher, smoothing the surface (as shown previously from the SEM pictures) and, most importantly, generating a Te rich layer which facilitates formation of a good electrical back contact in CdTe [28].

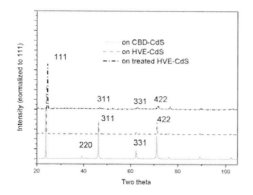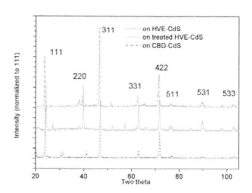

Figure 7. XRD Spectra of as deposited (left) and CdCl₂ treated (right) CdTe layers deposited on vacuum annealed HVE-CdS and CBD-CdS and on CdCl₂ treated HVE-CdS layers [9].

As already mentioned, the morphology of the CdTe layers is also dependent on the substrate on which is grown, such as different types of CdS layers. So due to the different structures of CBD-CdS and HVE-CdS layers, the as-deposited CdTe layers grow with different morphologies. The grains of CdTe on HVE-CdS are in the range of 0.5-1 μm.

The CdTe grains grown on CBD-CdS are between 1 μm and 5 μm width, while the grains on CdCl₂ treated HVE-CdS are larger compared to those on vacuum-annealed HVE-CdS but smaller than those of CdTe on vacuum-annealed CBD-CdS.

If we apply the CdCl₂ recrystallization treatment to the CdTe layers deposited on different CdS films, the change in the morphology of the CdTe is different depending on which substrate is deposited. For the CdTe on vacuum annealed HVE-CdS, a big enlargement of the grain size and the formation of wide grain boundaries are observed. Also for the CdTe on CBD-CdS there is an enhance in grain size but moreover an entirely different microstructure and morphology of CdTe is created. On the contrary, for CdTe grown on the CdCl₂ treated HVE-CdS, there is a minimum change in the grain size but the morphology is different: the layer is more compact and the grain boundaries are reduced.

Considering the XRD patterns[16] of CdTe layers (not shown here) grown at a substrate temperature of 150°C on vacuum annealed CdS layers grown by CBD and HVE [9]. The as-deposited CdTe layers exhibit a strong (111) preferred orientation in both cases. Figure 7 (right) shows the XRD of as-deposited CdTe layers grown at 300°C on different CdS/FTO/glass samples (vacuum annealed HVE- and CBD-CdS and CdCl₂ treated HVE-

---

[16] The crystallographic orientation of different types of CdTe/CdS stacks were investigated with X-ray diffraction. The measurements were performed with a Siemens D-500 diffractometer and Cu-Kα source.

CdS). CdTe layers on vacuum annealed and CdCl$_2$-treated HVE-CdS are (111) oriented. In contrast, the XRD pattern of CdTe on vacuum annealed CBD-CdS exhibits (111), (311) and (422) peaks of similar intensities, indicating that the layer is not (111) oriented.

There is a correlation between the size of the CdS grains and the orientation of the CdTe grown on top of it. The CdTe layers are generally more (111) oriented if the CdS layers are highly crystallized with large grain size, as for the case of the CdCl$_2$ treated CdS.

Loss in texture increases with the application of the CdCl$_2$ treatment on the CdTe (see figure 7, left). The loss of the (111) preferred orientation is common to all the layers, but becomes very strong for the CdTe deposited on the vacuum annealed CBD-CdS where the (311) peak is predominant. While in the case of CdTe on vacuum annealed HVE-CdS the (111) orientation is still stronger, but the (311) and (422) peaks have now a strong intensity so that the (111) preferred orientation is lost, and there is a more random orientation.

The crystallographic rearrangements in CdTe are so related to the stress in the layer and to the application of the CdCl$_2$ sintering flux and high temperature annealing.

The creation of new grains as a result of the disintegration of some large grains is due to the relaxation of the excessive strains in the lattice while the coalescence of small grains into bigger ones is caused by the recrystallization treatment.

THE BACK CONTACT

CdTe requires, because of its high electron affinity, metals with a work function greater than 5.7 eV to form an ohmic contact. Unfortunately such metals are not available and the formation of a Schottky barrier at the back contact would be unavoidable. To overcome this problem generally a heavily p-doped CdTe surface is created by chemical etching and a buffer layer of high carrier concentration is often applied [29]. Subsequent post-deposition annealing diffuses some buffer material into CdTe where it changes the band edges and interface states. The contact barrier is lowered, resulting in a quasi-ohmic contact.

Commonly used buffer layer/metallization combinations are Cu/Au [29,30], Cu/graphite [31] or graphite pastes doped with Hg and Cu, ZnTe doped with Cu [32-34] and Au or Ni metallization, Cu/Mo [35]. Alternatively, Cu free back contacts such as Ni:P, ZnTe [36], Au [37] or Sb$_2$Te$_3$/Ni [38] contacts have also been investigated.

Metals such as Cu, Au, Al or Ni have high diffusivity in CdTe layers, typically they accumulate at the CdS/CdTe or CdS/TCO interface [36,39] causing a degradation of the cell performance.

However, a PVD-deposited Sb buffer layer with Mo metallization has yielded high efficiency and low degradation in long-term performance [39]. Best cell stabilities have been achieved with RF-sputtered Sb$_2$Te$_3$ buffer layer with Mo metallization as introduced by N. Romeo et al. [40]. Recently the same group has introduced an alternative buffer layer, As$_2$Te$_3$:Cu, which has the advantage of incorporating Cu in a stable compound in order to improve the ohmic contact without any copper diffusion [41]. A similar solution was introduced by Wu et al. using Cu$_x$Te$_{1-x}$ [42]

A different approach introduced by Tiwari et Al. [43], applied on low temperature processed solar cells, is to apply a transparent back contact on CdTe solar cells: a thin layer of transparent and conducting ITO.

Due to the transparency of the ITO back contact and FTO front contact, solar cells can be illuminated both from the front and rear sides resulting in a bifacial solar cell.

ITO back contact is applied by RF sputtering after a bromine methanol etching: the layer has a sufficient transparency, more than 85%, together with a good conductivity about 10 Ohm/square. As deposited solar cells with pure ITO back contact perform typically very low efficiencies around 2.5% in particular V_{oc} and fill factor are poor, respectively, around 350mV and 35%.

However, if an annealing treatment to the ITO back contacted CdTe cells is applied with different temperatures from 200 up to 400 °C, the performance improves up to 4–5% absolute efficiency which reaches, under light soaking treatment, an upper efficiency above 6%; best results were given with annealing made of around 350 °C.

The interesting feature of these solar cells is the stability under accelerated lifetime test [45]. Tests on these devices (solar cells were kept under 1 sun illumination and at 80 °C temperature in open circuit conditions) demonstrated an excellent stability. The performance actually improves instead of degrading due to the light soaking effect: the electrons fill the recombination centers at the junction. In the annealed devices, there is a constant increase in an almost 10 years estimated working time.

The transparency of the back contact can still be maintained if a very thin layer of copper is deposited on the CdTe layer before deposition of the ITO film. The idea behind is that a small amount of copper would still leave the back contact transparent and at the same time might not affect the stability of the solar cell especially if it is bound to the tellurium in excess. Indeed prior to the copper deposition, a standard bromine–methanol etching is applied to form a Te rich layer that will react with the copper deposited by PVD afterwards.

ITO layer is deposited by RF sputtering as previously presented.

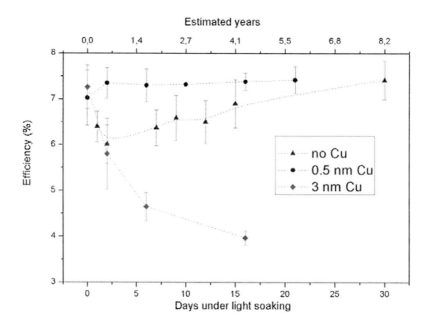

Figure 8. Accelerated lifetime stability tests at 80°C and one sun of an ITO/CdTe/CdS/FTO solar cell and of ITO/Cu/CdTe/CdS/FTO solar cells with 3 nm copper deposition and less than 0.5nm copper deposition [44].

In Figure 8, accelerated lifetime stability tests of cells with two different copper layer thicknesses are shown. Both start with very similar efficiency but the stability trend is opposite: the 3-nm-Cu cell has a degradation that tends to stabilize after 4 years estimated time, on the other hand the 0.5 nm-Cu cell starts from the same efficiency but instead of degrading it improves slightly in the performance. The behavior of the 3-nm-Cu cell is similar to the one observed for Cu/Au [45] but still the stability is higher. In the back side illumination, the solar cells perform efficiencies in the order of 0.5% with Voc of about 300mV, Jsc of 3–4 mA/cm^2 and fill factor of 45%. The low current in the back side illumination is due to the long distance between the hole–electron pair generation and the junction. To improve it, it is than necessary to reduce the CdTe thickness in order to reduce the path of the carriers from the generation zone to the junction.

However, the 0.5 nm-Cu cell has a very similar behavior of Sb_2Te_3/Mo cell with a rapid increase at the beginning of the light soaking followed by a stabilization. This suggests that the copper would give instability only if the back contact is not well designed: if a stable compound with Cu is formed (such as copper telluride as suggested by Zhou et al [42], then the back contact is stable.

Quantum efficiency measurements, not shown here, compared with the one of the standard Cu/Au contacted cell show that the latter has a very similar response in the red part near the absorption edge, but has a better response in the 400–550nm region most probably due to the presence of copper in the bulk of CdTe which passivates the grain boundaries increasing the quantum efficiency [46]. This is an indirect proof that copper is not penetrating the CdTe.

As already mentioned, in order to improve the performance in the back side illumination, CdTe thickness must be reduced.

The solar cell fabrication process has been adjusted to the different CdTe thickness; in particular the $CdCl_2$ treatment has been processed with only 20% of the standard $CdCl_2$ quantity and by reducing the annealing time. The first cells were made with about 2.5 μm CdTe thickness and the optimized quantity of copper (less than 0.5 nm). The solar cells started with a very good performance from the beginning (9.5%) and increased after light soaking to reach an efficiency exceeding 10%. The back side illumination performance improved considerably compared to the standard cell, after 3 days light soaking we have around 2% in performance [44].

Finally, if only one micron CdTe is used to prepare solar cells, with a consequently reduced $CdCl_2$ treatment, the current density is reduced, most probably due to possible formation of micro pin-holes, and the overall efficiency drops to 7–8%, but with an increased performance in the back side illumination of up to 3% [44].

FLEXIBLE CdTe CELLS IN SUPERSTRATE CONFIGURATION

The CdTe solar cell can be fabricated in a "superstrate" or a "substrate" configuration (Figure 9).

High-efficiency CdTe solar cells are generally grown in a superstrate configuration where the CdTe/CdS stacks are deposited on transparent conducting oxide (TCO) coated glass substrates. Generally, the fabrication of CdTe solar cells on substrate configuration raises a series of different problems such as the choice of a stable substrate (metals would induce diffusion of impurities into the cell), the difficulty of preparation of the p^+ surface for the back contact, the different growth of CdTe and the different way to form the CdTe/CdS junction.

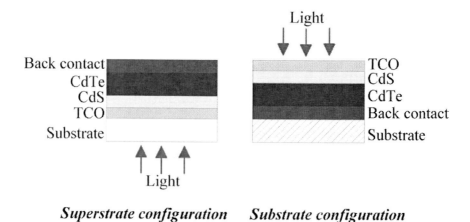

Figure 9. Schema of a CdTe solar cell in superstrate and substrate configuration.

For this reason, in order to have a high efficiency flexible CdTe solar cell the superstrate configuration still remains the better choice. However, efforts to develop flexible CdTe superstrate solar cells were, for long time, not successful, because of the incompatibility of the polymeric substrate with high-temperature processing steps.

For superstrate configuration the choice of an appropriate substrate is crucial; the substrate should be optically transparent and should withstand a high temperature during deposition and processing steps. Most of the CdTe solar cell fabrication processes require temperatures of about 450–550 °C, while most of the transparent polymers are not stable at such high temperatures. However, some polyimides are "stable" at temperatures of up to 450 °C with an optical transparency sufficient for solar cell applications. If the thickness of the polyimide film is reduced, the absorption loss in the substrate can be reduced. The average transmission of the polyimide film is more than 75% for wavelengths above 600 nm, while there is a strong thickness-dependent absorption of photons in the wavelength range of 400–600 nm. CdTe solar cells on 50–100- µm-thick-polyimide films will yield a low current due to a large optical absorption loss in the substrate.

For this kind of application, low temperature deposition process like high vacuum evaporation is the ideal solution since the processing temperature is not higher than 450°C.

One method to prepare a flexible cell is to prepare an in-house "specific" type of polyimide film: a thin buffer layer of NaCl can be evaporated on a glass substrate [46], and then a polyimide layer is spin coated and cured at about 430 °C. The thickness of the polyimide film can be controlled by the spin-coating process, typically about 10 µm-thin polyimide films are used. A layer of ITO is then deposited by RF magnetron sputtering on top of the polyimide, followed by the standard HVE-CdS and CdTe deposition and the $CdCl_2$ annealing treatment. After processing, the device is rinsed in water in order to dissolve the NaCl layer and the solar cell can be removed from the glass carrier.

The CdS and CdTe layers are deposited with substrate temperatures of respectively 150°C and 300°C.

The morphology and microstructure of CdTe on polyimide are similar to those on glass substrates [46]. The as-deposited CdTe layer is homogeneously compact with grains of up to 0.1 μm size. The CdCl$_2$-annealed layers are crack-free and consist of large grains of up to 0.5 μm. The solar cells grown on the spin-coated polyimide with copper-gold back contact exhibit efficiencies of 11% with ITO as front contact under AM1.5 illumination. High values of open circuit voltage (842mV) and fill factor (70.9%) are obtained, despite current density is lower (18.5 mA/cm^2) than the ones of CdTe/CdS solar cells on FTO-coated glass because of the absorption loss in the polyimide [46].

A slightly different method to prepare a CdTe flexible cell is to take into account commercial polyimide films. For example, the commercially available Upilex^{TM} film with a thickness of about 10 μm is sufficiently transparent and suitable for CdTe solar cells. The average transmission of the polyimide films is more than 75% for wavelengths above 550 nm. There is a strong absorption of photons in the wavelength range of 350–600 nm.

The ITO layers are than deposited with an RF-magnetron sputtering system on the Upilex polyimide foils. An annealing of the ITO/Upilex stacks prior to the deposition of the subsequent layers improves the stability of the ITO to further high-temperature processes. Moreover the annealing of these stacks in air at 450 °C, improves transmission (in the range of 500–900 nm) from 69% to 72% but also increases the sheet resistance from 5 to 12 Ohm/square. Repeated annealing cycles in the same conditions increased the ITO sheet resistance by only 0.8–0.9 Ohm/square, with no degradation in transparency. The quantum efficiencies of cells made with annealed and not annealed ITO/polyimide structures show that the difference in quantum efficiency is partly due to low transparency of non-annealed ITO/polyimide layers in the spectral region of 0.6–0.8 μm. The overall lower quantum efficiency for both solar cells is mainly caused by the lower transparency of the polyimide substrate. However efficiencies exceeding 11% (AM1.5 measurement conditions) are obtained with flexible solar cell on annealed ITO/Upilex-polyimide film (Voc = 765mV, Jsc = 20.9mA/cm^2, FF= 71%). Those are comparable to the ones obtained on the in-house polyimide solar cells (Voc ¼ 842 mV, Jsc =18:5mA=cm^2, FF = 70.9%).

Morevover if an intrinsic buffer layer is introduced between the TCO and CdS, it is possible to reduce the thickness of CdS and enhance its transparency improving the overall efficiency. Lately Perrenoud et al. have improved the flexible CdTe solar cell by introducing ZnO:Al as TCO and ZnO as insulating buffer layer, reducing the thickness of CdS obtaining an efficiency of 12.4% [47].

FLEXIBLE CDTE CELLS IN SUBSTRATE CONFIGURATION

In this process the CdTe/CdS/ITO stacks are not deposited on the polyimide substrate but directly on a thin layer of NaCl deposited by vacuum evaporation on soda-lime glass. After the complete processing of the solar cells, a metal foil can be laminated on the back of the device and the front glass is removed by dissolving the NaCl in water. This results in solar cells in the substrate configuration. The fabrication process is similar to the standard one, presented before. A TCO layer can be deposited on NaCl/glass substrate. The CdS and CdTe layers are grown by HVE and the back contacting is done by Cu/Au deposition. In this way no polyimide layer is present during processing and there is no absorbing layer, except of TCO, between the incident light and the photovoltaic layers.

The structural properties of CdTe on CdS/TCO/NaCl stacks are similar to that on glass substrates. The as-deposited CdTe layer is very compact with a small grain size. The CdCl2-annealed layers are crack-free and consist of large grains of up to 5 um. It is interesting to observe that the presence of NaCl on the substrate does not affect the morphology of the CdTe layer since TCO is a good barrier for NaCl [4].

Cells in the substrate configuration exhibit efficiencies of 7.3% with FTO and 6% with ZnO:Al front contacts. The fill factor has very low values (49% and 40%), perhaps it is related to the low stability of the TCO on NaCl layer, which means that the electrical properties of the front contact tend to degrade after processing. Despite lower values of open circuit voltage (692 and 743 mV, respectively, for FTO and ZnO:Al as front contact) and fill factor the current density is higher than in the superstrate case (21.6 and 20.3 mA/cm^2, respectively, for FTO and ZnO:Al as front contact) because of the absence of the polyimide on the front of the cell.

SUBSTRATE CONFIGURATION

An easier way to fabricate flexible solar cells would be to deposit the single layers directly in a substrate configuration, not as shown before. Substrate configuration would allow to use a large variety of different substrates that do not have to be transparent, for example metal foils. For the growth of solar cells in this configuration the choice of an appropriate metal or semiconductor surface is critical, because this layer has to form a low resistance ohmic contact with CdTe. Generally Cu-Au, Mo, Te-Mo, Sb_2Te_3-Mo are evaluated as back contact layers on glass substrates [48].

First a layer of 0.5 μm thick molybdenum is deposited with a DC magnetron sputtering system either on a soda lime or alkali free glass. Thin layer of Sb_2Te_3 or Te were deposited in an ultra high vacuum chamber with a substrate temperature of 200 °C. CdTe layers are grown at 300 °C in the same chamber without breaking the vacuum. For the $CdCl_2$ treatment, vacuum evaporation is used for the deposition of $CdCl_2$ layers on CdTe and the stacks are annealed at 430 °C for 30 minutes in air.

After washing CdTe in water, CdS is then deposited in a high vacuum chamber at a substrate temperature of 150 °C. The CdTe thickness is about 3 to 4 μm and CdS is about 0.5 μm thick in a standard deposition. After the growth of CdS layer, the CdS/CdTe stack is annealed in air at 430 °C or is treated with $CdCl_2$ at 430 °C. For the front contact, a layer of ZnO:Al is deposited on the CdS surface by RF sputtering at room temperature.

We have previously shown that the growth and morphology of CdTe depend strongly on the substrate on which it is grown. After the $CdCl_2$ treatment a large increase in the grain size is observed. The $CdCl_2$ acts like a sintering flux in CdTe as small grains grow and coalesce together, but it also causes the widening of grain boundaries, which are not desired since they affect the parallel conduction across the grain boundaries as well as they may also cause the shunting of solar cells by providing a conducting link between the top and bottom electrodes.

The as-deposited CdTe layer on Mo is compact and it consists of grains of 0.5 μm to 2 μm size, the morphology is rough due to irregular shape and size of grains. This condition is similar to the as deposited CdTe on CdS/FTO/glass, even if, in this latter case, the shape and size of grains are more randomly distributed. After $CdCl_2$ treatment, the CdTe layer on Mo/glass has grains with well defined shape but with grain size going from 3 μm to10 μm.

The layers on Sb_2Te_3/Mo/glass instead exhibit a different microstructure. In the as deposited case the grains have homogeneous shape and size of about 1 to 2 μm, and they are very compact. After $CdCl_2$ treatment, the grains enlarge up to 8 μm with wide grain boundaries.

The morphologies of CdTe, grown under identical conditions, on alkali free and soda lime glasses are substantially different. On alkali free glass, in the as-deposited case the grains have a size of about 0.1 μm-0.5 μm and they are homogeneous in shape. After $CdCl_2$ treatment, the layers are very compact (narrow grain boundaries) and the size of the grains is much smaller compared (1 to 5 μm) to the other treated layers on soda lime substrates.

The influence of substrate on the growth of CdTe is also evident from the X-ray diffraction (XRD) measurements.

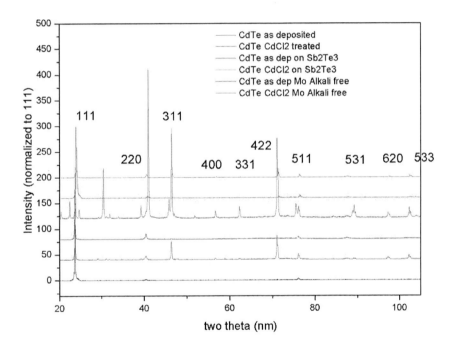

Figure 10. XRD Spectra of CdTe in substrate configuration deposited on different substrates [48].

As shown in figure 10, CdTe in the substrate configuration has a pronounced (111) preferred orientation, which is not the case for CdTe in superstrate configuration, as shown before. The (111) preferred orientation is still maintained after the $CdCl_2$ treatment, while in the superstrate configuration the (111) preferred orientation is almost lost. In figure 12 are also shown the XRD patterns of CdTe grown on alkali free glass, a comparison with the samples deposited on soda lime glass clearly indicates that the sodium has a very strong influence on the orientation. The as-deposited layer has a strong (111) orientation and the $CdCl_2$ treatment does not change this preferred orientation.

From the diffractograms of CdTe grown on Sb_2Te_3/ glass, it is clearly seen that the $CdCl_2$ treatment has a different effect on this layer; in the as-deposited case the CdTe is (111) oriented and almost no other peaks are present. For the $CdCl_2$ treated layer the (111) orientation is completely lost, and the (311) and (422) peaks are the stronger. Once again the observations suggest that the growth of CdTe is affected by the substrate. Therefore different back contacts provide different template for the growth of CdTe and the microstructure of the layer is influenced.

Heat treatment of the CdS/CdTe stack is necessary for interfacial intermixing and junction activation. In case of the superstrate solar cells, CdS-CdTe interdiffusion already takes place during the deposition of CdTe on CdS at 300 °C.

However, in the substrate configuration CdS is grown at a low temperatures of 150 °C. Therefore high temperature annealing is still required for an optimum intermixing even if the CdTe is already recrystallized in $CdCl_2$. The intermixing and photovoltaic properties are better when CdS/CdTe substrate cells are annealed in presence of $CdCl_2$.

However substrate configuration CdTe solar cells still perform low efficiency since it is very difficult to prepare the CdTe surface to the back contact and the junction between CdS and CdTe is difficult to control [47]

CONCLUSIONS

Low temperature processes open up a wide variety of new possibilities for CdTe solar cells.

Flexible cells with commercial polymide as substrate have delivered an efficiency of more than 12%. If a type of polimiyde would be specifically developed in order to have higher transparency with some optimization, even higher efficiencies can be obtained.

Moreover the study and understanding of preparation of solar cells in substrate configuration would give even more possibilities for preparation of flexible cells by using thin metal foils.

Transparent back contacts not only deliver solar cells that convert light from both sides but, even more interesting, give the possibility of reducing considerably the CdTe thickness reducing the production costs. Cells with less than one micron thick CdTe have performed efficiencies above 8%.

New recrystallization treatments can be applied maintaining high efficiency but getting rid of really poisonous materials like $CdCl_2$, if a deep understanding of the CdTe activation treatment is made.

If on one hand with low temperature it is possible to prepare high efficiency flexible cells on the other hand improving efficiency by reducing CdS thickness and optimizing the recrystallization treatment and the back contact would considerably improve the production costs and the overall emissions (see figure 11) compared to the high temperature processes like vapor transport deposition that are now typically used in production.

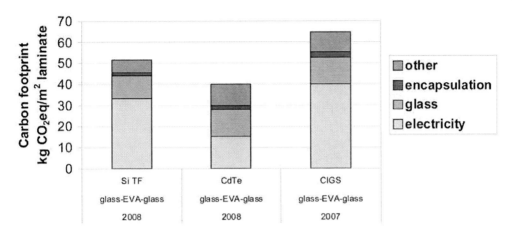

Figure 11. Carbon footprint for different thin film technologies: reducing production electricity would considerably reduce greenhouse gas emissions (Mariska Wild de Scholten, TIF Berlin 2009).

REFERENCES

[1] S. Seyrling, J. Perrenoud, A. Chirila, S. Buecheler, D. Guettler, Y. Romanyuk and A.N. Tiwari, Proceedings at the "2nd International Symposium on Innovative Solar Cells", Tsukuba, Japan, Dec. 7-8, 2009
[2] Wu X. Solar Energy 77 (2004) 803–814
[3] Bonnet D, Oelting S, Harr M, Will S. Proceedings of the 29th IEEE Photovoltaic Specialists Conference, Anaheim, 2002; 563–566.
[4] Romeo A, Tiwari A.N., Zogg H., Wagner M, Guenter J.R. Proceedings of the 2ndWorld Photovoltaic Solar Energy Conference, Vienna, 1998; 1105–1108.
[5] Gupta A, Compaan AD. Proceedings of the 2003 MRS Spring Meeting, Francisco, 2003; B3_9_1–6.
[6] Klein A. Applied Physics Letters 2000; 77(13): 2009–2011.
[7] Alamri SN, Brinkman AW. Journal of Physics D 2000; 33(1): L1.
[8] Massalski TB. Binary alloy phase diagrams. ASM International: Materials Park, 1990.
[9] Romeo A, Baetzner DL, Zogg H, Tiwari AN, Vignali C. Solar Energy Materials and Solar Cells 2001; 67: 311–321.
[10] Moutinho HR, Al-Jassim MM, Abulfotuh FA, Levi DH, Dippo PC, Dhere RG, Kazmerski LL. Proceedings of the 26th IEEE Photovoltaic Specialists Conference, New York, 1997; 431–434.
[11] Durose K, Cousins MA, Boyle DS, Beier J, Bonnet D. Thin Solid Films 2002; 403–404: 396–404.
[12] Oleinik GS, Mizetskii PA, Nuzhnaya TP. Inorganic Materials 1986; 22: 164–165.
[13] Terheggen M, Heinrich H, Kostorz G, Ba"tzner D, Romeo A, Tiwari AN, Romeo N, Bosio A. Thin Solid Films 2003; 431–432: 262–266.
[14] McCandless BE. Proceedings of the 2001 MRS Spring Meeting, San Francisco, 2001; H1_6_1–12.

[15] McCandless BE, Birkmire RW. Proceedings of the 16th European Photovoltaic Solar Energy Conference and Exhibition, Glasgow, 2000; 349.
[16] Ohata K, Saraie J, Tanaka T. Japanese Journal of Applied Physics 1973; 12: 1198–1204.
[17] Lane DW, Conibeer GJ, Wood DA, Rogers KD, Capper P, Romani S, Hearne S. Journal of Crystal Growth 1999; 197: 743–748.
[18] Tyan YS, Vazan F, Barge TS. Proceedings of the 17th IEEE Photovoltaic Solar Energy Conference, New York, 1984; 840–845.
[19] Rose DH, Levi DH, Matson RJ, Albin DS, Dhere RD, Sheldon P. Proceedings of the 25th IEEE Photovoltaic Solar Energy Conference, New York, 1996; 777–780.
[20] Rohatgi A, Sudharsanan R, Ringel SA, MacDougal MH. Solar Cells 1991; 30: 109–122.
[21] Jahn U, Okamoto T, Yamada A, Konagai M. Journal of Applied Physics 2001; 90: 2553–2558.
[22] Galloway SA, Durose K. Microscopy of Semiconducting Materials 1995, Institute of Physics Conference Series 1995; 146: 709–712. Romeo, M. Terheggen, D. Abou-Ras, D.L. Bätzner, F.-J. Haug, M. Kälin, D. Rudmann, A.N. Tiwari, Prog. Photovolt.: Res. Appl. 12 (2–3) (2004) 93.
[23] N. Romeo, A. Bosio, A. Romeo, S. Mazzamuto, V. Canevari, Proceedings of the 21st Eu-PVSEC, Dresden, Germany, 2006, p. 1857. Romeo, G. Khrypunov, S. Galassini, H. Zogg and A.N. Tiwari, N.Romeo, A.Bosio, S. Mazzamuto. Proceedings of 22nd European Photovoltaic Solar Energy Conference, 3-7 September 2007, Milan, Italy, pp 2367-2372
[24] Amirtharaj P.M., Pollack Fred H., Appl. Phys. Lett. 45 (1984) 789.
[25] Brajesh K. Rai, H.D. Bist, R.S. Katiyar, K.-T. Chen, A. Burger, J. Appl. Phys. 80 (1) (1996).
[26] Romeo A., Buecheler S., Giarola M., Mariotto G. and Tiwari A.N., Romeo N., Bosio A., Mazzamuto S. Thin Solid Films, Volume 517, Issue 7, 2 February 2009, Pages 2132-2135
[27] Ferekides CS, Viswanathan V, Morel DL. Proceedings of the 26th IEEE Photovoltaic Specialists Conference, Anaheim, 1997; 423–426.
[28] Bonnet D. International Journal of Solar Cells 1992; 12: 1–14.
[29] Suyama N, Arita T, Nishiyama Y, Ueno N, Kitamura S, Murozono M. Proceedings of the 21th IEEE Photovoltaic Specialists Conference, Kissimimee, 1990; 498–503.
[30] Tang J, Mao D, Ohno TR, Kaydanov V, Trefny JU. Proceedings of the 26th IEEE Photovoltaic Specialists Conference, Anaheim, 1997; 439–442.
[31] Gessert TA, Sheldon P, Li X, Dunlavy D, Niles D, Sasala R, Albright S, Zadler B. Proceedings of the 26th IEEE Photovoltaic Specialists Conference, Anaheim, 1997; 419–422.
[32] Rohatgi A. International Journal of Solar Cells 1992; 12: 37–49.
[33] Romeo N, Bosio A, Canevari V. International Journal of Solar Energy 1992; 12: 183–186.
[34] Dobson KD, Visoly-Fisher I, Hodes G, Cahen D. Solar Energy Materials and Solar Cells 2000; 62(3): 295–325.
[35] Niemegeers A, Burgelman M. Journal of Applied Physics 1997; 81(6): 2881–2886.

[36] Abken AE, Bartelt OJ. Thin Solid Films 2002; 403–404: 216–222.
[37] Baetzner DL, Romeo A, Zogg H, Wendt R, Tiwari AN. Thin Solid Films 2001; 387(1-2): 151–154.
[38] Romeo N, Bosio A, Tedeschi R, Romeo A, Canevari V. Solar Energy Materials and Solar Cells 1999; 58(2): 209–218.
[39] Romeo N., Bosio A., Mazzamuto S., Romeo A., Vaillant-Roca L.. Proceedings of 22nd European Photovoltaic Solar Energy Conference, 3-7 September 2007, Milan, Italy, pp 1919-1921
[40] Zhou J., Wu X., Duda A., Teeter G., Demtsu S.H., Thin Solid Films 515 (2007) 7364–7369
[41] Tiwari A.N., Kryphunov G., Kurdzesau F., Bätzner D.L., Romeo A., Zogg H.. Progress in Photovoltaics: Research and Applications 2004; 12:33-38 Romeo, G. Khrypunov, S. Galassini, H. Zogg, A.N. Tiwari, Solar Energy Materials & Solar Cells 91 (2007) 1388–1391
[42] Bonnet D., Meyer P., J. Mater. Res. 13 (10) (1998) 2740.
[43] Tiwari A.N., Romeo A., Baetzner D.L., Zogg H., Prog. Photovoltaics: Res. Appl. 9 (2001) 211.
[44] J. Perrenoud, S. Buecheler, A.N. Tiwari, Proceedings of the SPIE Solar Energy + Technology Symposium, 2-5 August 2009, San Diego, USA (in press during the preparation of this chapter).
[45] Romeo A., Bätzner D.L., Zogg H., Tiwari A.N.. Proceedings of 16th European Photovoltaic Solar Energy Conference and Exhibition, May 1- 5,2000, Glasgow, Scotland, pp 843-846

In: Thin Film Solar Cells: Current Status and Future Trends ISBN 978-1-61668-326-9
Editors: Alessio Bosio and Alessandro Romeo © 2010 Nova Science Publishers, Inc.

Chapter 7

POLYCRYSTALLINE CdTe THIN FILMS SOLAR CELLS

Alessio Bosio[1], Alessandro Romeo[2] and Nicola Romeo[1]
[1]University of Parma, 43124-Parma, Italy;
[2]University of Verona, 15-37134-Verona, Italy.

ABSTRACT

We report the state of the art of the second generation of solar cells, based on CdTe thin film technology. This type of cells reached on laboratory scale, photovoltaic energy conversion efficiencies of about 16.5% higher than those obtained with the bulk materials. In particular, we describe the materials, the layers sequence, the characteristic deposition techniques and the devices that are realized by adopting this semiconductor as absorber material. Particular emphasis will be placed on major innovations that have made possible to achieve high efficiencies with polycrystalline materials by showing how the thin-film technology is mature enough to be easily transferred to industrial production. We will focus our attention on a technological project, completely developed in Italy, with the aim to transfer the CdTe/CdS polycrystalline thin film photovoltaic module process to an industrial production.

1. INTRODUCTION

Solar cells made with silicon (Si) and gallium arsenide (GaAs), have reached energy conversion efficiencies near the theoretical limits for their respective band gaps. In addition, the small remaining losses are reasonably well understood, and any additional effort would appear to offer only a slight development. Conversely, during the last 15-20 years the photovoltaic world has been enhanced with other interesting materials such as CdTe (cadmium telluride) and $CuInSe_2$ (copper indium diselenide). Both these materials are considered very suitable for the fabrication of solar cells because of their direct band gap; this implies that the absorption edge is very sharp and more than 90% of the incident light is absorbed in a few micrometers of the material. The theoretical maximum efficiency of CdTe is over 27% and the maximum photocurrent available from a CdTe cell under the standard global spectrum normalized to 100 mW/cm^2 is 30.5 mA/cm^2.

One of the best characteristics of this semiconductor is that it is possible to fabricate a complete photovoltaic device using only thin film technology. This amazing fact has been well known ever since 1972 when Bonnet and Rabenhorst [1] published an interesting paper on CdTe/CdS thin film solar cells reporting an efficiency of 6%. There followed a long period during which several research groups tried to develop a solar cell fabrication process based on related thin film deposition techniques. However, it was only in the 1980s that the 10% efficiency value was overcome by Tyan and Albuerne [2]. Subsequently an efficiency of 15.8% was reached by Ferekides et al. [3] and most recently a research group of NREL (National Renewable Energy Laboratory- Golden- Co) reported a record efficiency of 16.5% [4].

Table 1. Summary of the best results obtained so far with p-type CdTe thin film and single crystal CdTe solar cells

Type of cells	Open circuit voltage V_{oc} (mV)	Short-circuit current J_{sc} (mA/cm^2)	Energy conversion efficiency (%)	Ref.
Thin films				
All thin film CdTe solar cells	---	---	6.0	[1]
CdS and CdTe by low temperature CSS	750	17.00[a]	10.5	[2]
CdS by CBD and CdTe by high temperature CSS	843	25.10[b]	15.8	[3]
CdS by CBD and CdTe by high temperature CSS	845	25.88[b]	16.5	[4]
CdTe single crystal				
Buried homojunction:n-ITO/p-type CdTe	890	20.00[b]	13.4	[5]
Heterojunction:n-ZnO/p-type CdTe	540	19.50[b]	8.8	[6]
CdTe homojunction:p-type CdTe by CSVT on n-type CdTe single crystal	820	21.00[b]	10.7	[7]

[a] Under simulated AM 1.5 solar illumination at 100 mW/cm^2.
[b] Under simulated AM 2 solar illumination at 75 mW/cm^2.

Even more surprising is the fact that CdTe/CdS solar cells fabricated using thin film technology exhibit higher efficiencies than those fabricated from single crystal materials. In fact, solar cells with an efficiency around 10% or higher have been made as hetero-junctions, homojunctions, buried homojunctions and MIS junctions, using CdTe single crystal. The highest 13.4% efficiency concerns an *n*-ITO/*p*-CdTe single crystal buried homojunction [5]. Best performances are shown in Table 1. All photovoltaic devices involving CdTe as an absorber material contain a highly transparent and *n*-type partner, which promotes the creation of a depleted region in the *p*-type CdTe film. In highly efficient CdTe thin film solar cells this partner is CdS (cadmium sulfide) and, despite the lattice mismatch which is 9.7%, the interface shows a good behavior without the typical recombination losses associated with junction interface states [8-9]. The electrical resistance of the CdS film may become an important factor that affects the whole solar cell behavior. Generally the thickness of the CdS film is minimized having in mind the need to preserve the best properties of the cell such as the open-circuit voltage.

The dark resistivity of as-deposited CdS films is usually in the range of about 10^4-10^6 Ωcm. These CdS films are suitable for solar cell purposes because they increase their conductivity under illumination. In order to increase the photocurrent, the CdS film thickness has to be as small as possible. For this reason, an additional transparent electrode must be used. This electrode called "Transparent Conducting Oxide" (TCO) is generally a semiconducting material such as SnO_2 or ITO (indium-tin-oxide). When the transparent layer is on top of the device, the cell is in the so-called "frontwall" configuration and can be fabricated on a non-transparent substrate. Otherwise, if the TCO is deposited directly on the substrate, the device is said to be in the "backwall" configuration and the solar cell fabrication needs a transparent substrate. Nowadays, highly efficient solar cells based on CdTe as an absorber material coupled with CdS as a window layer are made in the backwall configuration and soda-lime glass is used as a transparent substrate. The schematic diagram of all the stacked layers of the CdTe/CdS system is shown in Figure 1. Despite the good performance and efficiency, the preparation of these thin film solar cells based on CdTe/CdS heterojunction still exhibits quite a few open problems and it is therefore subject to a margin of uncertainty in its progress. One of the major open questions is certainly the back contact, which is crucial for the time stability of the solar cell. In fact, in order to obtain a low resistance or possibly ohmic contact with a *p*-type CdTe film, use is made of various metals like Cu, Hg, Pb, Ag or Au which, due to their ability to diffuse into the different layers may deteriorate the device [10-14]. All the devices described in Table 1 were completed with these kinds of back contact and nothing is reported about their time stability. The back-contact problem was solved, several years ago, by using a different type of material such as Sb_2Te_3 or As_2Te_3 as described in the patent filed in 2006 by the authors [15-18]. These semiconductors demonstrate ohmic behavior and a very low contact resistance when coupled with *p*-type CdTe. Commonly, in efficient solar cells, CdS films are prepared by RF sputtering, Close-Spaced Sublimation (CSS) or Chemical Bath Deposition (CBD) [19-21]; CdTe films are deposited by CSS, electrodeposition or metal-organic chemical vapor deposition (MOCVD) [9,22-24].

Since any mature solar cell technology tends to the stage where costs are determined by those of constituent materials, this means that highly efficient processing operations that produce solar cells with high-energy conversion efficiency are favored. Each technique, as described above, has its own merits but in order to produce a device that is easily scalable for industrial needs (i.e. the costs of the equipment, the deposition rate and the final cell performance) the CSS and sputtering techniques seem to be the most advantageous processes.

A big challenge for thin film photovoltaics is the development of large area semiconductor technology. In fact, one of the advantages of the thin film technology is the potential increase in the manufacturing unit from a silicon wafer ($\approx$100-200 cm^2) to a glass sheet ($\approx 10^4$ cm^2) that is about 50-100 times larger. In order to achieve this goal, high quality materials and high throughput on large areas have to be obtained.

As it can be observed from table 1 the progress of the technology was slow but the continuity in the improvement of the performances and the solution of typical problems such as long-term stability and reproducibility, have brought to the realization of productive facilities at industrial level.

In the last years photovoltaic modules production continued to be one of the industrial sectors in more rapid growth, with an increase well in excess of 40% per year. This growth is driven not only by the progress in materials and technology, but also by incentives to support the market in an increasing number of countries all over the world. Besides, the increase in price of fossil fuels in 2008 highlighted the necessity to diversify the provisioning for the sake of energy security and to emphasize the benefits of local renewable energy sources such as solar energy. The high growth was achieved by an increase in production capacity based on the technology of crystalline silicon, but in recent years, despite the already very high industrial growth rates, thin film photovoltaics has grown at an increasingly fast pace and its market share is increased from 5% in 2005 to over 10% in 2008. However, the majority of photovoltaic modules installed today are produced by the well-established technology of monocrystalline and polycrystalline silicon, which is very close to the technology used for the creation of chips for electronics. The high temperatures involved, the necessity to work in ultra-high vacuum and the complex cutting and assembly of silicon "wafers", make the technology inherently complicated and expensive. Other photovoltaic devices based on silicon are produced in form of "thin film" or in silicon ribbon; these devices are still in experimental stage. Amorphous silicon is a technology that has been on the market for decades and it is by now clear that it does not keep the promises of change and development that were initially launched.

Without resorting to sophisticated photovoltaic devices such as multi-junction solar cells, where the cost of production is high, thin film silicon modules were generally poor in conversion efficiency and demonstrate low stability. On the other hand, silicon is not an appropriate material for the implementation as thin film, both for the difficulties of processing (necessity of high temperatures) and the inherent characteristics of the semiconductor which, being an "indirect gap" has a low absorption coefficient in the visible radiation region. Because of this, silicon must be deposited in thick layers or it is necessary to use complex techniques of light trapping. Beyond the use of silicon, the thin film technology has the advantage to provide a large-scale productions, in which the panel is the final status of in-line processes and not the assembly of smaller cells, as in the case of crystalline silicon or polysilicon wafer based modules. The highest rates of production (in terms of square meters of modules per minute) have assumed since the '70s, that in the future, in order to compete with traditional sources, there will be just thin film modules. What has not allowed until 2000 the effective start of industrial production of thin film modules is that this technology had some problems, among which the reproducibility of the results, the stability in time and scalability on large areas of the layer deposition. Overcoming these problems, the photovoltaic modules that use $CuInGaSe_2$ (CIGS) and CdTe thin film technology are already being produced with high quality and conversion efficiency (12%), with expected values up to 14% for the near future. The cell interconnection integrated into modules of large area (0.6 x 1.2 m^2), with very limited use of raw materials, can minimize the production cost, so that the thin film modules will soon be able to compete with conventional modules based on silicon wafer.

In this chapter we want to analyze the typical thin film techniques that are able to produce highly efficient CdTe/CdS solar cells and especially those that are ready for industrial production. In particular, we want to show an all in-line process, starting from a laboratory scale and based on classical thin film technology, in order to demonstrate that it is possible to produce large area photovoltaic modules with high throughput.

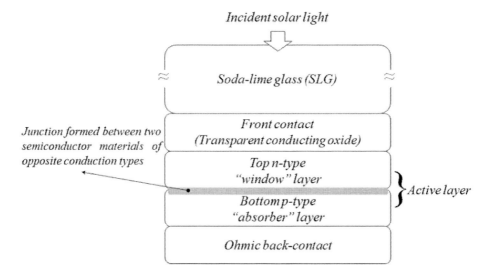

Figure 1. CdTe/CdS thin film solar cell structure in the backwall configuration.

2. THE ACTIVE LAYERS: CDTE AND CDS

2.1. The CdTe Layer

CdTe exhibits a forbidden gap of 1.45 eV very close to the maximum for solar energy conversion. In addition, its gap is direct and its absorption coefficient is in the range of 10^4-10^5 cm^{-1} for photon energies larger than the forbidden gap. This means that only a few micrometers of material are enough to absorb all the light. A theoretical maximum efficiency over 27% and a practical efficiency of 18.5% could be expected for this material with an open-circuit voltage of 880 mV and a short-circuit current density of 27 mA/cm^2 (with a negligible thin layer of CdS).

CdTe is one of the few II-VI compounds that can be prepared both *p* and *n*-type, with typical doping like B, Al, In and Ga for *n*-type and Cu, Ag or group I and V elements for *p*-type. Doping is rather simple for single crystal samples. The main problem arising with polycrystalline thin films is that all metallic doping can very easily segregate in the grain boundaries giving rise to metallic, highly electrical conducting phases that can shunt the CdTe layer. This segregation problem is still present during preparation of the CdTe film and is present during exposure to light of the CdTe/CdS based solar cell. Photo-induced diffusion is a very dangerous phenomenon for solar cells, because light produce an enhanced electrical field, which is able to move ions within the lattice at relatively low temperatures. For these reasons, it is usually preferred not to dope these materials using external atomic species. Fortunately, in contrast, CdTe can be grown self-doped by both Cadmium and Tellurium vacancies. The former (inducing an electrical *p*-type behavior) is the most favored during CdTe growth at high temperatures ($\geq$ 500°C). Thus, CdTe and CdS thin films employed in solar cells are usually grown undoped and it is necessary to take particular care with respect to the purity and stoichiometry of the starting materials.

Several methods have been used to deposit CdTe, namely vacuum evaporation and atomic layer epitaxy (ALE) [26,27], Electro-deposition [18,22,28], Chemical Spray Pyrolysis [10], Screen Printing [29,30], Chemical Vapor Deposition or Metal-Organic Chemical Vapor Deposition (MOCVD) [22] and Close Spaced Sublimation (CSS) [9,22,23,24,31,32].

High Vacuum Evaporation: With this technique, CdTe films are made by evaporation from a heated crucible and subsequent condensation of CdTe vapor on a heated substrate. Grain size enhancement and re-crystallization by thermal annealing in an ambient of chlorine salt have led to moderate device efficiencies of about 12-13%.

Atomic Layer Epitaxy (ALE): This is a thin film growth method, which is based on sequential chemical reaction on the surface. ALE offers the possibility to grow both CdS and CdTe in a single process. The main characteristics of thin film deposited by this method are the good crystalline quality, very high uniformity over a large area and very low pinhole density. A low-pressure type ALE reactor involves reactant transport and shuttering using computer controlled inert gas flow. The reactors are typically equipped with four controlled sources for the solid reactants and a certain number of external sources for gas and liquid reactants. The reaction zone normally consists of a 1-3 mm space between 10x10 cm^2 substrates. Cadmium sulfide films are grown using elemental Cd and S as reactants. The process temperature for the CdS deposition is in the range of 300-500°C. Using elemental Cd and Te atoms as reactants the process temperature is in the range 350-450°C. In order to obtain high efficiency devices a graded layer is processed between the CdS and the CdTe in which the Cd-S-Te proportion is gradually changed from pure CdS to pure CdTe. Solar cells entirely processed by means of the ALE technique have shown efficiencies up to 14%.

Electrodeposition: in electrodeposition, films of CdTe and CdS are formed from aqueous solutions of $CdSO_4$ and Te_2O_3 at temperatures of around 90°C. Two steps may represent the film deposition:
- tellurium reduction $HTeO_2^+ + 3H^+ + 4e^- = Te + 2H_2O$
- reaction of deposited tellurium with Cd^{2+} ions in the solution $Te + Cd^{2+} + 2e^- = CdTe$

The critical variables of this deposition method are the type and the concentration of species present in the solution, the design of the deposition system, the solution temperature and the flow geometry. In order to produce high photoelectronic quality films, a thermal annealing at high temperature for several minutes is needed.

Even though electrodeposition uses relatively cheap equipment, the low deposition rate and the health hazards of the toxic components used are aspects that do not correspond to the industrial criteria for large area production.

The best device on a laboratory scale (0.02 cm^2) exhibited an efficiency of 14.2%.

Chemical Spray Pyrolysis: In chemical spraying an aerosol of water droplets containing heated decomposable compounds of Cd, Te and S is sprayed onto a heated substrate forming CdTe or CdS films. The deposition temperature is about 480-520°C. A CdS film with a thickness of 6 μm is deposited on top of a commercial conductive soda-lime glass and subsequently 6 μm of CdTe are deposited on top of CdS.

Due to the poor optoelectronic quality of the very small grains typical of the sprayed films, device efficiencies produced with these layers are quite low. Since the CdS film thickness is rather high, over 20% of the possible photocurrent is lost because 20% of the incident light is absorbed through the window layer. For this reason the maximum solar cells efficiencies obtained by spraying CdS and CdTe are in the range 8-10%.

Screen-Printing: this process consists of an initial screen-printing with subsequent drying and sintering of first a CdS film and then a CdTe film. In comparison with other technologies, screen-printing normally leads to thick layers (10-20μm instead of some μm) and requires very high temperatures for sintering. The sintering is carried out in a furnace in an inert gas atmosphere at a pressure of several bars and at a temperature in the range of 550-750°C. The CdTe paste is obtained by mixing CdTe powders milled in water, $CdCl_2$ powder and propylene glycol. In a similar way the CdS paste is prepared. During the sintering process $CdCl_2$ is evaporated away from the film.

In CdS/CdTe solar cells prepared by this method, after the sintering step, a certain amount of intermixing between CdS and CdTe layers occur as a consequence of the high temperatures (up to 600-700°C). This is strongly connected with the presence of liquid $CdCl_2$, which is used as a sintering flux. Massive intermixing has to be avoided in order to obtain a good photo-response in the short wavelength region of the visible spectrum and then a high photovoltage and photocurrent. The morphology and the crystal structure of the screen printed-sintered solar cells benefit from the particular properties of sintering, such as the high temperature and the presence of liquid $CdCl_2$. Despite the micrometers sized grains in contact in the junction region, the final efficiency of the device is quite low and, on a laboratory scale, scarcely reaches 12%. This is principally due to the high thickness of both the layers constituting the junction, which causes losses in photocurrent and a high series resistance inside the active layer.

Metal-Organic Chemical Vapor Deposition: In MOCVD, volatile, thermally decomposable organic compounds of Cd and Te are transported in gaseous form by a carrier gas, in general at ambient pressure (1 bar), toward the heated substrate where they decompose and react to form the CdTe film. In general, MOCVD is a complicated process not easily scalable to large area production. Anyway, on a laboratory scale a device efficiency of 11.9% was obtained.

All these techniques are able to produce cells with efficiencies larger than 10%. However, the highest efficiency solar cells have been prepared by closed space sublimation (CSS).

2.2. CdTe by "Close Spaced Sublimation" (CSS)

In our laboratory, the deposition is carried out at high temperature in an inert or reactive gas at a pressure in the range 1-100 mbar. This type of deposition is possible because CdTe dissociates into its elements ($2CdTe(s) \leftrightarrow 2Cd(g)+Te_2(g)$), which can recombine on the substrate to form the CdTe film. The CdTe deposition can be carried out with the use of a sintered sputtering-like target in a graphite crucible (see Figure 2). This target is disk shaped 3.0 inches in diameter and is manufactured using 99.9999% purity CdTe powder (eg provided by 5N Plus) which is placed in a graphite crucible inside an oven. Then under a 10 bar N_2 pressure, the temperature is raised up to 1200°C for few minutes and then slowly lowered to room temperature.

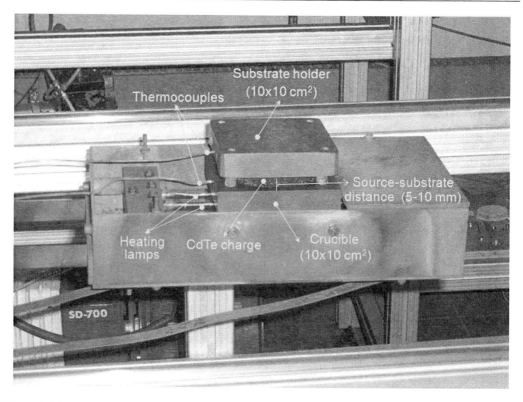

Figure 2. Photograph of the Close-Spaced Sublimation (CSS) system on laboratory scale.

As an encapsulant, B_2O_3 is used on top of the CdTe. The use of a sintered target as a CdTe source inhibits spitting of the CdTe so that the substrate can be directly faced to the source. In addition, a 3 inches diameter dimension ensures a uniform deposition because of its wider size with respect to the substrate (1 inch^2). The distance between source and substrate is typically 2 to 7 mm and the temperature of source and substrate are 500 and 600 °C respectively. With these parameters, a 6 μm thick CdTe film can be deposited in 2 min.

As mentioned, a CdS/CdTe junction can never work as a photovoltaic device after deposition of both the layers without further treatment. The lattice mismatch between the two materials is 9.7% and the interface defects can capture the majority of free carriers crossing the junction. The only way to effect the formation of a good device is to create in the junction region a mixed compound, namely CdS_xTe_{1-x}, between the two active materials. In this way, the lattice mismatch is gradually adapted, and the number of defects at the interface can be greatly reduced.

Some of the main advantages of the CSS method for junction formation with respect to the other mentioned techniques are related to the high quality of the as-deposited films, characterized by large crystalline grain size, which means a low defect density. During the CSS high temperature (500-600 °C) deposition of the CdTe layer on top of CdS film, the beneficial intermixing between these two layers begins to take place.

The absorption coefficient of CdTe, in the wavelength range corresponding to the sun light spectrum, is higher than 10^5 cm^{-1} thus a thickness of the film of 1-2 μm could be enough for converting all the visible light completely. Such a thickness would be optimal for both optical and electrical reasons, but in practice, it is very hard to achieve without side problems.

The grain size of CSS-deposited CdTe films is in the range of 5-10 µm (Figure 3(a)). Large grain size and columnar morphology are desirable in order to reduce transverse grain boundaries. On the other side, with the increase of the grain size there is also an increase in the dimensions of voids between the grains, and the film has to be thicker in order to avoid the presence of pinholes.

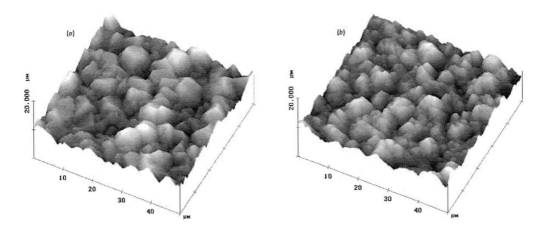

Figure 3. AFM (Atomic Force Microscope) picture of two CdTe layers deposited at 500 °C substrate temperature by CSS technique with 650°C source temperature; (a) ambient: pure Ar; total pressure: 1 mbar; average grain size ≈10 µm. (b) ambient: Ar+O$_2$; total pressure: 1 mbar (Ar) + 1 mbar (O$_2$); average grain size ≈ 2-3 µm.

The typical thickness of CdTe films deposited in a pure Ar atmosphere is around 8-10µm and, in spite of the thickness, the problem of pinholes is present. The problem is partly solved by increasing the Ar pressure up to 100 mbar. The deposition rate can be kept at high levels, around 20-30 nm/s, by simply raising the temperature of the CdTe source. The higher pressure allows a more uniform heat distribution, a greater diffusion of the Cd and Te vapors and a smaller re-evaporation of the growing film from the substrate. The resulting CdTe films are flatter, denser, with smaller grain size and with less problems of shunting. Since the substrate temperature is fixed to a maximum value of 520 °C, if soda-lime glass is used, one of the most important parameters which can control the formation of the CdS$_x$Te$_{1-x}$ layer, is the re-evaporation from the substrate at the beginning of the deposition. If the CdTe film grows on top of the CdS layer too quickly, (very low re-evaporation) the intermixing between CdS and CdTe has not time to be effective. For this reason the resulting devices show a poor performance, which are characterized by low photovoltages and low fill factors with maximum efficiencies below 10%.

Many research groups have found that CdTe films grown by CSS enhanced their crystalline and electro-optical quality if the deposition is carried out in Ar+O$_2$ atmosphere. The effect of oxygen on the CdTe growth is very strong: it increases the CdTe conductivity and reduces the grain size making the film more compact allowing the use of thinner films about 4-6 µm thick. By varying the oxygen percentage from 1 to 100% with respect to the total Ar+O$_2$ atmosphere it is possible to adjust the interaction between the CdS surface and the arriving Cd and Te atoms at the beginning of the deposition. In other words, independently of the total pressure, in presence of oxygen there is a change of the equilibrium between the CdTe sticking coefficient and the surface diffusion coefficient. Oxygen makes more stable the presence of Cd and Te atoms on the CdS surface due to its tendency to form compounds with both Cd and Te. These compounds, such as CdO and TeO_2, have chemical stability intermediate between that of CdTe and CdS and they play an important role on the surface diffusion of the impinging atoms. In fact, in presence of O_2, the reaction between Cd and Te atoms is lowered and Cd and Te have more time to diffuse along the substrate before recombining to produce a less defective CdS_xTe_{1-x} intermixed layer. This starting point affects the nucleation process, increasing the number of nucleation sites and promotes a denser growth for the whole CdTe film. This can be seen also from the AFM analysis made on the surface of CdTe film as it is shown in Figure 3 (b). In particular, a fast Fourier analysis has indicated that the average grain size of the CdTe films changes from 10 µm for the films grown in pure Ar, to 2-3 µm for the films grown with Ar+O_2. XRD spectra acquired with a diffractometer using Cu K_α radiation, show that the as-deposited CdTe films (Figure 4 (a)) always exhibit the (002) hexagonal preferred orientation. CdTe films prepared in Ar+O_2 generally contain a very small amount of $CdTeO_3$, which crystallizes with a monoclinic structure, probably by segregating inside the CdTe grain boundaries (Figure 4 (a)).

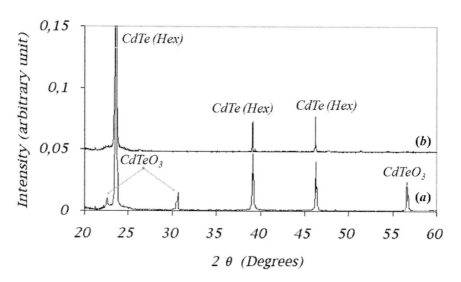

Figure 4. X-Ray spectra of CdTe layers grown by the CSS technique. (a) CdTe films grown with and (b) without O_2 in the CSS deposition chamber. In this picture, in order to make the spectra more legible, the CdTe peaks belonging to the (002) orientation have been removed.

2.2.1. The Heat Treatment in a Chlorine Ambient

Treatment in presence of Cl_2 is necessary, independently from the technique used to deposit CdTe, in order for the solar cell to exhibit high efficiency. If the $CdCl_2$ treatment is not performed, the short circuit current of the solar cell is very low and the efficiency is very low. This treatment is generally carried out by depositing a $CdCl_2$ film on top of CdTe by evaporation or by dipping the CdTe layer in a solution of $CdCl_2$-methanol and with a subsequent annealing at 400 °C in air or in an inert gas such as Ar [33,34]. One of the most serious problems encountered by using this material is that it is very hygroscope (and toxic), thus it is better not to keep films in air at room temperature for a long time. During the annealing, the small CdTe grains re-crystallize giving a better-organized CdTe matrix following possibly the process:

$$CdTe(s) + CdCl_2(s) \Rightarrow 2Cd(g) + {}^1/_2Te_2(g) + Cl_2(g) \Rightarrow CdCl_2(s) + CdTe(s)$$

The presence of Cl_2 could favor the crystalline growth of CdTe by means of local vapor phase transport. In this way, the small grains disappear and the CdS/CdTe interface is re-organized. This reaction seems to be strongly temperature dependent, with a marked sensitivity over some tens of degrees around 400 °C. It is independent from the thickness of the $CdCl_2$ layer (over 100 nm), from its uniformity and from the duration of the annealing (over ten minutes). If the process is carried out in air, some oxides are formed on the CdTe surface. These oxides need to be removed before making the back contact by using an acid etching or by cleaning the film surface in a solution of Br-methanol. The introduction of oxygen, into the CSS chamber during the deposition of CdTe layers, enables one to carry out the $CdCl_2$ thermal treatment at the same temperature but in an inert gas atmosphere such as Ar or N_2. This fact is very important because the presence of oxygen during the CdTe deposition or during the $CdCl_2$ treatment can increase the number of *p*-type majority carriers. Generally, the presence of oxygen results in CdTe films having a lower resistivity compared with the films prepared in a pure Ar atmosphere, typically with a carrier density of the order of 10^{14}-10^{15} cm^{-3}. Moreover, in case of annealing in Ar, the films do not show any oxide layer and thus no chemical etching is needed. In addition, the rinsing in methanol can be avoided, because the $CdCl_2$ residues can be easily removed by evaporation by evacuating the annealing chamber. This is a very remarkable innovation for the CdTe/CdS based solar cells fabrication process since a wet stage is eliminated. In fact, in an in-line large area photovoltaic module production, any use of acids or liquid solutions should be avoided in order to not interrupt the process and for safety considerations also.

From an industrial production perspective, we discovered another method for making the Cl_2-treatment, which does not use any $CdCl_2$ [35]. The CdTe/CdS structure is put in an ampoule, which is evacuated. A mixture of 100 mbar of Ar and 20 mbar of a non-toxic gas containing Cl_2 such as HCF_2Cl (difluoroclorometane) is introduced into the ampoule. The temperature of the ampoule is raised to 400 °C, an annealing of ≈ 5-10 minutes is carried out and then the ampoule is evacuated again. After the treatment, the CdTe morphology is completely changed due to an increase in the size of the small grains. Since HCF_2Cl is decomposed at 400°C and CdTe starts to decompose at around 400 °C, we suppose that the following process happens especially for the small grains that are the first-ones to decompose:

$$CdTe\ (s) + 2\ Cl_2\ (g) \Rightarrow CdCl_2\ (vap.) + TeCl_2\ (vap.) \Rightarrow 2Cl_2\ (g) + CdTe\ (s)$$

By keeping vacuum conditions for a few minutes at a temperature of 400°C, the $CdCl_2$ formed on the CdTe surface by the above treatment re-evaporates leaving a clean CdTe surface ready for the back contact. The Cl_2-treatment can be made with any gas of the Freon© family. The only need is that the gas contains chlorine.

The behavior of CdTe films treated with Freon© or with any other chlorine sources is quite the same from many points of view since the reaction mechanism is similar. Actually, during thermal treatment in chlorine ambient the smallest and more unstable CdTe grains are vaporized and, when they re-crystallize, larger and more stable CdTe grains are formed. The effect is very evident when CdTe is deposited by high vacuum evaporation in view of the fact that the average grain size is smaller than one micron. If we treat CdTe films produced by CSS (Close-Spaced Sublimation), starting grains are generally big, more than some microns, and only a re-crystallization of the grain edges is appreciated. The surface of the as-deposited CdTe film is shown in Figure 5 (a). The single crystals on the surface are very angled and between the grains it is possible to see very deep grain boundaries.

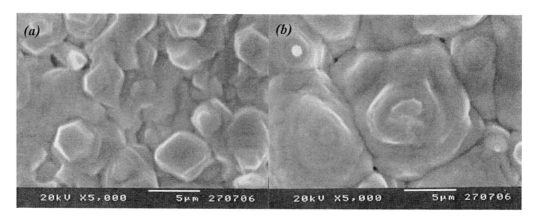

Figure 5. SEM photos of the surface of CdTe films. (a) morphology of an untreated CdTe film deposited by CSS method; (b) morphology of the same film after thermal treatment in Ar+HCF$_2$Cl atmosphere at a temperature of 400°C for 5 min.

The CdTe polycrystalline film grows in the zinc-blende structure. The (111) plane exhibits a hexagonal symmetry. This is clearly observed in Figure 5(a). After Freon treatment, the edges of the grains on the surface of CdTe layer are more round and the morphology appears to be smoother than before treatment. In Figure 5(b) one still can see the (111) hexagonal crystal structure but, after treatment, one observes a pyramid-like crystal with a hexagonal base. This indicates that some CdTe evaporates and gets lost during the treatment process leaving an etched-like surface. This behavior is typical of all the CdTe films heat-treated in a gas-containing chlorine atmosphere. In general, non-homogeneous distribution of the grain size is found, while larger grains are observed in the treated films.

Also for Freon© based process, the introduction of oxygen into the CSS chamber during the deposition of CdTe films, permits to execute the thermal treatment in chlorine ambient in an inert gas such as Ar or N$_2$ instead of an oxidizing atmosphere, such as air. This fact is very important because, as said before, the presence of oxygen during CdTe deposition or during Cl$_2$ treatment can increase the number of p-type carriers.

A clear dependence of the radiative recombination mechanism on the increase of HCF$_2$Cl partial pressure is observed by studying the cathode-luminescence (CL) spectra acquired on solar cells treated with different Freon© partial pressure. All the samples show the near band edge (NBE) emission, centered at 1.57 eV. The CL spectra acquired on HCF$_2$Cl treated CdTe (see Figure 6) show a broad CL band (1.4 eV), whose intensity increases by increasing the HCF$_2$Cl partial pressure from 6% to 10% in respect to the total Ar + HCF$_2$Cl pressure. The origin of this band could be attributed to the incorporation of Cl (or F) impurities from the annealing gas, which creates a V$_{Cd–Cl}$ complex.

Typical I-V characteristics are reported in Figure 7. We have taken the I-V characteristics by illuminating the cells by means of an Oriel™ solar simulator with 100mW/cm² incident power and AM 1.5 solar light spectrum. The measurement system has been calibrated with a 14% efficient CdTe/CdS thin film solar cell previously certified by an authorized laboratory. The most efficient device obtained for this series, corresponding to 8% HCF$_2$Cl partial pressure treatment, exhibits the following parameters: I$_{sc}$= 26.2 mA/cm², V$_{oc}$=820 mV, ff=0.69, η=14.8%. A pressure increase to 10% of the reactive gas in the annealing chamber yields to a degradation of the device reverse saturation current that is increased of various orders of magnitude, showing the high reactivity of the treatment and the impact of an annealing excess on the device electrical performance. At the same time, the quality diode factor N shows a variation of transport mechanism depending on the treatment conditions (inset of Figure 7).

For the not-treated sample N=1.8 indicates that recombination currents dominate the junction transport mechanism, which is representative of a high density of interface defects. An increase of the HCF$_2$Cl partial pressure gives rise to a situation in which diffusion and recombination currents are present, until a 8% HCF$_2$Cl partial pressure is reached.

At the same time, the quality diode factor N shows a variation of transport mechanism depending on the treatment conditions (inset of Figure 7).

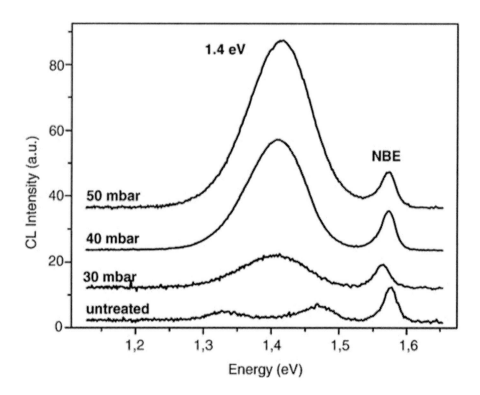

Figure 6. Comparison among the CL spectra acquired on the CdTe surface of the 6%, 8% and 10% HCF$_2$Cl treated CdTe/ZnO films and the untreated sample; Argon + Freon© total treatment pressure of 500 mbar, temperature of 77 K and beam energy of 25 keV, corresponding to 2.8 μm of maximum penetration depth, are used in all the CL analyses.

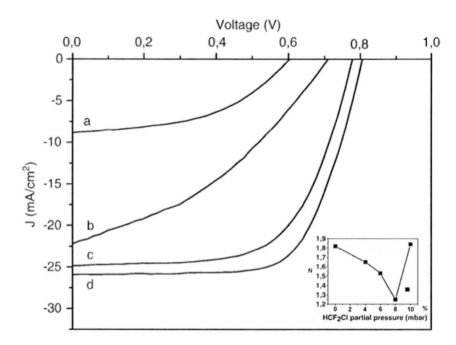

Figure 7. I–V characteristics under AM1.5, 100 mW/cm² illumination conditions of 0% (a), 10% (b), 6% (c) and 8% (d) HCF₂Cl partial pressure treated solar cells at room temperature. The inset shows the diode quality factor n as a function of the treatment partial pressure.

A minimum value of N=1.2 appoints to a predominant diffusion current, which suggests that the treatment is promoting the decrease of interface recombination centers density. The cell treated with 10% Freon© partial pressure shows again a sharp behavior, by increasing the diode factor n up to 1.8, giving indication of degradation of the junction. The increase of the diode reverse saturation current is responsible for a drastic reduction of fill factor. The evolution of the I–V light curves while comparing not treated, 6% and 8% HCF₂Cl partial pressure treated cells evidences an increase of the efficiency. We assume that this behavior is strongly related to the incorporation of Cl (or F) impurities into CdTe matrix. The increment of the photocurrent collection should be essentially due to an increase of the photo-generated minority carriers lifetime in the CdTe layer, which suggests the passivation of defects that, in absence of Cl, would contribute as non-radiative recombination centers. We consider the 10% HCF₂Cl treated cell an over-treated sample, where the intermixing process is so strong that CdS is locally consumed. The presence of shunt paths through the junction can explain the high reverse current and low fill factor values.

Depth-dependent cathode-luminescence analyses demonstrate that the impurities introduced during the Freon© heat treatment are not homogeneously distributed along the deposition direction of CdTe. In particular, chlorine concentration is higher near the CdTe surface and the CdS/CdTe interface. This is probably due to the fact that the penetration of Cl into the grains is smaller, in comparison with the migration through the grain boundaries. In this way Cl atoms could reach the CdS/CdTe interface and be incorporated as donors in the S-rich part of CdS_xTe_{1-x} alloy region, while in the CdTe-rich part, chlorine can form complexes with Cd-vacancies which have acceptor behaviour. The value of X and the extension of the intermixed layer could depend both on the CdTe deposition parameters and on Freon© heat treatment temperature.

The spatial distribution of the Cl-related defects is studied by acquiring monoCL images at the emission energies of the bands observed in the CL spectra, 1.57 eV and 1.4 eV respectively. The 1.57 eV monoCL (Figure 8b) is concentrated inside the CdTe grains, while 1.4 eV shows a not-uniform intensity distribution with clear increase at the grain edges (Figure 8c).

Figura 8. Luminescence mapping of a CdTe film treated with 8% HCF$_2$Cl; a) SEM image of the CdTe surface morphology; b) monoCL acquired at E=1.57 eV, NBE emission; c) monoCL at E=1.4 eV.

This treatment method is very effective in producing high efficiency CdTe/CdS thin film solar cells. It avoids the use of CdCl$_2$ that could be dangerous and instead it uses a gas that is stable, inert and not toxic at room temperature. Besides, it eliminates the CdCl$_2$ deposition step and, as a consequence, it is much more suitable for an industrial production. This process has been patented [36]. We think that this way of treating the CdTe film is quite general. For example, the treatment can be performed in the presence of any gas of the Freon family on condition that chlorine and fluorine are present in gaseous state. This will allow us to easily overcome to the fact that starting from 2010 the production and importation into Europe of R22 (HCF$_2$Cl) will be banned.

2.3. The CdS Layer

The CdS film in the CdTe/CdS solar cell is the so-called window layer. Since it is *n*-type it enables the formation of a *p-n* junction with *p*-type CdTe. With an energy-gap of 2.42 eV, CdS is transparent in the visible part of the solar light spectrum and therefore the solar light can penetrate into the CdTe layer thus giving rise to the photovoltaic effect. In order to have a highly efficient solar cell, an excellent *p-n* junction with a very good contact on the *p*-type CdTe is needed. Since the *p-n* junction formed between *n*-type CdS and *p*-type CdTe is strongly dependent on a proper interaction between these two layers, the deposition technique used to prepare these materials becomes very important. In particular for CdS the most suitable deposition techniques are: R.F. Sputtering [21], Close-Spaced Sublimation (CSS) [20], High-Vacuum Thermal Evaporation (HVTE) [37,38] and Chemical Bath Deposition (CBD) [19,39]. Although the highest energy conversion efficiency was obtained by using a CdS layer prepared by CBD, it is normally preferred to use the sputtering or CSS deposition method since CBD is not so suitable for large-scale production.

2.3.1. CdS by Chemical Bath Deposition (CBD)

Thin film semiconductors of metal chalcogenides, such as sulfides or selenides, may be deposited on metal foils, glass and polymer substrates. This happens when these substrates are dipped into a solution containing metal complex ions and sulfide or selenide ions. This technique is the so-called Chemical Bath Deposition (CBD) method [19, 39] and it is often demonstrated that CBD is a low-cost simple technique to achieve good quality CdS films, suitable for obtaining high efficiency CdTe/CdS based solar cells.

In CBD, CdS films can be prepared by exploiting the decomposition of thiourea in an alkaline solution of cadmium salts, following the reaction:

$$Cd(NH_3)_4^{2+} + SC(NH_2)_2 + 2OH^- = CdS + CH_2N_2 + 4NH_3 + 2H_2O$$

According to this reaction a chemical bath for the deposition of CdS could contain an aqueous solution of cadmium acetate ($Cd(CH_3CO_2)_2$) as a cadmium ion source, thiourea (N_2H_4CS) as a sulfur ion source, ammonium hydroxide (NH_4OH) as a complexing agent and ammonium acetate ($CH_3CO_2NH_4$) as a stabilizing buffer agent for the reaction.

The previous reaction implies that the growth of CdS films can occur either by ion-by-ion condensation of Cd and S ions on the surface of the substrate or by adsorption of colloidal particles of CdS. The behavior of the growth process is schematically summarized in these three remarkable points:

- *Nucleation period.* First, the chemical reactions in the bath are completed and an initial monolayer of the metal chalcogenide is formed on the surface of the substrate (the CBD technique requires extremely careful substrate cleaning).
- *Growth phase.* When the surface of the substrate is entirely covered by the initial monolayer, this can act as a catalytic surface for the condensation of metal and of chalcogenide ions resulting in film growth.
- *Terminal phase.* During growth the chemical reactions change and, as a consequence, the deposition rate changes. In fact, the growth rate assumes a maximum value after a certain time depending on the solution parameters and finally, when the ions species diminish, it achieves a terminal phase at which the film stops growing.

Taking these considerations into account, it is possible to assert that, if the aim of the deposition is a high film thickness, it is necessary to combine a relatively high initial concentration of metal ions in the bath with a relatively low temperature. In contrast, if the objective is the deposition of a thin layer, typically ≤ 50 nm, it is more appropriate to use a very dilute concentration and a relatively high deposition temperature.

Since CdS exists in two crystalline forms, the hexagonal (wurzite) and the cubic (zincblende) phase, with CBD technique it is possible to deposit CdS films in both these crystalline structures depending mainly on the chemical composition of the bath, the temperature and the pH of the solution. It was found out that the as-deposited CdS films contain both cubic and hexagonal structural forms as a mixture. It was also seen that the percentage of hexagonal structured grains increases by annealing the film at temperatures above 400°C. This is explained by considering that the wurzite form is more stable than zincblende at high temperatures in CdS films. Another confirmation of the transformation in the crystalline phase comes from the absorption spectrum; in fact, the energy gap of 2.42 eV is observed only for films annealed at a temperature above 400°C for 1 hour in a Nitrogen atmosphere. As-deposited films exhibit a lower value of the energy gap typical of the cubic phase.

Although the CBD technique allows the deposition of CdS thin films that are very dense, compact, smooth and without pinholes, the as deposited films are not suitable for photovoltaic device fabrication when used as an *n*-type partner with *p*-type CdTe. An annealing of the film at high temperature is needed for the reasons explained before. The annealing promotes some beneficial changes in the film such as the reorganization in the crystalline structure or the re-evaporation of excess sulfur atoms, but some detrimental effects, like the self-oxidation of the CdS film, were also observed.

Even though the annealing is always carried out in an inert atmosphere, one of the possible oxidation agents is the absorbed water in the film from the bath solution. Another possible mechanism able to explain the oxidation may be the following reaction:

$$Cd(OH)_2 = CdO + H_2O$$

Here some $Cd(OH)_2$ precipitates in the film during the CBD process and it can change to CdO at high temperature giving water vapor that can cause further oxidation. When the oxidation produces more than 30% of CdO in the CdS film we see a sharp decrease in the energy gap value. In other words, in a mixture of CdS + CdO one measures always the energy gap of CdO. In spite of this aspect it is important to notice that, by using a CBD CdS thin film in a CdTe based solar cell, an energy efficiency record was obtained, but an important step of this process consisted in the annealing of the CdS film in an atmosphere containing Ar + 20%H_2 (Forming Gas). Despite this remarkable result, it must be considered that CBD is not suitable for large-scale production since the deposition process is not fast and gives a waste that needs to be recycled.

2.3.2. CdS by Close-Spaced Sublimation (CSS)

In Close-Spaced Sublimation (CSS) CdS is sublimed from a solid source (see Figure 4) [20,21]. The deposition of CdS films by CSS method is based on the reversible dissociation of CdS at high temperature in an inert gas ambient at a pressure in the range 1 to 100 mbar. The CdS source dissociates into its elements, which recombine on the substrate surface depositing a CdS film. Since the rate of sublimation depends strongly on the source temperature and the gas pressure in the reaction tube, the rate of CdS deposition varies similarly.

The most critical CSS process parameters for the deposition of the CdS film are the substrate/source temperature and the ambient conditions. For an inert gas such as Ar, it was found out that the films contained pinholes for temperatures as low as 400 °C and high deposition rates. Considerable changes in film structure were observed when the depositions were carried out in the presence of O_2. Oxygen may act as a transport agent as well as it can be incorporated in the film to form defect complexes. The use of Oxygen was found out to be beneficial since it yields pinhole free films over a wide range of deposition conditions. The presence of O_2 in the deposition chamber however reduces the deposition rate and the mean grain size.

As an example, we describe the behavior of several CdS films prepared in our laboratory by CSS both, in pure Ar or in an Ar containing 50% of O_2 at a substrate temperature in the range of 480-530 °C. The CdS film deposition is carried out by putting in a graphite crucible a sintered sputtering-like CdS target. This target is disk shaped and manufactured with 99.999% pure CdS powder. Using a sintered target of CdS as a source any spitting of CdS powder is inhibited and the substrate can be placed directly in front of the source. During the CdS deposition, the Ar+O₂ total gas pressure was 50 mbar and the distance between the source and the substrate was around 2 to 7 mm. The typical deposition rate was 50 nm/min and the time for film preparation was a few minutes for 10 to 150 nm thick CdS films.

Films prepared in pure Ar even though they exhibit high transparency and a good absorption spectrum, were found not to be suitable for the preparation of efficiency CdTe/CdS solar cells.

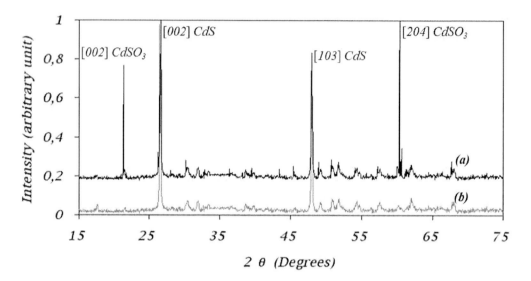

Figure 9. X-Ray spectra of the CdS layers grown by a CSS technique. (a) CdS films grown with and (b) without O₂ in the deposition ambient. In this picture, in order to make the spectra more legible, we eliminated the reflections belonging to the TCO layers.

Films prepared in Ar+O₂ generally contain a large amount of O₂, both on the surface and in the bulk as it has been seen by X-ray Diffraction (XRD). The XRD spectra were acquired with a diffractometer using Cu K_α radiation and CdS, CdSO₃ and In₂O₃ standards were used to index the peaks. All the films were found to be of the hexagonal phase which is the stable structure of CdS films deposited at high temperatures (more than 500 °C). Figure 9 shows an XRD scan in the 15-75° (two-theta) range for films prepared in Ar+O₂ (top) and pure Ar (bottom). The spectrum of the CdS films prepared in an ambient containing more than 50% of oxygen indicated that a second phase was present, in fact the 21.2° and 60.1° peaks could be associated with monoclinic CdSO₃. The presence of oxygen seems to affect the nucleation process, increasing the number of nucleation sites and promoting a denser growth of the CdS films. These results from the AFM analysis on the surface of a CdS film are shown in Figure 10. In particular, a fast Fourier analysis has indicated that the average grain size of the CdS films changes from 80 nm for the film grown in pure Ar, to 10 nm for the film grown in Ar+O₂. The O₂ content reduces to less than 10% in the surface when the film is annealed for half an hour at 400-420°C in 400 mbar of Ar containing 20% of H₂. The annealed CdS film is more transparent than the as-deposited one. Solar cells made with as-deposited CdS films exhibit low short-circuit currents indicating the presence of an insulating layer at the interface between the CdS and the CdTe which could be the $CdSO_3$ layer seen by X-rays. Solar cells made with CdS annealed in $Ar+H_2$ always exhibit a good performance with efficiency higher than 14% and a maximum efficiency approaching 15% (solar cell behavior is included in a later section). The beneficial effect of O_2 in preparing CdS by CSS has been also reported by Ferekides et al [20].

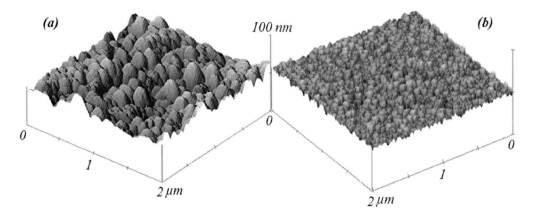

Figure 10. Atomic force micrograph of the CdS film, deposited by CSS in (a) pure Ar and (b) Ar + 50% O_2 atmosphere. Both the films were 100 nm thick.

2.3.3. CdS by R.F. Sputtering

The CdS layers were grown by radio-frequency (R.F.) sputtering at a substrate temperature of about 220 °C, with a deposition rate up to 10 Å/sec in an Argon atmosphere of 10^{-3} mbar by using a 99.995% pure CdS vacuum-cold pressed target supplied by Cerac Inc.. After the deposition, the soda-lime glass covered by the TCO-CdS bi-layer was heat treated in a vacuum chamber at 500 °C for 30 min.

The efficiencies of solar cells prepared with this kind of CdS layer are quite poor, in the range of 8-10%. This result is due to the fact that these devices exhibit a diode reverse saturation current too high. One possible explanation is that the interaction between CdS and CdTe is anomalous in that the grain boundaries in the CdS film are active and can channel the diode reverse current.

It is possible to get over this problem by introducing in the sputtering chamber, during the CdS deposition, Argon containing 3% of CHF_3. This gas is decomposed and ionized in the sputtering discharge freeing F^- ions which, being strongly electronegative, are directed to the substrate that is the positive electrode; here two different events can happen:
- The presence of energetic F^- ions near the substrate favors the formation of a fluorine compound such as CdF_2 during the growth of the CdS film [21].
- The F^- ions, accelerated by the electric field present in the discharge, hit the film surface during the deposition with energy sufficient to sputter back the more weakly bonded Cd or S atoms.

This effect leaves a CdS film with high quality optical and the structural properties. We can see in Figure 11 that the CdS films deposited in Argon+CHF_3 have an energy gap greater than that of the films deposited in Argon alone! Despite the beneficial role of fluorine in the growth of sputtered CdS films, it was demonstrated that fluorine is not an effective dopant in CdS, since no change in the resistivity of the CdS layers was observed. However, solar cells fabricated using 80 nm CdS thick films deposited by sputtering in the presence of fluorine showed a very high efficiency of the order of 15%-15.8%. We explain this fact by considering that CdS(F) may contain CdF_2 probably segregated in the grain boundaries. While the CdF_2 segregated in the grain boundaries can be useful to passivate them, the CdF_2 layer grown on the CdS surface may adjust the interaction between CdS and CdTe during the CdTe deposition by CSS.

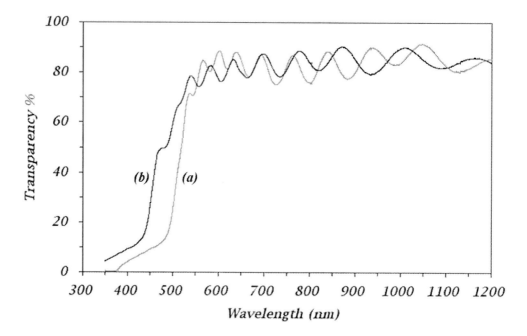

Figure 11. Transmission spectrum of an 80 nm thick sputtered CdS film: (a) deposited in pure Argon and (b) deposited in Argon+CHF_3. The shift towards shorter wavelengths of the absorption edge proves the beneficial effects of deposition in the presence of CHF_3.

These remarkable results are shown in Figure 12 and demonstrate that CdS(F) can be used as deposited without any other treatment. Probably this fact makes the CdS films deposited by sputtering in presence of fluorine the best candidates for the industrial production of photovoltaic modules based on the CdTe/CdS thin film technology.

2.3.4. CdS by "High Vacuum Thermal Evaporation" (HVTE)

Good photovoltaic devices are also obtained by using evaporated CdS layers. In fact it is possible to evaporate CdS thin films in a high vacuum evaporation chamber at a substrate temperature in the range of 150-200°C. Graphite crucibles are normally used as evaporation sources in the range of temperatures between 700-900°C. As-deposited, these CdS layers are not suitable for producing high efficiency CdTe/CdS solar cells, since they are not sufficiently stable. For this reason, before the growth of the CdTe layer, CdS films are annealed in vacuum or in a hydrogen atmosphere at 400-450°C for 30 min. The beneficial effects of this thermal treatment are well known in terms of grain growth, re-crystallization and re-evaporation of stoichiometric excess such as not well bonded Cd and S atoms [37,38]. The as-deposited CdS films change their average grain size from 100-300 nm to 500 nm after the annealing and it appears that the heat treatment re-crystallizes the CdS layers in such a way that some of the smaller grains can coalesce together to form bigger grains some of which are 500 nm in width. Despite this, the resulting CdS layer is not so dense as the films obtained by sputtering or CBD and the average thickness of the CdS films used to prepare CdS/CdTe based solar cells is never below 300 nm. Actually, in HVTE CdS films, it is easy to find pinholes if their thickness is below 200 nm. Since it has been widely shown that the CdS layer contributes a negligible photocurrent, it constitutes a photocurrent loss for the wavelength range from 300 nm to 520 nm. This poor photo-response is probably due to the low lifetime of the holes and to a high recombination despite the high absorption coefficient of CdS in the wavelength range of interest ($>10^5$ cm^{-1}). The parasitic optical absorption in CdS layer is the predominant loss in photocurrent, making it necessary to minimize the CdS thickness. The highest efficiency CdTe/CdS solar cells are distinguished by their high short circuit current (>25 mA/cm^2). This has been achieved using CdS film with a thickness below 100 nm deposited by CBD. It is possible to reduce the CdS film thickness only down to 100 nm whilst maintaining a uniform interface throughout processing in order to avoid the formation of parallel junctions between CdTe and the transparent conductive oxide (TCO), which have a much higher recombination current than CdTe/CdS. In practice, HVTE CdS films are suitable as *n*-type partners in high efficiency CdTe/CdS based solar cells only if their thickness is greater than 200-300 nm in order to avoid any complication due to pinholes. The use of thin CdS is also complicated by a further reduction in the thickness of the CdS film, which occurs during CdTe film growth or during post deposition heat treatment. Anyway, devices prepared using HVTE CdS films, thermally treated as described above, with a thickness of about 300 nm show a fairly good performance, exhibiting efficiencies up to 12-13%.

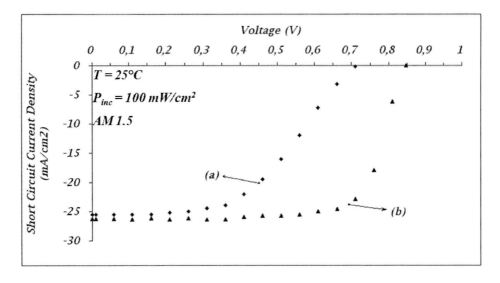

Figure 12. J versus V characteristics of the CdTe/CdS solar cells taken in standard conditions in our laboratory: (a) behavior of a device obtained by using a CdS layer deposited by sputtering in pure Argon and then annealed in vacuum at a 500°C substrate temperature for 30 min. (b) behavior of a device obtained by using a CdS layer deposited by sputtering in an Argon+CHF$_3$ atmosphere.

The use of the HVTE technique could find a role in all the processes in which the temperature regimes don't exceed 450°C. For this reason the HVTE deposition method is attractive for a very simple in-line production of large area CdTe/CdS based solar modules on polymer films [39], facilitating the roll-to-roll manufacturing of flexible solar modules.

3. THE ELECTRICAL CONTACTS

3.1. Front Contact

In recent years there has been a great interest in metallic oxides thin films due to their many industrial applications [25, 40, 41]. Thin films of these materials (TCOs) are produced by several deposition techniques. The most studied TCOs are: SnO_2:F (FTO), ZnO:Al (AZO), In_2O_3:Sn (ITO), and Cd_2SnO_4 (CTO) [42]. These metallic oxides exhibit very good optical transparency nearly or more than 90% for visible light and near infrared radiation and very high n-type conductivity. For these reasons TCOs are generally nearly degenerate semiconducting materials with a free carrier concentration between 10^{18} cm^{-3} and 10^{20} cm^{-3}. The high transparency and the high electrical conductivity make the TCOs suitable for a great variety of applications. In fact, they are used in optoelectronic devices and as transparent electrode in photovoltaic modules. In addition, they have been employed in glass coatings, for example as transparent heating elements for planes and car windows. Because of their high reflectivity in the IR part of the spectrum they could also be used as transparent heat-mirror coatings for buildings, cars and energy saving light bulbs. Since it is not possible to obtain both high electrical conductivity and optical transparency in any intrinsic material, one way to reach this aim is to create electron degeneracy in a wide band-gap oxide. This could be made in two different ways:
- Introducing donor elements into the oxide matrix.
- Exploiting deviation from correct stoichiometry by using, for example, structural defects and/or oxygen vacancies.

The first point is explained by considering that the substitution of a higher valence cation by a donor impurity in the oxide, e.g. tin or antimony in indium oxide or fluorine in tin oxide, increases the electron concentration and so the n-type conductivity. On the contrary, the replacement of a lower valence cation by an acceptor impurity generates a hole (broken bond) that works like a trap (deep level in the energy gap) in the n-type metallic oxide decreasing its n-conductivity. Since the mean grain size of TCO thin films is in the range of 10-100 nm depending on the deposition method, the high electrical conductivity of doped and undoped films depends mainly on carrier (electron) concentration and not on their mobility. This is due to the fact that the mobility in these films is considerably lower than that in the bulk materials, because it is limited by grain boundaries.

In the last few years a lot of new TCOs have been developed starting from multicomponent oxides such as: $GaInO_3$, $ZnSnO_3$, $Cd_2Sb_2O_6$:Y, Zn_2SnO_4, $MgIn_2O_4$, In_4SnO_{12} [43]. All these metallic oxides can exhibit high n-type conductivity following the behavior described above. In addition, a new p-type TCOs has been intensively studied in recent years in order to make a p-n junction. In 1997, it was reported for the first time that a $CuAlO_2$ thin film exhibits p-type conductivity. After that, a new series of materials based on copper was discovered such as: CuGaO$_2$ and SrCu$_2$O$_2$ [44,45]. In 2000, a UV-emitting diode based on a p-n heterojunction composed of p-SrCu$_2$O$_2$ and n-ZnO was successfully fabricated using heteroepitaxial thin film growth. Anyway, the major area of interest is in n-type TCOs due to their utilization in industrial applications. One of these applications is in photovoltaic (PV) module fabrication. In this case, it is necessary to reach a very low resistivity. This direction has been strongly accelerated by the rising demand for enlargement of the module size. In thin film PV module production, the specification needed for TCOs not only concerns the very high electrical conductivity and very high optical transparency but also their chemical and physical stability [46].

3.1.1. The In$_2$O$_3$ (IO) Family

The In$_2$O$_3$ (IO) films are normally polycrystalline with cubic structure and a typical grain size of about 10-50 nm depending on the deposition technique. The most common techniques used for deposition of IO thin films are the following: reactive R.F. sputtering, chemical vapor deposition, spray pyrolysis, glow discharge and activated reactive evaporation. IO films often exhibit superior electrical and optical properties with respect to the other transparent conductors; this fact is principally due to the higher mobility in IO. In fact In$_2$O$_3$ films prepared using various deposition methods have mobilities in the range 10-75 cm^{2}V^{-1}s^{-1} that is very high considering that they are polycrystalline thin films. IO films exhibit a direct optical band gap, which lies between 3.55 and 3.75 eV, which was shown to increase with increasing carrier concentration owing the Burstein-Moss shift. For films with thicknesses below 1 µm the optical transparency in the visible and near IR regions is about 80-90%. For this reason IO has been widely used as a transparent conductor.

3.1.2. The Tin Doped In$_2$O$_3$ (ITO) Films

ITO films prepared in our laboratory were obtained by R.F. magnetron sputtering from several targets with different stoichiometry; in particular, we studied four kinds of ITO: In$_2$O$_3$ containing, respectively, 1, 2, 4, and 10 weight% of SnO$_2$. Some ITO films were deposited in a sputtering gas mixture containing Ar + O$_2$; the O$_2$ partial pressure was varied between 2 and 20% in respect to Ar + O$_2$ total pressure. All these films exhibit a very low resistivity (2x10^{-4} Ωcm); the resistivity of the ITO films does not depend on the stoichiometry of the target and on the sputtering deposition parameters, such as the substrate temperature, deposition rate, Ar pressure and power density. This probably means that Sn is an effective dopant and the doping level does not depend on the typical sputtering parameters. Since the optical transparency of the ITO layers is more than 80%, the films are sufficiently transparent to be used as front contacts for CdTe/CdS solar cells. However, we found out that the efficiency of the solar cells, when using an ITO film thicker than 1.2 µm, could be quite high (above 14%) but not very reproducible. We explain this by the fact that all ITO targets modify their surface after several runs forming some In-rich nodules which can cause some occasional discharges during sputtering deposition [47]. This discharge instability produces a nonstoichiometrically uniform ITO film. In order to improve the reproducibility of the ITO film we introduce in the sputtering chamber, during the film deposition, a small amount of H$_2$ and trifluoromethane (CHF$_3$). The H$_2$ partial pressure was changed in the range of 1-10% and the CHF$_3$ in the range of 1-10% with respect to the total Ar+H$_2$+CHF$_3$ pressure. With the maximum quantity of CHF$_3$ we did not observe any indium-rich nodule formation on the surface of the ITO target. This is probably due to the fact that the free indium atoms can react with fluorine on the surface of the target forming a stable compound. Therefore, this reaction prevents the segregation of In into superficial nodules and, as a consequence, the sputtering discharge is more stable producing stoichiometric and more uniform ITO films. From the point of view of the stability of the whole device, the quality of the ITO layers is better for films prepared in Ar+4%H$_2$+10%CHF$_3$ gas. This kind of film, being very compact and dense, offers a better shield against Na diffusion from the soda-lime glass. Since the presence of Na atoms favors the reactivity of all the present elemental species, this can cause the formation of undesirable compounds. For example, the formation of CdSO$_4$ has been observed, when a CdS film is deposited in the presence of oxygen on top of ITO layer if ITO is not sufficiently dense to avoid the diffusion of alkaline atoms from the soda-lime glass [29].

Table 2. Summary of the best reults for different types of transparent conducting oxides (TCO) used in polycrystalline thin films solar cells

TCO	Deposition Temperature [°C]	Film thickness [nm]	Resistivity (Pure Ar) [Ωcm]	Resistivity (Ar+H$_2$+CHF$_3$) [Ωcm]	Resistivity (PureAr +O$_2$) [Ωcm]	Chemical stability	Transparency [%] λ(400-800nm)
In$_2$O$_3$:SnO$_2$ (ITO)	300-450	400	2.0x10^{-4}	1.8x10^{-4}	2.0x10^{-4}	Poor w/out O$_2$ / Good with O$_2$	≥ 80
In$_2$O$_3$:GeO$_2$ (IGO)	350-450	500	2.2x10^{-4}	1.8x10^{-4}	1.8x10^{-4}	Very good	85
In$_2$O$_3$:SiO$_2$ (ISO)	300-450	400	2.0x10^{-4}	2.0x10^{-4}	2.0x10^{-4}	Very good	85
In$_2$O$_3$:F (IFO)	300-420	500	1.0x10^{-2}	2.8x10^{-4}	---	Excellent	88
ZnO:Al (AZO)	200-350	1000	8.0x10^{-4}	---	---	Poor (Al diffusion)	89
ZnO:F (FZO)	200-350	1000	1.0x10^{3}	8.5x10^{-4}		Excellent	92
Cd$_2$SnO$_4$ (CTO)	100	300	---	---	1.2x10^{-4}	Poor (hygroscopic)	≥ 90
Zn$_2$SnO$_4$ (ZTO)	100	200	---	---	1.0x10^{-2}	Poor (hygroscopic)	≥ 90
SnO$_2$:F (FTO)	350-450	1000	1.0x10^{-1}	8.0x10^{-4}	---	Excellent	≥ 90

3.1.3. Germanium and Silicon Doped Indium Oxide (IGO-ISO) Films

Germanium doped indium oxide (IGO) films are normally prepared by R.F. sputtering from a target of In$_2$O$_3$ containing 4% weight of GeO$_2$. A typical deposition rate is larger than 1 nm/s. The surface of this target did not present any formation of In-rich nodules and the film deposition is much more stable and reproducible. The preparation of an IGO film is carried out by using pure argon or a mixture of Ar, H$_2$ and CHF$_3$ as a sputtering reactive gas. In the first case it exhibits a resistivity on the order of 8.8x10^{-4} Ωcm and a very good transparency in the visible region of the spectrum. This means that germanium has the same behavior as tin in an In$_2$O$_3$ matrix and acts as an effective dopant for In$_2$O$_3$. The IGO films deposited by reactive sputtering by using a mixture of Ar+H$_2$+CHF$_3$ as sputtering gas presents a resistivity of 2.2x10^{-4} Ωcm (Table 2). This result shows that hydrogen can create some oxygen vacancies in the In$_2$O$_3$ matrix giving more possibility for fluorine to substitute oxygen and therefore to better dope the IGO film during the deposition. The H$_2$ and CHF$_3$ gas pressures were both 5% with respect to the total Ar+H$_2$+CHF$_3$ gas pressure. The IGO film stability and its very low resistivity permit the use of a thickness as low as 400 nm for the TCO layer. With this TCO we fabricated solar cells with an efficiency up to 15% and with very good reproducibility.

Under the same conditions of deposition of IGO we deposited films of In$_2$O$_3$ doped with Si (ISO) from sputtering targets obtained by hot-pressing technique of In$_2$O$_3$(96% weight)+SiO$_2$(4% weight) powder. With this type of material we have obtained films with a resistivity of 1.8×10^{-4} Ωcm and this value does not change if we use Ar+H$_2$+CHF$_3$ as process gas in the deposition chamber, which suggests that the doping of Si is very effective following the same behavior of Sn. The excellent electrical conductivity, the very high transparency in the range of wavelengths 400-800 nm, the very good and chemical stability of the material, the ability to shield sodium from soda-lime glass and the observation of no formation of indium-rich nodules on the surface of the target, put this material as one of the most suitable for use as a transparent contact for solar cells.

3.1.4. Fluorine Doped Indium Oxide (IFO) Films

Another TCO of the IO family is fluorine doped indium oxide (IFO). In$_2$O$_3$ can be sputtered at a relatively high deposition rate (≥10 Å/s) without any change on the target surface. When In$_2$O$_3$ is not intentionally doped during sputtering deposition it grows with a resistivity of the order of 1×10^{-2} Ωcm. In this case the conductivity is due to native defects, such as oxygen vacancies (Table 2). It is possible to prepare IFO films with a resistivity of ≈ 2.8×10^{-4} Ωcm by introducing Ar containing 5% of H$_2$ and 5% of CHF$_3$ in the chamber during sputtering deposition (Table 2). The IFO films obtained in this way are very smooth and transparent. Besides, we found out that 100 nm of this material are sufficient to passivate the diffusion of sodium atoms into the film from the soda-lime glass. For this reason and for its intrinsic stability, IFO is, perhaps, one of the best-suited material for solar cell production. Depositing in sequence 400 nm of IFO and 100 nm of CdS onto 1 inch^2 soda-lime glass, we were able to obtain a 14% efficient CdTe/CdS solar cell.

3.2. Aluminun and Fluorine Doped Zinc Oxide (AZO-FZO)

ZnO thin films can be prepared by a variety of thin film deposition techniques, such as D.C. reactive and R.F. magnetron sputtering, chemical vapor deposition (CVD), reactive thermal evaporation and Chemical Bath Deposition (CBD) [48-50]. The sputtered films are normally polycrystalline with an average grain size of about 5-30 nm with a wurtzite-type structure with a strong c-axis preferred orientation perpendicular to the substrate. Pure and stoichiometric zinc oxide films show a very high resistivity and a direct band gap of about 3.2 eV. Furthermore ZnO has one of the largest electromechanical coupling coefficients. Due to this property, zinc oxide is a well-known piezoelectric material, which has been used as a transducer for surface acoustic wave devices. In the last 10 years, ZnO has been intensively used as a TCO for photovoltaic applications and as a gas sensor device. The primary need for a ZnO film is a high transparency in the visible region and a high conductivity. In order to reach a good conductivity it is common practice to dope ZnO with trivalent cations such as indium or aluminum. In our laboratory we obtained ZnO:Al (AZO) thin films, deposited by R.F. sputtering, with a resistivity of the order of 8×10^{-4} Ωcm and with a very good transparency (over 90% in the range 450-850 nm of the visible spectrum). The starting target was a hot-pressed powder mixture of 98 weight % ZnO and 2 weight% Al_2O_3 supplied by Cerac Inc. Since the ZnO:Al thin films are not stable at high temperature with respect to Al diffusion, as we observed by making use of these films in our CdTe/CdS solar cell, we tried to dope ZnO films with fluorine. In order to do that we used reactive R.F. sputtering technique starting from a pure Cerac ZnO target and introducing into the sputtering chamber a gas mixture containing Ar, H_2 and CHF_3. The H_2 partial pressure was fixed at 5% with respect to the total Ar+H_2 pressure, while the CHF_3 pressure was varied in the range 1-10% of the total Ar+CHF_3 pressure in order to change the fluorine doping level in the ZnO:F film. The presence of H_2 in the sputtering chamber causes the creation of some oxygen vacancies in the ZnO matrix giving more possibility for fluorine to substitute oxygen atoms and, as a consequence, to better dope the ZnO film during its deposition. We observed that the presence of hydrogen into the sputtering chamber made the deposition of ZnO films almost independent of substrate temperature in the range 200-350°C. In fact, low resistivity, of the order of 8.5×10^{-4} Ωcm and high transparency has been achieved by making use of the maximum quantity of CHF_3, independent of the substrate temperature in the above-mentioned range (Table 2). It was also demonstrated that this doping level did not change with heat treatment of the ZnO film in air at a temperature higher than 500 °C. This means that fluorine substitutes oxygen atoms into the ZnO matrix, making a stable chemical bond with Zn suggesting that it is an effective dopant for ZnO.

3.3. Zinc and Cadmium Stannate (Zn_2SnO_4 - Cd_2SnO_4)

Zn_2SnO_4 and Cd_2SnO_4 thin films are prepared by reactive R.F. sputtering in a mixture of Ar+50%O_2. Both the materials are very transparent exhibiting a transparency larger than 90% in the wavelength range between 400 and 850 nm. Zn_2SnO_4 thin films showed a resistivity of the order of 10^{-2} Ωcm. This resistivity is too high for Zn_2SnO_4 films to be used in solar cell production. We tried to dope Zn_2SnO_4 by reactive sputtering with CHF_3 but, unfortunately fluorine did not act as an effective dopant in this material; probably, the presence of oxygen during the sputtering deposition, inhibits fluorine atoms from entering the Zn_2SnO_4 matrix. The preparation of Cd_2SnO_4 thin films was carried out with the same sputtering parameters as used with the Zn_2SnO_4 deposition and the resistivity of these films was on the order of 1.2×10^{-4} Ωcm that is comparable with the resistivity of the best ITO, IGO and ISO films. Due to its high transparency and conductivity, this material seems to be quite suitable for achieving high efficiency CdTe/CdS solar cells [51]. However, the sputtering target supplied by Cerac Inc., since it is made by hot pressing a mixture of CdO and SnO_2 powders, and since CdO is hygroscopic, is difficult to handle. Due to this fact, the target is not sufficiently stable and in order to avoid some damage, the sputtering power density must be kept very low. Consequently, the deposition rate was $\approx$ 2 Å/s. Since this material is highly hygroscopic, we discovered some problems when the soda-lime glass covered by Cd_2SnO_4 was exposed to air. Despite this, we were able to obtain highly efficient solar cells.

3.4. Fluorine Doped Tin Oxide (FTO) Films

SnO_2 (TO) is the first transparent oxide to have received relevant commercialization. Nowadays TO films are used in products like "low-emissivity" windows, photovoltaic modules, flat-panel displays, heated windows, etc. In general, commercial TO thin films are produced by Chemical Vapor Deposition (CVD) [52-54]. In CVD, a solid film is deposited onto a substrate, typically glass, from gas or liquid reactants that are contemporaneously supplied and pre-mixed close to the deposition zone. Many inorganic and metalorganic precursors are used for the deposition of pure TO films. The most industrially used are tin tetrachloride ($SnCl_4$, TTC), dimethyltindichloride (($CH_3)_2SnCl_2$, DMTC), tetramethyltin ($Sn(CH_3)_4$, TMT) and O_2 or O_2 containing about 3-10 mol.% O_3. When TO films are used as transparent conducting oxides (TCO) such as in solar cells, a very high electrical conductivity is needed. In TO films the n-type conductivity is primarily due to O-vacancies, and often a mixture of two phases: SnO and SnO_2 are involved; $Sn^{4+}_{1-x}Sn^{2+}_{x}O^{2-}_{2-x}$ oxide can be formed where x is the fraction of SnO in the mixture. This oxide contains x oxygen vacancies and x Sn^{2+} atoms that donate 2 electrons for conduction. But this mechanism doesn't provide films with sufficient conductivity suitable for solar cells. In this case, the CVD process is provided with precursors containing HF (hydrogen fluoride)-acid. In fact, it is well known that fluorine can dope TO films producing very low resistivities.

Some useful properties of the CVD TO films are:
- High transparency in the visible part of the light spectrum (more than 90%)
- High reflectivity for infrared light
- Low electrical resistivity (on the order of 10^{-4} ohm·cm for fluorine doped TO)
- Good environmental steadiness
- High mechanical hardness

Despite these remarkable features, most on-line coated glass is still produced by sputtering techniques. Only about 25-30% of all coated float glass is produced by CVD. In part this is due to the fact that the sputtering techniques were developed for large area coating before the CVD technique, and furthermore the development of a successful on-line CVD process is very difficult. CVD is a complicated process since it involves liquid and gas phases and surface chemistry as well as the hydrodynamics of the whole reactor system. For this reason, it is not possible to plan an industrial CVD process based only on theoretical considerations or physic-chemical models, but rather on technological results and experience. This means that optimal conditions of film deposition are not always achieved and, as a consequence, low process yields and high product rejection rates are usual (principally due to optical non-uniformities) in the glass coating production process. Besides this, the utilization efficiency of reagent species in an industrial CVD reactor is around 10%; this implies the need for very expensive chemical scrubbing units that can create more than one hundred thousand of tons of waste per year.

Meeting these problems, one can readily appreciate that CVD methods not only need a lot of expertise in chemical reactions, fluid dynamics, mass and heat transfer and material science, but also many years of development to implement more reliable techniques for the glass coating industry. Thus it is clear why sputtering is the most utilized technique for the deposition not only of TO films, but also for many other TCO. For this reason, typical sputtering parameters were studied in order to prepare several TCO thin films with very high quality for photovoltaic application. In this section, we briefly described the main characteristics of the TCOs tested in our laboratory when used as a front contact in photovoltaic devices. Pure TO thin films, prepared by sputtering in an Ar atmosphere, starting from an oxide target, exhibit a resistivity of 10^{-1} Ωcm and an optical transparency up to 90% in the visible region of the light spectrum. The films were doped by mixing the Ar sputtering gas with CHF_3 in the range 1-10% with respect to the total $Ar+CHF_3$ gas pressure.

The minimum resistivity that we were able to obtain by making use of the maximum quantity of CHF_3 in the sputtering chamber during the SnO_2 deposition was about $8 \cdot 10^{-4}$ Ωcm. A disadvantage of the use of TO thin films in comparison with the other TCOs is that the resistivity of the SnO_2:F is 3-4 times greater. This means that, in order to obtain the same sheet resistance as that of the TCOs films, one has to deposit a SnO_2:F film 3-4 times thicker with a loss in the transparency. As a consequence, we didn't prepare any solar cell with SnO_2:F alone but we used this material as a buffer layer against Na diffusion from soda-lime glass. For example we deposited a SnO_2:F film, 100-500 nm thick on top of an ITO film prepared as will be described before. The final performance of the solar cell didn't change if compared with that of the device fabricated by making use of ITO alone.

3.5. Back Contact

Most researchers make the contact on *p*-type CdTe films by using Cu-containing compounds, such as a Cu-Au alloy, Cu_2Te, ZnTe:Cu or Cu_2S [10-14]. It is believed that Cu is necessary to make an ohmic contact on *p*-type CdTe. In fact, copper, by diffusing into CdTe lowers its resistivity and for a while it gives a higher solar cell performance. Moreover CdS/CdTe solar cells made with contacts not containing Cu behave as if they have a high series resistance. A possible explanation of this behavior is that the series resistance does not come from the contact but from CdTe that possibly is more conducting at the interface than in the bulk. The higher conductivity close to the interface could come from the fact that CdTe mixes with CdS and lowers its gap. It must be considered that the highest efficiency solar cell so far reported has been made with some copper at the back contact.

Before depositing Cu, an etching in Br-methanol or in a mixture of HNO_3/HPO_3 acids is carried out in order to enrich the CdTe surface with Te. There are two reasons why the CdTe surface is etched before the deposition of the back contact. The first one is to remove the oxygen compounds that form on top of the CdTe film due to the fact that normally the chlorine treatment is carried out in air. The second reason is to form a Te-rich surface that could be highly *p*-type conducting thus facilitating the formation of a good contact.

In this manner a Cu_2Te thin film, or some other copper-tellurium compound can be formed, which could limit the copper diffusion into the CdTe film. The back contact is then completed with graphite paste.

This type of back contact works quite well but it introduces another drawback that is the possibility for copper atoms to diffuse through the grain boundaries to the junction where they could create shunting paths. For this reason, if the solar cell has to live 20-30 years, copper must be avoided. This problem was completely solved at the University of Parma (Italy) where a new contact was developed. The novel contact is constituted by a layer of As_2Te_3 deposited directly onto the CdTe surface without any etching of CdTe, followed by a thin layer of Cu deposited onto As_2Te_3 at a substrate temperature of about 200°C. As_2Te_3 is a *p*-

type semiconductor that has a forbidden gap of about 0.6eV and exhibits a room temperature resistivity of $10^{-3}\Omega\cdot cm$. It melts at 360°C and can evaporate at a temperature higher than 250°C in vacuum. The As_2Te_3 thickness can be varied between 100 and 300 nm while the Cu thickness can be varied between 2 and 20nm. As_2Te_3 can be deposited at a substrate temperature between RT and 200°C, while Cu has to be deposited between 100 and 200°C substrate temperature in order to get a good contact. Actually the back contact is completed by depositing, by D.C. sputtering, 100 nm of Mo or W.

If both As_2Te_3 and Cu are deposited at room temperature, the contact is rectifying as one can see in Figure 13, curve a. Here the "roll-over" is clearly visible in the first quadrant of the I-V characteristic. If Cu is deposited at a substrate temperature around 200 °C, the "roll-over" disappears and the fill factor is much higher (see Figure 13, curve b).

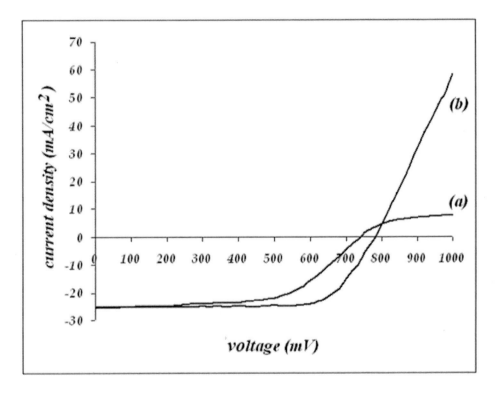

Figure 13. J-V characteristic of a CdTe/CdS solar cells in which the back contact is deposited with different substrate temperatures: (a) As_2Te_3 + Cu both deposited at room temperature; fill factor = 0.57, (b) As_2Te_3 + Cu both deposited at 200°C ; fill factor = 0.7

In order to understand the behaviour of the double layer As_2Te_3+Cu, some samples were prepared by depositing As_2Te_3+Cu directly on glass with Cu deposited at 200 °C substrate temperature.

Besides, some samples were prepared by depositing onto As_2Te_3 a Cu layer up to 20nm thick, while other samples were prepared by depositing a 50nm thick layer. In all cases the thickness of As_2Te_3 was 200nm. These samples were analyzed by X-rays and were compared with samples containing only As_2Te_3.

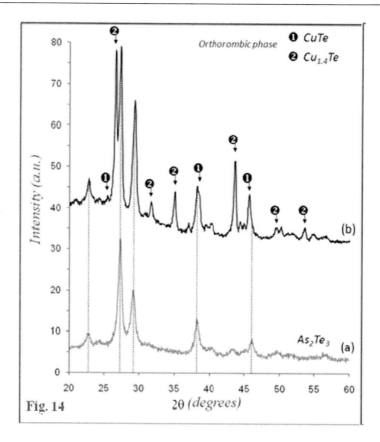

Figure 14. XRD analysis of (a) 200nm thick As_2Te_3 deposited on glass at 200°C substrate temperature, (b) 200nm thick As_2Te_3 film deposited on glass at 200°C substrate temperature on which 20nm thick Cu is deposited at the same temperature.

As we can see in Figure 14 and 15 curves (a) and (b), the samples containing Cu up to 20nm exhibit several Cu_xTe phases with $1 \leq x \leq 1.4$, while the samples containing 50 nm of Cu exhibit also the Cu_2Te phase. Back contacts on CdTe/CdS thin film solar cells made by using 50nm of Cu resulted to be not much stable, while cells made with 20nm or less of Cu resulted to be very stable and more efficient. This is in accord with what is reported by Wu et al. [55], that is, the Cu_xTe phases with $x > 1.4$ are not stable. A CdTe/CdS thin film solar cell, in which the back contact has been done by depositing, in sequence, at a substrate temperature of 200 °C, 200nm of As_2Te_3 and 20nm of Cu, is shown in Figure 13, curve b. No etching of CdTe has been done. Fill factor of this cell is ~ 0.7. Furthermore the quality of the As_2Te_3+Cu back contact has been investigated by studying the behavior of the current-voltage-characteristics of the CdTe/CdS based solar cells over a period of six months by keeping the devices at 60°C in a dry ambient, under 5 suns and in open-circuit conditions. Under these conditions no appreciable degradation of cells performance has been noticed apart from a slight increase in the open-circuit voltage (10-30 mV) while the fill-factor suffered a decrease which, after repeated tests, was never greater than 1%. So we can conclude that it is possible to make very stable high efficiency CdS/CdTe polycrystalline thin films solar cells by using As_2Te_3 + Cu as back contact on *p*-type CdTe films.

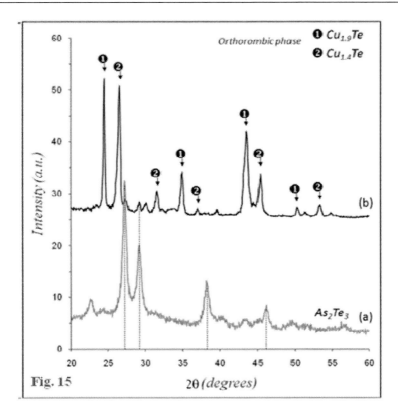

Figure 15. XRD analysis of (a) 200nm thick As$_2$Te$_3$ deposited on glass at 200°C substrate temperature (b) 200nm thick As$_2$Te$_3$ deposited on glass at 200°C substrate temperature on which 50nm thick Cu are deposited at the same substrate temperature.

In order to further understand the As$_2$Te$_3$+Cu double layer behaviour, some experiments have been done by substituting As$_2$Te$_3$ with Sb$_2$Te$_3$. In this case, the substrate temperature was varied between 200 and 350 °C both for Sb$_2$Te$_3$ and Cu. In all the finished cells the back contact was not stable even if the Cu thickness was only 2nm. The cell efficiency is quite high at the beginning, but decreases very fast.

These results can be explained in the following way. When Cu is deposited onto As$_2$Te$_3$ at a sufficiently high temperature, namely 200 °C, Cu makes a solid-state reaction with As$_2$Te$_3$ in which Cu substitutes As forming Cu$_x$Te. Depending on both As$_2$Te$_3$ and Cu thickness, Cu$_x$Te phases with $x \leq 1.4$ or $x > 1.4$ can be formed. The stable phase with $x \leq 1.4$ can be formed by using an As$_2$Te$_3$ thickness of 200nm and a Cu thickness of 20nm or less at a substrate temperature around 200 °C. This type of reaction cannot happen if Sb$_2$Te$_3$ is used instead of As$_2$Te$_3$, since Sb$_2$Te$_3$ is a much more stable material than As$_2$Te$_3$. With Sb$_2$Te$_3$, Cu remains free and diffuses into the CdTe/CdS structure, damaging the cell. This novel method of making the back contact on CdTe is similar, in some aspects, to the one commonly used in which first a Te-rich surface is created by means of a chemical etching of CdTe and than a very thin layer of Cu is deposited in order to form the Cu$_x$Te with $x \leq 1.4$. However, a substantial difference is in the fact that, in our case, no etching of the CdTe is done since we do not need a Te- rich surface (this is substituted by As$_2$Te$_3$) and that a Cu thickness ten times higher can be used. This renders less critical the formation of the non-rectifying contact giving a good stability to the CdTe/CdS thin film solar cell. This novel contact was patented [56].

4. THE CdTe/CdS SOLAR CELL

The schematic structure of a typical CdTe/CdS polycrystalline thin film solar cell is shown in Figure 16. The cells are in the super-strate configuration, with light coming to the junction through the substrate [1,57,58].

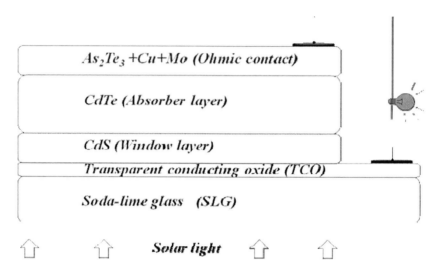

Figure 16. Sequence of the layers constituting the solar cell. The structure of CdTe/CdS solar cells is of the superstrate arrangement, which means that the solar light enters the cell through the glass substrate.

Afterwards it is possible to draw up the list of all the layers that constitute the cell including the substrate:
- The substrate can be soda-lime glass (SLG, the common window glass) or special alkali-free glass. The choice of alkali-free glass assures no diffusion of pollutants species into the overhanging films and allows more freedom in cell processing, having a higher softening temperature. On the other side it is quite expensive, thus research oriented to large-scale production employs SLG, despite the problem of Sodium or Potassium diffusion and that of the lower softening temperature.
- The front-contact layer is commonly made of a TCO such as ITO (indium tin oxide), IFO (fluorine doped indium oxide) or FTO (fluorine doped tin oxide). On top of this electrical conductive layer are often deposited a few nanometers (50-200) of a buffer layer such as pure TO (tin oxide), ZnO (zinc oxide) or Ga_2O_3 (gallium oxide) which has the role of a shield against the probable diffusion of Na and K atoms. Generally, these layers are rather resistive showing electrical resistivity in the range of 10^2-10^5 Ωcm; for this reason they are also effective in preventing an increase of the junction reverse saturation current (J_0) if some pinholes are present in the subsequent very thin CdS film [59].

- The window layer, that is the CdS film, represents the *n*-type part of the junction. In highly efficient CdTe based solar cells the CdS films are deposited either by CSS (CdS:O) or by sputtering (CdS:F). In order to maximize the solar cell photocurrent it is necessary to minimize the CdS film thickness and this is helped by the buffer layer presence between the CdS and the TCO films. Typical CdS layer thickness is in the range of 70-120 nm.
- In efficient CdTe/CdS solar cells, CSS (close-spaced sublimation) is the most popular technique used to deposit 4-7 μm thick CdTe films. This deposition method is particularly suitable for large-scale application since it is very fast (only 1 or 2 minutes are needed). Best cell performance is obtained if the CdTe films are grown in an oxidizing atmosphere.
- A heat treatment is usually carried out for all the system, as described before, at 380-420°C in a medium containing chlorine (typically chlorine-containing gas or vapor). The treatment involving Cl_2 is typical for CdTe/CdS based solar cells in order to re-crystallize the nanograins, if they are present, and to remove structural defects inside the CdTe and the CdTe-CdS intermixed layer.
- If the goal to be reached is a long-term stability, as well as very efficient devices, the best suited back-contact has so far been made by depositing 200 nm of As_2Te_3 on top of the CdTe layer followed by the deposition of 20 nm of Cu in order to obtain a stable and ohmic contact. If sputtering technique is used for the deposition of these two layers, a substrate temperature of 200° C has to be used. The contact is well finished by depositing a metallic electrode that is a 100 nm thick Mo or W film, in the same sputtering chamber in which the As_2Te_3 and Cu layers are deposited. On average, the efficiency of the cells made under similar conditions, by using the layers described above, is in the range of 15-15.8%.

At Parma's laboratories, this efficiency has been measured with an Oriel™ solar simulator under 100mW/cm² incident power and AM 1.5 solar light spectrum. The measurement system has been calibrated with a 14% efficient CdTe/CdS thin film solar cell previously certified at the Renewable Energies Unit of the Joint Research Centre, Ispra, Italy. A J versus V characteristic of a CdTe/CdS polycrystalline thin film solar cell fabricated in our laboratory is depicted in Figure 17. Typical parameters of this device are: an efficiency of 15.8% with a V_{oc} (open circuit voltage) of 862 mV, a J_{sc} (short circuit current density) of 25.5 mA/cm² and a ff (fill factor) of 0.72.

In order to obtain high efficiency solar cells with a high fill factor, as we have described in a previous section, it is necessary to deposit CdTe by CSS in the presence of oxygen. Moreover, the efficiency of the solar cell depends on the amount of oxygen in the CSS chamber, on the substrate temperature during the deposition and on the chlorine- treatment temperature. A high content of O_2 or a high chlorine annealing temperature tends to give a higher fill factor and a lower open circuit voltage. The short circuit current is generally not affected except in the case in which a very large amount of O_2 and/or a high Cl_2 annealing temperature is used. In extreme cases, the short circuit current density could be much lower than 25 mA/cm². When a high substrate temperature, namely 520 °C, is used during the CdTe deposition, the fill factor tends to decrease since, at this temperature, less O_2 is incorporated in the film. In order to interpret these results, we made the hypothesis that O_2, mixing with CdS and CdTe reacts to form at the interface a mixed compound containing Cd, S, Te and O with a forbidden gap close to that of CdTe which is *n*-type. In this way, a pseudohomojunction is formed between this compound and *p*-type CdTe and this can explain the strong increase in the fill factor. This hypothesis is corroborated by several facts.

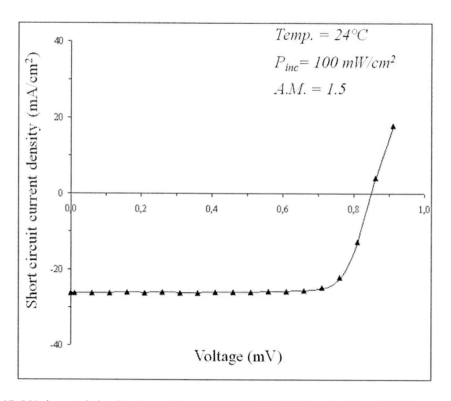

Figure 17. J-V characteristic of the best all dry-processed CdTe/CdS based solar cell fabricated at Parma University (Italy).

First the I-V characteristic in the dark exhibits an *N* diode factor close to 1 (where *N* is given in the diode formula J=J₀ exp [(eV/*N*kT) -1]). An *N* factor close to 1 is typical of a homojunction.

When a high content of O_2, namely more than 10%, and/or a high chlorine-annealing temperature, namely more than 420 °C, are used, the short circuit current density is much lower than 25 mA/cm² despite a high open circuit voltage and principally a high fill factor (0.74-0.75). Probably due to a strong intermixing, favored by the oxygen presence, a thick *n*-type layer, larger than the absorption length of the visible light, is formed and a good fraction of the carriers created by the photon absorption, do not reach the junction region. Consequently, the amount of O_2 used during the CdTe deposition is very important in order to control the formation and the thickness of this presumed *n*-type layer. The formation of an *n*-type thick layer has been confirmed by measurements of the photocurrent as a function of wavelength: when the solar cell exhibits a J_{sc} much lower than 25 mA/cm² the photocurrent response drops very quickly for wavelengths shorter than that corresponding to the gap of CdTe.

It has also been reported that oxygen in presence of chlorine enhances the concentration of Cl_{Te} which is a donor for CdTe and can form an *n*-type CdTe(S,O) layer at the interface [60].

5. MANUFACTURING OF CDTE/CDS BASED PV MODULES

The techniques, which are used to make CdTe/CdS based solar cells, are CSS and sputtering, both fast and easily scalable. A possible in-line process is shown in Figure 18 [61, 62]. Modules of 0.6 x 1.2 m² can be covered with a cycle time of a few minutes.

From this schematic graph (Figure 18), we see that the production begins with the cleaning of the substrate, consisting of a soda-lime glass, with a special washing machine (1). The module (2) corresponds to the sputtering deposition of the TCO. This first film is cut with a laser beam in step (3) (see figure 19 (a) and (b)).

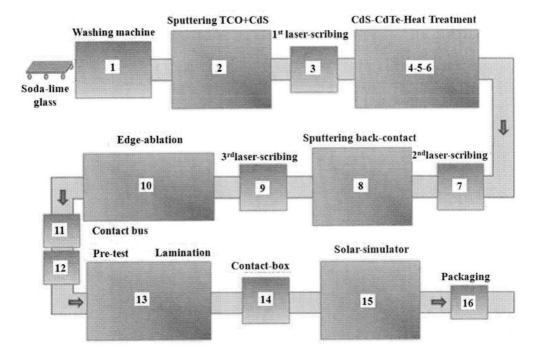

Figure 18. Schematic diagram of an industrial production line of CdTe/CdS thin-film solar modules.

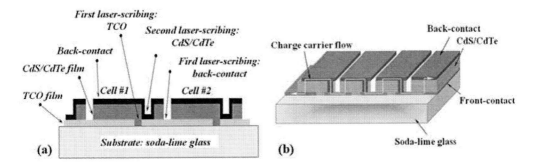

Figure 19. (a) Typical interconnects scheme for a CdTe/CdS based solar cell module. Laser scribing of the TCO film, the active layers (CdS/CdTe) and back contact/active layers are shown. (b) PV module in which the series electrical connections between all the solar cells constituting the module and the charge carrier path through the cell interconnections are visible.

Components (4-5-6) involve the deposition and the formation of the junction. A single stage machine with different process zone can overcome the sputtering deposition of CdS and the close-spaced sublimation of CdTe (see Figure 20 (a)). Immediately after, there is the process chamber for the chlorine heat-treatment for the entire stacked layer and for the activation of the CdS/CdTe junction.

Step (7) corresponds to the second laser scribing. Module (8) consists in the deposition of the back contact which is a film of As_2Te_3 followed by a film of Cu and Mo all deposited by sputtering (see Figure 20 (b)). A final laser scribing in step (9) serves to definitively separate the individual cells inside the module.

The removal of the film from the edges, made in step (10) serves to allow the perfect electrical isolation after encapsulation. In step (11) the electrical contacts, as a metal strip between the first and the last cell, are applied and simultaneously (12) a preliminary electrical test before the module is laminated with another glass in block (13) is carried out. In step (14) the metal strips are connected inside the contact-box and, when necessary, an external frame is applied. The modules are completed (15-16) with the test under solar simulator, classified according to their quality and then packed in order to be ready for shipment.

Figure 20. (a) Partial view of the CSS system, complete with CdS sputtering and the heating tunnel. (b) The back-contact sputtering plant complete with heating and cooling tunnel (Arendi SpA.).

Table 3. Some manufacturers of CdTe/CdS thin films photovoltaic modules, and their market status

Manufacturer	Production capacity MW/year	Glass Size (m x m)	Efficiency% max/average	Market
CdTe				
CTF Solar, Germania	50 (2010)	0.6 x 1.2	7/6	yes
First Solar, Malesia, Germania	475 (2008/2009)	0.6 x 1.2	11/9	yes
Arendi S.p.A., Italia	18 (2009)	0.6 x 1.2	-------	no
Abound Solar, USA	200 (2010)	0.6 x 1.2	------	no
Primestar Solar, USA	50 (2010)	0.6 x 1.2	------	no

Currently there is a real boom in the construction of factories for the production of thin film modules based on CdTe technology. Some new companies have recently announced the start of production or plan to do so in the future. Many of these plants are located in Germany, some in the U.S.A. In Italy, a new company (Arendi S.p.A.) has been constituted in 2007, with the aim of building a factory with a capacity of 18 MWatt/year. The production line, projected with the support of our group, is close to be completed and will enter into full production at the end of 2010.

6. CONCLUSION

In this paper we explain that high energy conversion efficiency of CdS/CdTe thin film solar cells cannot be achieved with heterojunctions where the space charge is about equally distributed between the window and the absorber layer. In fact, in CdTe based solar cells the materials and processing have to be self-adjusting in a manner that the conductivity type is inverted into the absorber layer surface giving the possibility to make a homojunction. In this way the photocurrent becomes a majority carrier current before passing through the metallurgical interface [63]. The type-inversion is obtained by considering the inter-diffusion between the window and absorber layers that is greatly influenced by the heat treatment in chlorine ambient. Besides, the grain boundaries of the CdTe polycrystalline film are self-passivated since, during the CdTe growth, a segregation of an insulating material takes place and this prevents the recombination of the photo-generated carriers. Taking into account these considerations we worked to make a simplified industrial in-line production of photovoltaic CdS/CdTe based modules and we made several process innovations regarding the solar cell fabrication.

These discoveries together with the intrinsic properties of the selected materials constituting the CdTe based PV module make possible the concept of a manufacturing plant as a single in-line thin film-processing unit. The core of the deposition line can be fully automated with minimal human interference. Only the module-finishing part components (lamination, external electrical contacting, support frames) will be manually handled.

Nowadays, considering these assets, CdTe/CdS based thin film technology is the only one that allows one to develop a so cheap (< 1€/W_P) mass production of PV modules as to make PV's competitive on the energy market. The last question that we want to point out is the environmental and health aspects linked to the cadmium presence in the CdS/CdTe based modules. Since CdTe PV's reached commercialisation, some problems were raised about potential cadmium emission from CdTe PV modules. First of all cadmium telluride is not cadmium alone. The basis of the stability of CdTe is its extremely high binding energy of 6 eV per atom. For this reason the melting point of CdTe is at quite high temperature, (around 1041 °C) and the material starts to evaporate at 1050 °C at atmospheric pressure. In case of fire, the covering glass melts sealing everything. This experiment has already been carried out and a great number of studies have shown that no environmental impact due to CdTe based PV is expected especially if proper life-cycle management system is provided by suppliers. In fact, recycling the modules at the end of their usual life would completely resolve any environmental concern [64-67].

From this point of view, we can conclude by asserting that during a useful life of 20-30 years, these modules don't produce any pollutant, and furthermore, by displacing fossil fuel and by the trapping of the Cd in a more stable form, they offer great environmental benefits!

REFERENCES

[1] Bonnet, D.; and Rabenhorst, H. *Proc. of the 9th Photovoltaic Spec. Conf.*, 1972, pp 129-131.
[2] Tyan, Y. S.; and Albuerne, E. A. 16th *IEEE Photovoltaic Spec. Conf.*, IEEE publishing, NY, 1982, p 794.
[3] Ferekides, C.; Britt, J.; Ma, Y. and Killian, L. *Proc. of 23th Photovoltaic-Spec. Conf.* IEEE, NY, 1993, p 389.
[4] Wu X.; Keane, J. C.; Dhere, R. G.; Dehert, C.; Albin, D. S.; Dude, A.; Gessert, T. A.; Asher, S.; Levi, D. H. and Sheldon, P. *Proc. of the 17th Eu. PVSEC*, Munich, Germany, 2001, Vol. II, p 995.
[5] Nakazawa, T.; Takamizawa K. and Ito K. *Appl. Phys. Lett.* 1987, 50, p 279.
[6] Aranovich, J. A.; Golmayo, D.; Fahrenbruch A. L.; and Bube, R. H. *J. Appl. Phys.* 1980, 51, p 4260.
[7] Mimila Arroyo, J.; Marfaing, Y.; Cohen-Solal, G.; and Triboulet, R. Sol. *Energy Mater.* 1979, 1, p 171
[8] Bonnet, D. *International Journal of Solar Energy*, 1992, 12, p 1.
[9] Chu T.L.; and Chu, S.S. *International Journal of Solar Energy*, 1992, 12, p 121.
[10] Albright, S.P.; Jordan, J.F.; Akerman, B.; and Chamberlain, R.C. *Solar Cells*, 1989, 27, p 77.
[11] Gessert, T.A.; Mason, A.R.; Sheldon, P.; Swartzlander, A.B.; Niles, D. and Coutts, T.J. *Journal of Vacuum Science Technology*, 1996, 14, p 806.
[12] Uda, H.; Ikegami, S. and Sonomura, H. Japanese *Journal of Applied Physics*, 1990, 29, p 495.
[13] McCandless, B.E.; Qu, Y.; Birkmire, R.W. *Proc. 1st World Conf. on Photovoltaic Energy Conversion*, Hawaii, USA, 1994, pp 107-110.
[14] Chou, H.C.; Rothagi, A.; Thomas, E. W.; Kamra, S.; Bath, A.K. *Journal of Electrochemical Society*, 1995, 142, p 254.
[15] Romeo, N.; Bosio, A.; Tedeschi, R.; Canevari, V.; *Thin Solid Films.* 2000, 361, p 327.
[16] Baetzner, D.L.; Romeo, A.; Zogg, H.; Wendt R.; Tiwari, A.N. *Thin Solid Films*, 2001, 387, pp 151-154.
[17] Zappettini, A.; Bissoli, F.; Gombia, E.; Bosio, A.; Romeo, N. *Nuclear Science Symposium Conference Record*, 2004 IEEE, 16-22 Oct., 2004, Vol. 7, pp 4518-19.
[18] Romeo, N.; Bosio, A.; Romeo, A. *PCT Int. Appl.*, 2008, p 22.
[19] Barker, J.; Binns, S.P.; Johnson, D.R.; Marshall, R.J.; Oktik, S.; Öznan, M.E.; Patterson, M.H.; Ransome, S.J.; Roberts, S.; Sadeghi, M.; Sherborne, J.M.; Turner A.K.; Woodcock, J.M.; *International Journal of Solar Energy*, 1992, 12, p 25.
[20] Ferekides, C.S.; Ceekala, V.; Dugan, K.; Killian, L.; Oman, D.M.; Swaminathan R.; Morel, D.L. *Proc. AIP Conf. 1995, vol 353*, pp 39-45.
[21] Romeo, N.; Bosio, A.; Canevari, V. Proc. of 3rd *World Conference on Photovoltaic Energy Conversion, WCPEC-3*, 11-18 MAY 2003,Osaka, Japan, 2003, 1, pp 469-470.
[22] Bonnet, D.; Henrichs, B.; Richter, H. *International Journal of Solar Energy*, 1992, 12, p 133.
[23] Li, X.; Albin, D.; Asher, S.E.; Mouthino, H.; Keyes, B.; Matson, R.J.; Hasoon, F.; Sheldon, P. Proc. *AIP Conference*, 1995,vol. 353, pp 376-383.

[24] Gordillo, G.; Florez, J.M.; Hernandez, L.C.; Teherán, P. Proc. *1st World Conf. on Photovoltaic Energy Conversion, Hawai, USA*, 1994, pp 307-310.
[25] Coutts, T.J.; Young, D.L.; Li, X.; Mulligan, W.P.; Wu, W. *J. Vac. Sci. Technol. A Vac. Surf. Films*, 2000, 18, p 2646.
[26] Skarp, J.; Koskinen, Y.; Lindfors, S.; Rautiainen, A.; Suntola, T. *Proc. 10th European Photovoltaic Solar Energy Conf., Lisbon, Portugal*, 1991, pp 567-569.
[27] Skarp, J.; Anttila, E.; Rautiainen A.; Suntola, T. *International Journal of Solar Energy*, 1992, 12, p 137.
[28] Morris G.C.; Das, S.K. *International Journal of Solar Energy*, 1992, 12, p 95.
[29] Clemmick, I.; Burgelman, M.; Casteleyn, M.; Depuydt, B. *International Journal of Solar Energy*, 1992, 12, p 67.
[30] Ikegami, S. *Solar Cells*, 1988, 23, p 89.
[31] Romeo, N.; Bosio, A.; Tedeschi, R.; Romeo A.; Canevari, V. Solar Energy Materials and *Solar Cells*, 1999, 58, pp 209-218.
[32] Britt, J.; Ferekides, C. *Appl. Phys. Lett.* 1993, 62 (22), p 2851.
[33] Zanio, K. Cadmium Telluride, *Semiconductors and Semimetals*, Academic Press, New York, NY, Vol. 13, p 148.
[34] Ferekides, C.S.; Marinskiy, D.; Viswanathan, V.; Tetali, B.; Palekis, V.; Selvaraj, P.; Morel, D.L. *Thin solid Films*, 2000, 361-362, pp 520-526.
[35] Romeo, N.; Bosio, A.; Mazzamuto, S.; Podestà A.; Canevari, V. *Proc. of 20th Photovoltaic Solar Energy Conference*, 6-10 June 2005, Barcelona, Spain, 2005, 1, pp 1740-1750.
[36] Romeo, N.; Bosio, A.; Romeo, A. Patent nr. WO 2006085348 A2
[37] Romeo, A.; Bätzner, D.L.; Zogg, H.; Vignali, C.; Tiwari, A. N. Solar Energy Materials & *Solar Cells*, 2001, 67, pp 311-321.
[38] Romeo, A.; Bätzner, D.L.; Zogg, H.; Tiwari, A. N. *Thin Solid Films*, 2000, 361, pp 420-425.
[39] Khrypunov, G.; Romeo, A.; Kurdesau, F.; Bätzner, D.L.; Zogg, H.; Tiwari, A. N. *Solar Energy Materials & Solar Cells*, 2006, 90, pp 3407-3415
[40] Sohn, M.H.; Kim, D.; Kim, S.J.; Paik, N.W.; Gupta, S. *J. Vac. Sci. Technol.* A, 2003, 21, p 1347.
[41] Biyikli, N.; Kartgoglu, T.; Artur, O.; Kimukin, I. ;Ozbay, E. *Appl. Phys. Lett.*, 2001, 79, p 2838.
[42] Chopra, K.L.; Major, S.; Pandya, D.K. *Thin Solid Films*, 1983, 102, pp 1-4.
[43] Minami, T.; Takeda, Y.; Takata S.; Kakumu, T. *Thin Solid Films*, 1997, 308-309, pp 13.
[44] Ueda, K.; Hase, T.; Yanagi, H.; Kawazoe, H.; Hosono, H.; Ohta, H.; Orita, M.; Hirano, M.*J. Appl. Phys.* 2001, 89, pp 1790.
[45] Kudo, A.; Hanagi, H.; Hosono, H.; Kawazoe, H. *Appl. Phys. Lett.* 1998, 73, p 220.
[46] Romeo, N.; Bosio, A.; Canevari, V.; Terheggen, M.; Vaillant-Roca, L. *Thin Solid Films*, 2003, 431-432, pp 364-368.
[47] Lippens, P.; Segers, A., Haemers, J.; De Gryse, R. *Thin Solid Films*, 1998, 317, pp 405-408.
[48] Peiró, A.M.; Ayllón, J.A.; Peral, J.; Domènech, X.; Domingo, C. *J. of Crystal Growth*, 2005, 285, pp 6-16.
[49] Smith, A.; Rodriguez-Clemente, R. *Thin Solid Film*, 1999, 345, p 192.

[50] Faÿ, S.; Kroll, U.; Bucher, C.; Vallet-Sauvain, E.; Shah, A. *Solar Energy Materials & Solar Cells*, 2005, 86, p 385.

[51] Wu, X.; Sheldon, P.; Coutts, T. J.; Rose D. H.; Moutinho, H. R. *26th, PVSC*, 29 Sept.-03 Oct.1997, Anaheim, CA.

[52] van Mol, A.M.B.; Chae, Y.; McDaniel, A.H.; Allendorf, M.D. *Thin Solid Film*, 2005, 502, 1-2, pp 72-78

[53] Rajaram, P.; Goswami, Y.C.; Rajagopalan, S.; Ghupta, V.K. *Materials Letters*, 2002, 54, pp 158-163.

[54] Vlahovic, B.; Persin, M. J. Phys. D: *Appl.Phys*. 1990, 23, pp 1324-1326.

[55] Wu, X.; Zhou, J.; Duda, A.; Yan, Y.; Teeter, G.; Asher, S.; Metzger, W.K.; Demtsu, S.; Wie, Su-Huai; Noufi, R. *Thin Solid Films*, 2007, 515,15, pp 5798-5803

[56] Romeo, N.; Bosio, A.; Romeo, A. Patent nr. WO 2009001389 A1

[57] Bube, R.H. In *Properties of Semiconductors Materials, Photovoltaic Materials*; Newman R.C.; Ed.; Imperial College, London, UK, 1998, vol. 1.

[58] Fahrenbruch, A. L.; Bube, R.H. In: *Fundamentals of Solar Cells*, Academic Press, New York, NY, 1983. Vol 1.

[59] McCandless, B.E.; Dobson, K. *Solar Energy*, 2004, 77, pp 839-856.

[60] Ikegami, S.; Nakano, A. Int. J. *Solar Energy*, 1992, 12, pp 53-65.

[61] Romeo, N. Bosio, A. Romeo, A. Bianucci, M. Bonci, L. Lenti, C. *PV in Europe from Pv Technology to Energy Solutions Conference and Exhibitions*, 7-11 October 2002, Rome, Italy, 2002.

[62] Bonnet, D. *Thin Solid Films*, 2000, 361-362, pp 547-552.

[63] Bosio, A.; Romeo, N.; Podestà, A.; Mazzamuto S.; Canevari, V. *Cryst. Res. Technol.*, 2005, 40, 10/11, pp 1048-1053.

[64] Moskowitz, P.D.; Zweibel, K. *Proc. of Conf. on Recycling of Cd and Se from PV Modules and Manufacturing Waste*, Golden, Co, 11-12 Mar, 1992.

[65] Patterson, M.S.; Turner, A.K.; Sadeghi M.; Marshall, R.J. *Proc. of 12th Eur. PVSEC*, 11-15 April, 1990, pp 950-952.

[66] Doty M.; Meyers, P. *AIP Conf. Proc*. New York, 1988, 166, pp 10-18.

[67] Ftenakis, V. M. Fuhrmann, M. Heiser, M. Lanzirotti, A. Fitts J. Wang, W.*Prog. Photovolt: Res. And Appl*. 2005, 13, 8, pp 713-723.

In: Thin Film Solar Cells: Current Status and Future Trends
Editors: Alessio Bosio and Alessandro Romeo
ISBN 978-1-61668-326-9
© 2010 Nova Science Publishers, Inc.

Chapter 8

THIN FILM SILICON SOLAR CELLS

Paola Delli Veneri and Lucia Vittoria Mercaldo
ENEA - Portici Research Centre, 80055 Portici, Naples, Italy.

ABSTRACT

Thin film silicon, in the amorphous and microcrystalline form, constitutes at present one of the most promising options for low-cost large-area applications of photovoltaics, especially suited for building integration. The technology has evolved considerably since the first amorphous silicon solar cells were made in the '70s, and is now mature: The last couple of years saw a rapid increase in the number of companies involved in thin film silicon solar cell production. This Chapter wants to be a complete, even if not exhaustive, survey on this topic, starting from material issues, in terms of physical properties and fabrication techniques, then moving towards an introduction to the devices on various grounds (geometries, operation principles, multi-junction approach, modelling, state-of-the-art overview), and closing with issues involved in module manufacturing and market prospects.

1. INTRODUCTION

About 82% of the current terrestrial photovoltaic production uses wafer-based crystalline silicon technology [1]. However, the market entry of companies offering turn-key production lines for thin film solar cells led to a massive expansion of investments into the thin film sector. Equally competitive technologies are amorphous/micromorph Silicon, CdTe and CIGS thin films. The growth of these technologies is accelerated by the positive development of the PV market as a whole.

The worldwide intense research on thin film solar cells finds its bases on the cost reduction potential of this technology (below US$1/watt). In the thin film approach the generally expensive semiconductor material is used in minor quantity and can be directly deposited on low-cost large-area substrates. The concept is to pursue cost reductions for device manufacturing rather than only high efficiency objectives.

Thin film silicon, in particular, both in the amorphous (a-Si:H) and microcrystalline (μc-Si:H) form, constitutes at present one of the most promising material options for low-cost large-area applications of photovoltaics [2–5]. Thin film silicon technology is an industrially mature technology: Modules on the market have efficiencies in the range 6–9% with a declared durability of 20 years. The main advantages of this technology are:

- silicon is abundant and non-toxic;
- low process temperatures, allowing for use of low-cost substrate materials (float-glass and plastics), are involved;
- a substantial cost reduction is potentially available, due to the lower amount of active material required and to the reasonable quantity of 'grey energy' for the fabrication of full solar modules;
- large-area deposition processes are feasible;
- monolithic series connection of cells to modules, and, hence, a variability of output voltages, is practicable;
- attractive modules can be realized to be used in architecture (flexible, semi-transparent, etc.);
- module performance at elevated temperature is superior with respect to the conventional counterpart.

Such attractive features are accompanied by lower efficiency values with respect to crystalline silicon, accentuated by the light-induced degradation (Staebler–Wronski effect). This effect is only mildly present in microcrystalline silicon solar cells, and can be reduced in amorphous silicon solar cells by adequate cell design (i.e., fabricating multi-junction devices, where the individual absorber layers can be kept suitably thin).

Except for their bandgaps and light-induced degradation behaviour, there are a number of similarities, both in deposition procedures and in transport and optical properties, that make a-Si:H and μc-Si:H well suited for use in solar cells. In this Chapter, after a short historical excursus, we introduce some of the physical properties of these materials and survey the principal methods that are employed in the fabrication of thin film silicon solar cells. Afterwards we illustrate the device p-i-n structure with its optical designs and describe how such diode works. Multi-junction devices and in particular "micromorph" tandems, the stacked combination of an amorphous silicon top cell with a microcrystalline silicon bottom cell are then introduced as the approach to overcome the efficiency limit of single junction solar cells and at the same time gain in stability. We conclude with the description of issues involved in module manufacturing and a survey on production and market.

2. HISTORICAL OVERVIEW

Crystalline silicon (c-Si) and crystalline semiconductors in general are used for many applications, but also non crystalline semiconductors can have sufficiently good optoelectronic properties. In crystals, the atoms are arranged in near-perfect regular arrays or lattices, consistently with the underlying chemical bonding properties of the atoms. In particular, a silicon atom forms four covalent bonds to neighbouring atoms arranged symmetrically around it. This "tetrahedral" configuration is perfectly maintained in the "diamond" lattice of crystal silicon (Figure 1a). In amorphous semiconductors the chemical bonding of atoms is nearly unchanged from that of crystals. However a random variation in the angles between bonds destroys the regular lattice structure (Figure 1b). Such noncrystalline semiconductors can have fairly good electronic properties – sufficient for many applications.

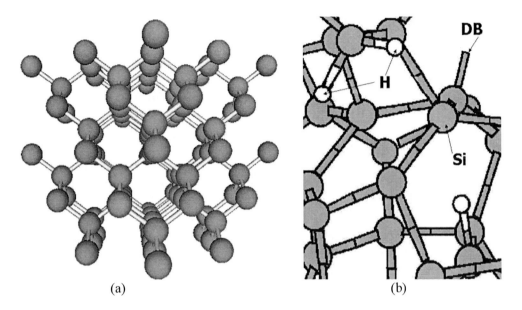

(a)   (b)

Figure 1. Model of c-Si structure seen along the <110> direction (a), and a-Si:H structure (b). DB stands for "Dangling Bond".

In contrast to the fabrication of crystalline silicon wafers, amorphous thin film technologies do not require the costly and delicate crystal traction and saving. The thin film silicon technology is based on a simple and inexpensive deposition process on glass or other low-cost substrate at deposition temperatures well below the melting point of c-Si. Nevertheless, the un-hydrogenated a-Si thin films, prepared by sputtering or by thermal evaporation, have a very high defect density that prevents doping and reduces the mobility and lifetime of free carriers — three desirable characteristics of a useful semiconductor.

In Scotland around 1973, Walter Spear and Peter LeComber, after earlier studies by R. Chittick et al., discovered that amorphous silicon prepared using a "glow discharge" in silane (SiH$_4$) gas had unusually good electronic properties [6]. After several years of uncertainty, it emerged that plasma-deposited amorphous silicon contained a significant percentage of hydrogen atoms bonded into the amorphous silicon structure and that these hydrogen atoms were essential to the improvement of the electronic properties of the plasma-deposited material [7]. As a consequence, the improved form of amorphous silicon has generally been known as *hydrogenated amorphous silicon* (or, more briefly, a-Si:H). Such improved material could also be doped: Spear and LeComber reported in 1975 [8] that amorphous silicon conductivity could be increased enormously either by mixing some phosphine (PH$_3$) gas or some diborane (B$_2$H$_6$) gas with silane. Just as for crystalline silicon, the phosphorus doping of the amorphous silicon had induced a conductivity associated with mobile electrons (the material was "*n*-type"), and the boron doping had induced a conductivity associated with mobile holes (the material was "*p*-type").

In 1976, David Carlson and Christopher Wronski reported a solar cell based on amorphous silicon [9] with a solar conversion efficiency of about 2.4%. In the first years after their introduction, amorphous silicon solar cells made rapid progress in reaching higher efficiencies. By 1982, a-Si:H solar cells with initial efficiencies over 10% had been obtained [10]. However, it became soon clear that a-Si:H solar cells suffer from a light-induced degradation effect (the so-called Staebler–Wronski effect) [11] and stabilize only at lower efficiency values.

In the 90's the University of Neuchâtel PV group pioneered a different form of thin-film silicon, microcrystalline silicon (μc-Si:H), only mildly influenced by the Staebler–Wronski effect, as a promising absorber material for solar cells [12]. Microcrystalline silicon was firstly produced by Veprek and Marecek in 1968 in Prague [13] by a low-temperature plasma-assisted deposition process, but it took several years before it could be successfully used for solar cell applications. Because of the poor quality (high defect density) of the first μc-Si:H layers and because of the n-type character of most layers, even without addition of doping gases, it was originally believed that it would not be possible to make reasonable solar cells with a μc-Si:H layer acting as the main photoconversion layer.

Amorphous and microcrystalline silicon thin films are currently applied in solar cells and numerous other electronic devices. A very successful example is the thin film transistor (TFT), a switching device that has already found application in commercially available liquid crystal displays, image sensor arrays, and printing arrays. Many applications benefit from the low processing temperatures and large area manufacturability of these thin films that facilitate the cost-effective production of large area devices. More recently the trend in thin-film silicon photovoltaics is toward the multi-junction approach, where amorphous silicon and/or amorphous silicon-based alloys and/or microcrystalline silicon are combined together in stacked structures. Overall, the advantages of the multijunction design are sufficiently sound to overcome the additional complexity of the structure. Both tandem and triple-junction devices are being manufactured today. The stabilized efficiency record (12.5%) within the silicon-based thin film photovoltaic device family is detained by a-Si/nc-Si/nc-Si triple junctions [14].

3. MATERIAL ISSUES

3.1. Atomic and Electronic Structure of Hydrogenated Amorphous Silicon

Structural disorder, in-homogeneities and presence of hydrogen bonded to the amorphous network are specific features of a-Si:H. Silicon atoms in amorphous silicon largely retain the same basic structure of crystalline silicon, where each silicon atom is connected by covalent bonds to four other silicon atoms arranged as a tetrahedron. However, bond angle distortions and coordination defects, called *D*-centers or dangling bonds, are found due to structural topological constraints in the amorphous network (Figure 1(b)). The dangling bonds can be monitored by electron paramagnetic resonance (EPR) [15], since unpaired electrons are paramagnetic and give a strong electron paramagnetic resonance signal. For hydrogenated amorphous silicon (a-Si:H), however, several percents of the silicon atoms make covalent bonds with three silicon neighbours plus one hydrogen atom. Thus, a lower amount of dangling bonds is found in the material. This crucial hydrogen is essentially invisible to X-rays, but is quite evident in non-destructive measurements (infrared spectroscopy [16] and proton magnetic resonance [17]) as well as destructive testing, like secondary ion mass spectroscopy [18] and hydrogen evolution during annealing [19]). Since the coordination defects are responsible for the formation of deep localized states in the forbidden gap, the dangling bonds can be monitored also by sub-gap absorption measurements, such as constant photocurrent method (CPM) [20] or photothermal deflection spectroscopy (PDS) [21].

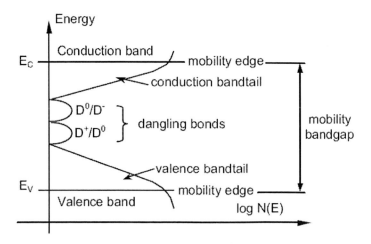

Figure 2. Schematic of the density of electronic states in hydrogenated amorphous silicon.

The bonding disorder determines most of the electronic properties of a-Si:H. Free carriers have very short scattering lengths, and consequently a mobility of only 10-20 cm^2/Vs compared to about 500 cm^2/Vs in crystalline silicon [22]. In Figure 2 the schematic of the density of electronic states N(E) is reported as it has emerged primarily from measurements of electron photoemission [23,24], optical absorption [25], and electron and hole drift mobilities [26]. Similarly to crystalline silicon there are two strong bands of extended states: one occupied valence band ($E < E_V$) and one unoccupied conduction band ($E > E_C$). Between the conduction and valence bands lies an "energy gap" where the density of states is very low. Any functional semiconductor, crystalline or noncrystalline, must have such an energy gap. For perfect crystals, the valence and conduction bandedge energies E_V and E_C are well defined, as is the *band gap* $E_G = E_C - E_V$. In disordered semiconductors the disorder causes an exponential tail of localized states near these band edges extending into the forbidden gap. In particular, in a-Si:H the conduction bandtail is narrower than the valence bandtail (typically the widths are ~ 20 meV and ~ 50 meV respectively) since electrons have higher mobility than holes. The energy dividing the extended and localized states at the band edges is known as the mobility edge, and the gap separating the valence and conduction band mobility edges is approximately the optical band gap.

Given the presence of exponential bandtails, for noncrystalline semiconductors there is no single, conclusively established procedure for locating the bandedges within the density-of-states. The band gap is thus difficult to determine without some ambiguity. By far the most common approach is to extract an "optical" band gap E_G from optical absorption coefficient $\alpha(E)$ measurements. Because of the lack of long-range order (structural uniformity is found only over tiny volumes), by the Uncertainty Principle, the strong spatial localization of charge carriers leaves their momentum largely undetermined. As a result, differently from crystalline silicon (indirect semiconductor[17]), where phonons are required for transitions between valence and conduction band states to satisfy the conservation of momentum, here the requirement on the conservation of momentum is relaxed, the distinction between direct and indirect optical transition disappears, and larger absorption coefficient is found. Three typical regions can be distinguished in the spectra: (a) the fundamental absorption region (high energy-above gap), (b) the exponential tail region (intermediate energy), ascribed to tail states-extended states transitions, characterized by the so-called Urbach energy (E_U), namely the logarithmic slope of the absorption distribution for which $\alpha \sim \exp(E/E_U)$, and (c) the weak absorption region (low energy-below gap), ascribed to defects-extended states transitions. Above gap, but sufficiently close to it, a dependence like:

$$\alpha(E) = \frac{A}{E}(E - E_G)^2$$

is usually assumed for the absorption coefficient, where A is a constant and E_G is known as the Tauc band gap [27]. This parameter is then determined through the so-called Tauc plot, by extrapolating the linear region of the experimental $\sqrt{\alpha \cdot E}$ curve down to the horizontal axis. The band gap obtained using this procedure is typically about 1.75 eV in a-Si:H, but varies substantially with deposition conditions, hydrogen percentage within the material, and alloying with elements like germanium or carbon. A simpler procedure is to give a rough estimate in terms of the photon energy corresponding to a particular optical absorption coefficient. Usually the energy corresponding to $\alpha = 10^4$ cm^{-1}, called E_{04}, is adopted to evaluate the material quality. However one has to keep in mind that there is a difference between the optical estimate of the band gap and the electrical band gap defined in terms of mobility edges [28]. Internal photoemission measurements [29] indicate that the electrical band gap is 50 to 100 meV larger than the Tauc band gap.

Between the bandtails, lie defect levels: the dangling bonds ("*D*-centers"). This kind of defects can be in the neutral state (D^0) when singly occupied and in +e or -e charged states (D^+ or D^- respectively) when empty or doubly occupied. They can be represented by two localized levels within the energy gap, a donor-like (D^+/D^0) and an acceptor-like (D^0/D^-) state (Figure 2). The actual level positions vary between doped and intrinsic a-Si:H [22], between intrinsic samples with varying defect densities [30], and possibly between dark and illuminated states [31]. The difference between the 2 levels is termed *correlation energy* of the *D*-center [32]. These defects control the trapping and recombination of carriers, and hence influence the carrier lifetimes. In device-quality intrinsic material the defect density is lower than 10^{16} cm^{-3} [22], while in doped materials the defect density increases by 2-3 orders of magnitude, greatly reducing the minority carrier lifetimes. For this reason, doped layers are primarily used for junctions rather than active layers in devices.

---

[17] In indirect semiconductors (like Si and Ge) the minimum of the conduction band and maximum of valence band occur at different values of the momentum, and excitations from the maximum of the valence band to the minimum of the conduction band are only possible with a change of momentum, i.e. the intervention of phonons.

Due to the continuous density of states distribution, the transport in amorphous silicon is much more complex than in its crystalline counterpart, where conduction of electrons and holes occurs in the extended states of conduction and valence bands respectively. In a-Si:H, beside the conventional extended-state transport, other two transport mechanisms are active: gap state multiple trapping, involving frequent trapping and release from the localized tail states, and hopping, involving the deep-levels within the gap. This last mechanism is a thermally activated process where electrons and holes tunnel from one localized state to another [33].

3.2. Metastability of a-Si:H

The most intense research about defects in a-Si:H has been focused on light-soaking effects. It was soon discovered that the electronic properties of a-Si:H are degraded by prolonged illumination. The first observation of the degradation of a-Si:H goes back to the work of Staebler and Wronski [11], who found in 1977 that the dark conductivity and photoconductivity of a-Si:H can be reduced significantly by prolonged illumination by an intense light. This degradation is called the Staebler-Wronski effect and has been observed in measurements of many properties of hydrogenated amorphous silicon. EPR measurements [34] have attributed the Staebler-Wronski effect to an increase of the dangling bond state density up to about $10^{17} cm^{-3}$ [22]. The observed changes were found reversible by annealing at temperatures above 150°C. The peculiarities of the Staebler-Wronski effect, as summarized by Stutzmann et al. in 1985 [35], are:
- The effect is intrinsic to hydrogenated amorphous silicon and does not depend on the concentration of the major impurities (like O, N, C) unintentionally incorporated in the material, below a critical value of about $10^{19} cm^{-3}$.
- It is a bulk effect involving the a-Si:H layer in its entirety; however, the probability of creating metastable defects can be larger in a surface/interface layer.
- The Staebler–Wronski effect has a self-limiting character, confining the densities of metastable dangling bonds to typically $10^{17} cm^{-3}$.
- The kinetic behaviour of the metastable dangling-bond density and of the photoconductivity, i.e., the dependence of these quantities on illumination time and light intensity, can be described quantitatively by stretched exponential behaviour.

To explain the metastability of a-Si:H, several microscopic models have been suggested. In ref [35] the breaking of weak Si-Si bonds is proposed as the mechanism forming extra dangling bonds. Although the appropriate incorporation of hydrogen is essential in reducing the defect density in hydrogenated amorphous silicon, it is now generally accepted that the ability of hydrogen to move in, out, and within a-Si:H is responsible for the degradation of the electronic properties of this material [22]. Since the hydrogen is sufficiently mobile and the activation energy for hydrogen motion decreases in the presence of intense illumination [36], both the light induced degradation and the reversal of the degradation can be explained by hydrogen motion within a metastable defect complex [37]. At present the most promising class of models explaining the Staebler-Wronski effect involves long-range diffusion of H, where the metastable H atoms weakly bounded in the three-center (Si-H-Si) configuration are the mobile species [38].

3.3. Microcrystalline Silicon

Hydrogenated microcrystalline silicon (µc-Si:H) is a complex material consisting of crystalline and amorphous silicon phases plus grain boundaries (Figure 3). It has to be kept in mind that µc-Si:H is not a single material, but exhibits a wide range of microstructures that depend both on the deposition conditions and on the underlying layer or substrate. Solar cells with the highest efficiencies are those using a material with a medium crystalline phase fraction, which means that only about half of the material volume is crystalline (regions of regular network of atoms), while the rest is amorphous (regions with no periodicity and no long-range order). Grain boundaries between large conglomerates of nanocrystals, as well as the nanocrystals surface itself, are source of defects, although the high content of hydrogen, as incorporated in the µc-Si:H material during its deposition, passivates most of the dangling bonds [39]. In addition, cracks and voids can easily be formed. Three main length scales for disorder can be identified in microcrystalline silicon:
1. local disorder due to the amorphous fraction;
2. nanometrical disorder due to the random orientation of the small crystalline grains of a few tens of nanometre size;
3. micrometrical disorder originating in the conglomerates, formed by a multitude of nanocrystals, often assuming a pencil-like shape (Figure 3).

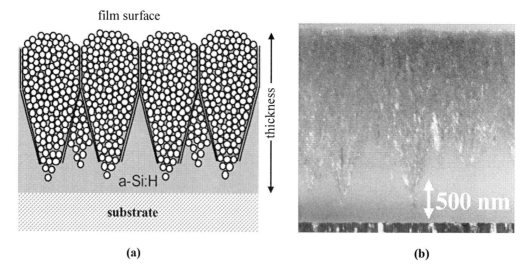

Figure 3. (a) Scheme of a µc-Si:H layer microstructure: pencil-like conglomerates formed by a multitude of nanocrystals (small white circles) are embedded in the amorphous tissue (grey background). (b) TEM dark field micrograph of the cross-section of a µc-Si:H layer deposited near the µc-Si:H / a-Si:H transition on sputtered ZnO [Reprinted from J. Non-Cryst. Solids 299-302, J. Bailat et al., "Influence of substrate on the microstructure of microcrystalline silicon layers and cells", pp 1219-1223, Copyright (2002), with permission from Elsevier].

The nanocrystals have been observed through Transmission Electron Microscopy (TEM) (Figure 3 (b)) [40], and the size of the conglomerates emerging at the free surface of the layers, as estimated from Atomic Force Microscopy (AFM) scans, is roughly half a micrometer. The nanocrystals are separated by grain boundaries and the conglomerates are embedded in the amorphous tissue. The crystalline volume fraction is obtained by Raman spectroscopy, from the deconvoluted integrated intensities of the narrow peak centred at about 520 cm^{-1}, which is the position of the transverse optic (TO) mode in crystalline silicon (c-Si), attributed to silicon crystallites, and the broad peak around 480 cm^{-1}, characteristic of the TO mode in amorphous silicon, attributed to the amorphous phase [41]. Usually a third component centered around 510 cm^{-1}, attributed to a defective part of the crystalline phase, has to be included in order to describe the shape of the spectra right below the crystalline peak [42]. In Figure 4 the Raman scattering spectra of a series of μc-Si:H films with different structural composition is shown, as an example: a clear change in the relative intensity of the peaks at 520 cm^{-1} and 480 cm^{-1} is present.

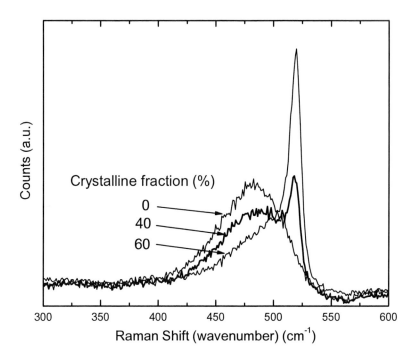

Figure 4. Raman spectra of a series of μc-Si:H films with different structural composition.

In Figure 5 the optical absorption coefficient spectrum of device-quality μc-Si:H is shown, after correction for light scattering due to surface roughness [43]. The α(E) spectra of device-quality a-Si:H and crystalline silicon are also shown for comparison. Three spectral regions can be distinguished [43]:
  (a) the range between 1.2 and 1.4 eV with parabolic behaviour close to the band edges from which the indirect band gap of silicon can be inferred:
  (b) the intrinsic absorption region above 1.5 eV;
  (c) the region below 1.1 eV governed by defect-related absorption.

The detailed study of region a on a large set of samples points to the conclusion that α(*E*) reaches values similar to crystalline silicon, hence the gap of typical device-quality microcrystalline Si has a value of approximately 1.1 eV at room temperature. In single crystalline silicon, the absorption coefficient around 1.1 eV is related to phonon assisted transitions. This mechanism is, however, masked in microcrystalline silicon by an exponential decay (Urbach tail). The appearance of an exponential decay in μc-Si:H may be interpreted as a result of the loss of translational symmetry at the grain boundaries.

As for the above-gap region b, α(*E*) is always higher than the absorption coefficient of crystalline silicon. The higher α(*E*) coefficient can be understood by the effective media approximation [44] taking the combination of three different material components into account: the crystalline silicon grains, the surface region of grains and the amorphous silicon tissue. However, in region b, α(*E*) for μc-Si:H is lower than α(*E*) for a-Si:H. For this reason thicker layers (usually 1-3 μm against ~ 0.5 μm for a-Si:H) are required in solar cells to absorb enough radiation. Finally, the defect-connected absorption (region c) is attributed to silicon dangling bonds mainly at the grain boundaries and in the amorphous tissue.

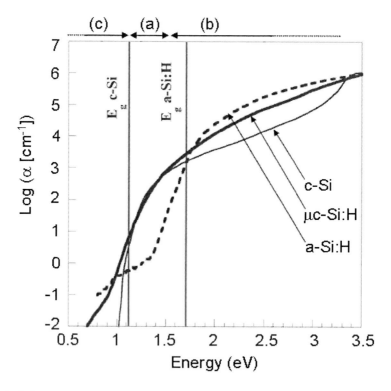

Figure 5. Optical absorption coefficient spectra for wafer-type crystalline silicon (c-Si) and device-quality a-Si:H and μc-Si:H layers on glass [2].

3.4. Alloys and Doped Materials

The structural and optical properties we have described can be varied substantially by changes in deposition conditions. In particular, films with larger hydrogen content have increased band-gap energy. Even bigger changes can be obtained by alloying with additional elements such as Ge, C, O, and N. The resulting alloys have very wide ranges of band gaps. Only some of these materials have proven useful for application in solar cells. In particular, a-SiGe$_x$:H alloys with optical gaps down to about 1.4 eV are employed as absorber layers in multijunction cells: the narrower band gap of a-SiGe$_x$:H compared to a-Si:H allows for increased absorption of photons with lower energies [45]. The bandgap E_G of the material decreases with increasing the Ge content in the alloy. However, when E_G is decreased below 1.4 eV the defect density becomes too high and the material can no longer be employed in solar cells.

Doping is the intentional incorporation of appropriate atoms in order to shift the Fermi energy of a material and change its electrical properties, creating n-type or p-type materials. For silicon, with valency 4, the most common dopants are boron (B) with valency 3 as acceptor, and phosphorous (P) with valency 5 as donor. Inhomogeneous doping in crystalline silicon is achieved at high temperatures by the diffusion of dopant atoms. In c-Si, P and B atoms substitute for Si atoms in the crystal lattice. For example, in the n-type material, P has five valence electrons, so in the fourfold coordinated sites of the Si lattice four electrons participate in bonding to neighbouring silicon atoms. The fifth free electron occupies a state just below the bottom of the conduction band, and the dopants raise the Fermi energy to roughly this level.

On the other hand, in a-Si:H most phosphorus atoms bond to only three silicon neighbours; they are in threefold coordinated sites. This configuration is actually advantageous chemically: phosphorus atoms normally form only three bonds (involving the three valence electrons in p atomic orbital). The final two electrons paired in "s" atomic orbital, do not participate in bonding, and remain tightly attached to the P atom. The reason why this more favourable bonding occurs in a-Si, but not in c-Si, is the absence of a rigid lattice. Occasionally, independent formation of both a positively-charged fourfold-coordinated P4+ and a negatively-charged dangling bond D− can occur instead of the more ideal threefold coordination [22]. Only these P atoms contribute a free electron and raise the Fermi energy.

Doping is then rather inefficient in a-Si:H. The Fermi-level cannot be pushed very close to the conduction and valence band edges, even with heavy doping. Additionally, the negatively charged dangling bonds induced by doping are very effective traps for holes, and the minority carrier lifetime is greatly reduced. Also boron tends to alloy with amorphous silicon rather than substitutionally dope it, leading to a strong reduction in band gap. The bandgap reduction is usually compensated by adding carbon to the lattice. On the other hand, μc-Si:H has much higher doping efficiency, with substitutional doping mainly happening within the nanocrystallites.

Beside the intentional introduction of different atoms to form alloys or doped layers, contaminant atoms (like O, N, C) are unintentionally incorporated in the films. Fortunately, because of the flexibility of the bonding network in an amorphous solid, the tolerance level for contaminants in a-Si:H is much higher than that of its crystalline counterpart. As long as the concentration of impurity atoms is below a critical value of about $10^{19} cm^{-3}$, solar cells using such material show good performance [35]. For higher contaminant level the reduced diffusion length of photogenerated carriers becomes an issue.

4. DEPOSITION TECHNIQUES

4.1. Plasma Enhanced Chemical Vapor Deposition (PECVD)

Amorphous and microcrystalline silicon are essentially still produced using the methods introduced by Chittick *et al.* [46] and Spear and LeComber [47] that first resulted in hydrogen incorporation in the material. The technique is a silane-based glow discharge induced by radio frequency (RF) voltage, commonly termed plasma enhanced chemical vapour deposition (PECVD). However, other thin film deposition methods, mostly attempting to get higher deposition rate or improved microcrystalline silicon, have been extensively explored.

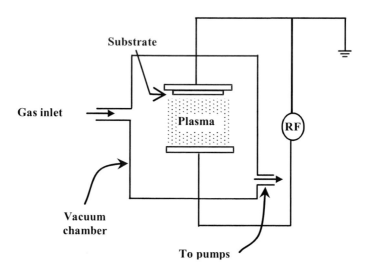

Figure 6. Simplified schematic of a PECVD chamber.

Figure 6 shows a simplified schematic of a PECVD chamber. Two parallel plates (electrodes) are placed in a vacuum reactor chamber, and appropriate substrates can be mounted on one or both of the electrodes. Controlled flows of silicon-containing gases, such as, typically, silane (SiH_4) or a mixture of SiH_4 and H_2, are admitted into the vacuum chamber[18], and a RF power is applied between the electrodes. The standard RF frequency for glow discharge deposition is $f = 13.56$ MHz, which is the frequency allotted for industrial processes by federal and international authorities. A glowing electric discharge, similar to that present in a fluorescent light bulb, is sustained between the electrodes by inelastic electron-impact processes initiated by electrons that have acquired enough energy from the electric field. The visible luminosity of the plasma glow region is the result of the de-excitation of emitting molecular and atomic species contained in the plasma.

The PECVD process includes four major steps: electron impact dissociation of the feed-gas, gas-phase chemical reactions, transport of the species to the substrate surface, and film growth. This final step includes the adsorption of species at the growing film surface and

---

[18] Alloys are deposited by adding other gases to the gas mixture, such as GeH_4, CH_4, NH_3, to obtain a-$SiGe_x$:H, a-SiC_x:H, a-SiN_x:H respectively; a-Si:H and μc-Si:H are doped by mixing into the feedgas typically phosphine (PH_3), to obtain n-type layers, and diborane (B_2H_6) or trimethylboron [$B(CH_3)_3$], for p-type layers.

reactions/re-aggregations (chemical and physical processes) that result in the final film. The weakly ionised plasma created between the two electrodes contains positive, negative and neutral species (radicals). As the electrodes (and the substrate) are slightly negative with respect to the plasma bulk, negative ions are trapped in the plasma. The positive ions and the radicals reach the substrate by drift and diffusion, respectively, and undergo surface and subsurface reactions during deposition. Radicals are considered to be the main precursors for the growth of both amorphous and microcrystalline silicon in a PECVD deposition process, and in particular it is believed that the SiH$_3$ radical, characterized by high surface migration thanks to its low sticking coefficient, is mostly responsible for the growth of high-quality a-Si:H films [48]. Ions, as long as their energy is not too high, help to improve the film quality.

The advantage of this deposition method is that externally applied electrical energy is used to provide the activation energy for gas decomposition in the glow discharge deposition process. The gas-phase reaction reduces the substrate temperature required for film deposition compared to thermal CVD, which, instead, depends on thermally-induced gas-surface interactions. During the process, the substrates are usually heated to achieve optimum film quality, thanks to the thermally activated diffusion of adsorbed species on the growing film surface.

Glow discharge is a highly complex process as physical and chemical interactions in the plasma and at the growing film surface are dependent on the various deposition parameters (rf power and frequency, substrate temperature, gas pressure and composition, gas flow rate and pattern, electrode geometry, etc.). To deposit good-quality a-Si:H films, the deposition conditions need to be controlled within certain ranges. Over the years, much research has been carried out to learn how plasma conditions relate to the film quality and the literature on this subject is vast. However, since there are numerous variations in deposition system design, the optimum process parameters vary from system to system and are often empirically optimized for the individual case. Moreover, the effects of the various deposition parameters are often interrelated and cannot be expressed as simple relationships. In the following section, we will shortly discuss some effects of the main process parameters.

4.2. PECVD Process Parameters

The electrode spacing in a plasma deposition reactor affects the deposited film quality because it limits the distance that radicals travel before reaching the growing surface and determines the discharge-sustaining voltage, which is a function of the product of process gas pressure and inter-electrode distance (Paschen's law) [49]. A small spacing is preferable to minimize the radicals residence time to avoid excessive gas-phase nucleation which causes powder formation (plasma polymerization). Inter-electrode distances of 1-5 cm are typically used.

For high-quality a-Si:H depositions, the supplied power is usually maintained at a level slightly above the minimum value needed to sustain the plasma. For pure silane or disilane glow discharge deposition, this value is usually less than 0.1 W/cm^2. Larger power density allows for higher deposition rate, but with too large values the rapid reactions in the gas easily create silicon polyhydride powder that contaminates the growing film. In fact, the deposition rate increases monotonically with power density until it is limited by the gas flow rate, when the deposition becomes gas-phase-limited instead of reaction-rate-limited. Also the gas pressure should be kept below the level that causes gas-phase polymerization. The pressure range is usually between 100 mTorr and 2 Torr, depending also on the electrode spacing, according to the Paschen's law.

During a-Si:H deposition, the substrate temperature affects the hydrogen elimination and surface rearrangement of the deposition precursors arrived on the growing film surface. The temperature is usually set between 150 and 350°C. With increasing substrate temperature, the total hydrogen content in the material decreases and films with lower optical gap are produced [50,51]. The deposition temperature is, as a matter of fact, used as a means to change the absorption of intrinsic amorphous silicon films [52]. However, films grown at temperatures above 350°C contain insufficient hydrogen to passivate dangling bonds defects.

As for the feed-gas composition, often silane is diluted in hydrogen or in inert gases like helium or argon. In case of the more common hydrogen dilution, it is believed that most of the hydrogen incorporated in the film network originates from SiH_x species and not from the hydrogen diluent. Hydrogen is a reactive diluent in the glow discharge whose role is mostly to scavenge deleterious radicals from the growing surface. In fact, amorphous silicon deposited with hydrogen dilution has enhanced stability against light soaking [53]. The improved quality is obtained at the expense of a lower deposition rate.

As the hydrogen dilution is increased to sufficiently large values, extensive etching of the growing film surface by atomic hydrogen is obtained. Since strained and weak bonds are selectively etched, this can result in the deposition of microcrystalline rather than amorphous silicon. Usually also higher pressure and power values are used to grow such material when compared to the amorphous case. By adjusting the various deposition parameters (commonly the hydrogen dilution, keeping the other parameters fixed), materials with different amorphous and crystalline phase fractions can be obtained going from completely amorphous to fully polycrystalline films. The best material for solar cell application is found within the transition region from crystalline to amorphous growth [54]. In any case, microcrystalline silicon deposition is an inhomogeneous growth process. The deposition generally begins with an amorphous phase [55]. It needs a minimum thickness, called *incubation layer*, before a localised phase transformation takes place, which is called nucleation. Once nucleation has occurred, the crystals grow around the seeds. Ion bombardment is considered deleterious for this process retarding nucleation.

As under these conditions the deposition rates are quite low (generally a few Å/s for a-Si:H and less than 1 Å/s for μc-Si:H obtained with heavy hydrogen dilution) much effort has been put in research on compatible methods to increase the deposition rate. Two methods attracting a lot of attention are the PECVD using a higher discharge frequency, called Very High Frequency PECVD (VHF-PECVD), and Hot Wire CVD (HWCVD), discussed in the following sections. Also with the standard PECVD, conditions for a better dissipation of the RF power at 2-5 times higher pressures are under exploration. For a more efficient deposition process, powder management is an important issue in achieving larger deposition rates while preventing the powders from being deposited on the substrates. At the same time, powder formation in silane plasmas, usually considered as a drawback, has recently been viewed as a source of new silicon thin film materials, called polymorphous silicon, resulting from the incorporation of small Si particles (with a few nanometres size) in the a-Si:H network [56].

4.3. VHF-PECVD

The standard RF frequency for glow discharge deposition is 13.56 MHz. A much larger frequency range has been explored, including DC, low frequency (kHz range), very high frequency (VHF) (20–150 MHz), and microwave frequency (2.45 GHz). The group at University of Neuchâtel [57,58] has pioneered VHF plasma as a route to higher deposition rates. One attractive feature of this technique, commonly thought to contribute to a lower production cost of a-Si:H based solar cells, is the possibility of producing device-quality material at 5-10 times higher deposition rate. The maximum achievable deposition rate is system dependent, where proper impedance matching of the RF power to the glow discharge is crucial. Usually a monotonic increase of the deposition rate with the excitation frequency is found, while the material properties are more dependent on other deposition parameters such as the pressure. The explanation of the increased deposition rate is not straightforward, and in particular cannot be simply related to the enhanced radical production, which has been found to increase only to a minor degree. As an alternative to the enhancement of precursor formation (gas phase effects) an enhanced surface reactivity of film forming species (plasma-wall interactions) has been proposed, thanks to an increased ion flux impinging on the substrate [59]. Ions also strongly contribute to the final film quality, as long as their energy does not exceed the threshold energy of defect formation. At VHF frequencies the increase in ion flux is accompanied by a decrease in ion energy. Therefore, a low-energy high-flux ion bombardment is active, enhancing the surface mobility of the adsorbed species, so that the properties of the material remain device-quality also at high deposition rate. This issue is particularly important in case of microcrystalline growth, as high-energy ions are believed to adversely affect the nucleation process and the defect density. Also the need of a higher deposition rate method to grow µc-Si:H is more stringent due to the thicker layers required in solar cells when compared to the a-Si:H case.

4.4. Hot-wire Chemical Vapor Deposition

The Hot-Wire Chemical Vapour Deposition (HWCVD) consists in the catalytic decomposition of silane-containing mixtures at the surface of resistively heated filaments (usually tungsten, tantalum or graphite) brought to temperatures significantly above 1500°C (preferably between 1750 and 2000°C) [60-63,55]. The obtained species, then, react and diffuse inside the deposition chamber and deposit onto the substrate placed a few centimetres away and heated to a temperature of 150 – 450°C. The success of HWCVD is based on the efficient decomposition of the feedstock gas into atomic fragments, which enables high deposition rates (much above 10Å/s). Moreover, in contrast to PECVD, no ions are created. This feature makes HWCVD a softer deposition technique, which is considered beneficial for microcrystalline growth. However, with large pressure or large substrate-filament distance significant gas phase polymerization occurs. In the opposite conditions the film properties would suffer from a high flux of Si atoms arriving at the surface leading to a large concentration of voids. In the absence of ions, the concentration of atomic hydrogen plays a role in creating the right radicals for deposition (again, mostly SiH_3) as well as in balancing the hydrogenation and etching of the growing surface. Good quality µc-Si:H films are obtained when the silane is diluted in hydrogen at relatively high pressure. The hot filament decomposes all hydrogen molecules into reactive atoms that effectively etch silicon from strained or disordered bonding sites, locally promoting the formation of a crystalline network. Especially for materials at the extremes of the amorphous-crystalline scale (purely amorphous or purely polycrystalline), superior properties have been reported in comparison with other deposition methods [55].

4.5. Large Area Deposition Techniques

When moving toward large-scale manufacturing, together with production throughput (also related to the deposition rate), key issues are machine maintainability, process automation, process reproducibility, feedgas utilization, and uniformity over large areas.

PECVD is particularly suitable for producing thin films for large area applications, in the photovoltaic field as well as in microelectronics. It is, in fact, adopted by the majority of the manufacturers. The basic concerns when going from small to large area are to maintain the material quality and the film uniformity on the enlarged scale. The most dominant sources for non-uniform deposition are local inhomogeneities in the RF impedance, loss of parallelism between the electrodes, local changes in the gas composition by depletion far from the injection side, local temperature variations. Moreover, parameters like RF power and gas consumption do not simply scale with the size of the deposition reactor. Also, the flux of ions towards the growing surface and the energy of the bombarding ions are affected when scaling up.

The principal challenge to applying VHF deposition in manufacturing is the difficulty to couple VHF power from the generator to large electrodes and the possible manifestation of standing waves, which would influence the uniformity of the deposited layers, when the electrode size is comparable to a quarter of the wavelength of the RF wave. However, significant progress has been made in this area. For example, multipoint feeding of the RF power improves the homogeneity. Recently, Mitsubishi Heavy Industries has proposed a ladder shaped electrode as well as a phase modulation method which controls the phase of the VHF electric charge supplied to the electrode [64].

There are several concerns about incorporating HWCVD processes in manufacturing. First, the uniformity of HWCVD films is still poorer than that of PECVD films, although some companies have worked on this and made significant improvement [63]. Second, the filament needs to be improved to reduce the maintenance time in production. Third, HWCVD-deposited solar cells have not yet achieved the same performance as cells prepared using PECVD and VHF PECVD.

5. THIN FILM SILICON SOLAR CELLS TECHNOLOGY

5.1. Thin Film Silicon Solar Cells Design: p-i-n Photodiodes

The operation principle of a solar cell is based on a sufficiently long lifetime of photoexcited electrons and holes such that they can become spatially separated and thus contribute to a net current. A "classical" solar cell consists of a p-n junction, whereas with amorphous and microcrystalline semiconductors a solar cell cannot be constructed by merely stacking a p-type and an n-type thin film. Although there are some "technological" reasons for using a p-i-n junction in the case of silicon thin films (difficulty of doping, decrease of layer quality with doping), the p-i-n structure also offers some basic advantages with respect to collection. For wafer-based c-Si solar cells collection is not a major problem as the minority carrier diffusion length is usually sufficiently large (over 200 µm) that minority carriers are transported by diffusion to the junction, where the actual separation between holes and electrons takes place, thanks to the internal field in the depletion region. This depletion region is typically much thinner than the total cell thickness and the electrical field extends only over this very thin region. Amorphous and microcrystalline silicon, in contrast, have relatively poor transport properties (low carrier mobilities μ and short carrier lifetimes τ) so that the diffusion length would be extremely small (around 0.1-0.2 µm for holes and about 2 orders of magnitude more for electrons) [65]. Therefore, collection is a major problem in thin film silicon solar cells. In order to avoid the recombination of the major part of generated free carriers, it is imperative to have a drift-assisted transport based on an internal electric field extending over most of the cell. For this reason an undoped layer with a low defect density must be incorporated between the p and n layer such that the built-in electric field, generated following the creation of the junction, extends over the whole i-layer.

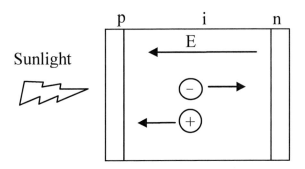

Figure 7. Scheme of a p-i-n photodiode. The electric field causes the drift of photogenerated carriers in the directions shown by the arrows.

In figure 7 the p-i-n photodiode schematic is shown. Sunlight enters the photodiode as a stream of photons that passes through the p-type layer, which is a nearly transparent "window" layer. The solar photons are mostly absorbed in the much thicker intrinsic layer; each photon that is absorbed will generate one electron and one hole photocarrier [66]. Simply speaking, the photocarriers are swept away by the built-in electric field to the n-type and p-type layers, respectively – thus generating solar electricity. A thorough description of the physics governing the device operation is given later in Section 6.

5.2. Superstrate and Substrate Solar Cells

One of the advantages of thin film silicon–based solar cells is that they absorb sunlight very efficiently: The total thickness of the absorbing layers is in the range 0.3 - 0.5 µm for amorphous silicon solar cells and 1-3 µm for microcrystalline devices. Consequently, these layers need to be supported on a much thicker substrate. Two totally different designs for solar cells have evolved corresponding to transparent and opaque supports: superstrate and substrate designs (figure 8).

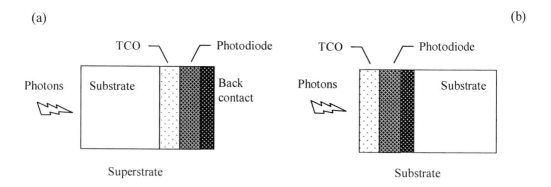

Figure 8. Different designs for thin film silicon solar cells: (a) superstrate and (b) substrate configuration. Generally, photons enters the photodiode through the p-layer in both cases.

In superstrate-type solar cells, the carrier on which the various thin film materials are deposited serves as a window to the cell: Usually, glass is used as the carrier. A cross section of the superstrate solar cell is shown in figure 8a. In this case the p–i–n deposition sequence is generally used. Light enters the cell through the glass substrate on which a transparent conductive oxide (TCO) layer has been deposited as contact layer. The p-layer is deposited as the first semiconductor layer on top of the glass–TCO combination. Being it the first layer crossed by the incoming light, it has to be relatively thin (about 10 nm) and needs to have a low absorption coefficient (for example for a-Si:H solar cells a-Si:C:H alloy layers are used as p-layers, so that the resulting optical gap is high, and light absorption in most of the visible range is correspondingly low). The light, then, crosses the much thicker i-layer, where the essential part of photogeneration takes place, the relatively thin n-layer (20 nm typically), and finally reaches (if not yet absorbed) the reflective back contact layer.

A more recent version of the superstrate configuration uses a transparent thin polymer foil as a substrate. This type of cell is bound to have performance limitations as the limited temperature resistance of most transparent plastics dictates the maximum temperature allowable in further process steps. As a matter of fact, deposition temperatures have often to be kept low (around 150°C or lower) a condition that is, indeed, generally possible with PECVD deposition, but could impose additional restrictions on semiconductor layer quality. Furthermore, most plastics are often easily damaged by ultraviolet light. Therefore, the present world production volume of superstrate solar cells consists primarily of p-i-n structures on glass. The superstrate design is particularly suited to building-integrated solar cells in which a glass substrate can be used as an architectural element.

In the substrate configuration the carrier, on which the various thin film materials are deposited, forms the back side of the cell: The sunlight enters the photodiode before reaching the substrate (figure 8b). The devices are usually made on lightweight, unbreakable, often opaque substrates, such as stainless steel, but also polyimide or PET (polyethylene terephtalate). In this case the deleterious UV-light effect, active on most plastics, is not an issue. The sequence used in this configuration is usually substrate/n–i–p, with light entering the device through the semiconductor layer deposited as the last layer, i.e., still through the p-layer. The fabrication issues of such substrate/n–i–p cells are quite different from those of the superstrate-type cells because of the modified deposition sequence.

The substrate configuration is being followed by one of the prime producers of a-Si-based modules, UNI-SOLAR. So far, it has yielded the highest stabilized cell efficiency [14,67]. The substrate material used here (stainless steel) lends itself to a roll-to-roll in-line deposition process. However, the electrical conductivity of the substrate material rules out, in this specific case, the monolithic series connection of cells to a module, which otherwise is one of the attractive features of thin film technologies (see Section 7.1). The use of plastic substrates (polyimide, PET) within the substrate configuration, would, on the other hand, allow for a combination of both roll-to-roll production and monolithic series connection of cells to modules; this combination may, in the medium term, prove to be economically very interesting.

5.3. Front Electrode Technology

The front electrode in thin-film silicon solar cells usually consists of a transparent conductive oxide (TCO) layer such as SnO_2:F, doped ZnO, or In_2O_3:Sn (ITO). The basic properties of TCO layers are: high optical transparency (preferably >85%) in the required spectral range (mainly the visible spectral range but also the near-infrared (NIR)), as well as sufficiently high electrical conductivity. The electrical properties are determined by the sheet resistance of the TCO film, that in turn is related to the resistivity ρ and the film thickness d, i.e., $R_{sheet} = \rho/d$. The sheet resistance is of practical importance in view of the integrated series connection of cells to a solar module (see Section 7.1): it should be preferably lower than 10 $\Omega/\square$ to minimize the series resistance. In principle, the reduction of R_{sheet} can be achieved by: (a) reducing the specific resistivity of the TCO material, which, in turn, should preferentially be achieved by raising the carrier mobility, rather than by raising the carrier concentration, because an increase in carrier concentration enhances the optical absorption in the near-infrared region (so-called free carrier absorption); (b) increasing the TCO thickness, which, unfortunately, increases the TCO absorbance, too. Thus, the TCO sheet resistance is inherently related to the absorption of near-infrared light within the TCO film: Increasing the carrier mobility is the only way to decrease the sheet resistance without increasing the optical absorption of the TCO layer in the near-infrared.

A further property of TCO layers used in thin-film solar cells is their ability to scatter the incoming light, particularly the long wavelength light (red and NIR). This scattering effect increases the path of the light inside the solar cell, and therefore also enhances the probability of photons to be absorbed by the cell (see figure 9). This property is especially important in the case of amorphous and microcrystalline silicon thin-film solar cells, because of their relatively low optical absorption coefficient in the red and NIR spectral range (Figure 5). Moreover, the use of thicker a-Si:H (>0.5 μm) and μc-Si:H (>3 μm) intrinsic(i)-layers within the corresponding solar cells is not convenient for several reasons. First of all, with thick i-layers the built-in electric field is reduced, and the probability of carriers recombination increases (see section 6.2). Moreover, the light-induced degradation effect, particularly strong in a-Si:H p–i–n solar cells (see section 5.6), becomes more pronounced as the thickness of the a-Si:H i-layer is increased. Finally, from a technological point of view, the deposition times needed for thicker silicon layers become so long that cell manufacturing becomes economically prohibitive (especially in the case of μc-Si:H solar cells).

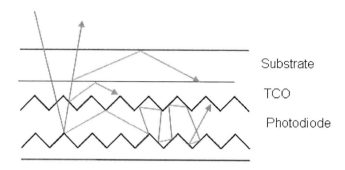

Figure 9. Light trapping effect caused by the surface morphology of the Transparent Conductive Oxide (TCO) layer.

The light scattering property is linked to the surface roughness of the TCO layer used in the cell. Indeed, a rough surface is required to efficiently scatter the light that enters the solar cells through the TCO layer (figure 10). At the same time the morphology should be such that shunting paths, pinholes, or local depletion are avoided.

The ability of a TCO layer to scatter light is often expressed in terms of the haze ratio H evaluated as the percentage of light diffused at a different direction than specular, $H=T_{diffuse}/T_{total}$, where $T_{total}= T_{diffuse}+T_{specular}$ and T indicates the transmittance. The measured value depends on the medium covering the TCO layer, but is usually measured in air. The used transmission data are a weighted average over the spectral region of interest. It should be noted that TCO layers with different morphologies can have the same numeric haze ratio value. Therefore, the haze value has only limited significance as a figure of merit.

In superstrate devices, it is important for the TCO as substrate material to exhibit not only 'optimal' electrical and optical properties and a suitable surface morphology, but also to remain chemically stable during plasma deposition (PECVD), as employed for subsequent fabrication of a-Si:H and μc-Si:H layers. The well-known indium-tin-oxide (ITO) is not at all suitable for use here, because it would be chemically reduced, even without any hydrogen dilution, i.e., already by the hydrogen formed by the decomposition of silane during the PECVD process. Therefore SnO_2:F and doped ZnO are used as TCO in thin film silicon solar cells.

Tin Oxide is usually deposited by Atmospheric Pressure CVD (APCVD) [68,69] which produces a natively textured coating. The deposition temperature for this process is 500 – 600 °C which basically limits the use of this TCO as front electrode in superstrate configuration. The use of SnO_2:F is quite common for a-Si:H solar cells, but is not recommended for single-junction μc-Si:H cells, as it is partly chemically reduced in the H_2-rich plasmas that are necessary in this case [70].

Current research is also focused on doped ZnO, deposited either by sputtering or by Metal Organic CVD (MOCVD). Sputtered ZnO has been widely used in R&D work on thin-film solar cells. Sputtering technique yields flat ZnO layers; they can be subsequently roughened (texturized) through an etching process in diluted HCl [71]. Incorporation of such sputtered and etched layers in entire a-Si:H and µc-Si:H solar cells has yielded excellent results [72]. However, the etching step could turn out to be a critical step when upscaling the manufacturing process for actual commercial module production (because of production hazards and material waste). In alternative it would be desirable to have a deposition process that yields as-grown rough ZnO layers.

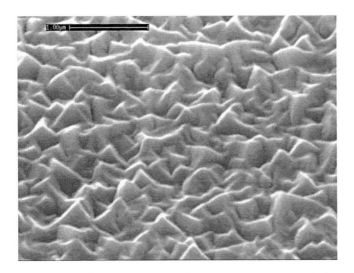

Figure 10. Scanning Electron Micrograph of the surface of a ZnO film grown by MOCVD.

Textured ZnO films are deposited by MOCVD. Films grown by this technique have been successfully used in thin film silicon devices as front TCO [73]. The high roughness of the MOCVD-grown ZnO material can make the surface morphology of the front TCO layer inappropriate to host the solar device. Many research groups [74,75] have been developing surface plasma treatments to fully adapt the surface morphology to subsequent growth of solar cells. This last task is challenging as the optimal morphology for the growth of amorphous or microcrystalline cells is not yet fully known.

As for the substrate configuration (n-i-p structure), where the transparent top electrode is deposited as the last layer, a restriction to the deposition temperature is imposed. The most suitable top contacts are evaporated ITO or ZnO deposited by MOCVD technique. The thickness of this layer is chosen in such a way to provide an antireflection effect. As the thickness is about 70-80 nm, the sheet resistance of the best ITO is still quite high (> 50 Ω/□), and therefore a metal grid is necessary to reduce the series resistance and thus enhance the fill factor of the device.

5.4. Back Electrode Technology

The back contact of the solar cell in the simplest configuration can be realized by depositing a single metal layer (generally Ag or Al) as first/last layer in the substrate/superstrate configuration respectively. In a superstrate p-i-n solar cell the roughness of the front electrode layer is largely replicated at the back contact (see Figure 9). Therefore the roughness of the interface between the silicon and the metal back reflector contributes to the light trapping. In order to further enhance the light trapping within the device, the reflectivity of the metal electrode is often increased by introducing a TCO layer (either ZnO or ITO) between the n-layer and the metal back contact. The refractive index value, intermediate between silicon and the metal, combined with the oblique incidence of a large fraction of the light at this interface, leads to total internal reflection for this fraction, and thus to high overall reflectivity.

An Ag/ZnO bilayer is often used as back contact also when substrate configuration is used. In this case, since the back contact is deposited on the substrate as the first layer, the ZnO may have the additional function of preventing the diffusion of Ag in the superimposed n-layer. Further, high temperatures can be used for the deposition of Ag and/or ZnO to obtain the desired texture, since no silicon layers are present at the stage of forming the back contact.

Finally, for both configurations, substrate and superstrate, a back electrode realized with an appropriate combination of a thin TCO layer with a thicker metallic layer (Al or Ag) provides both the desired optical properties (high value of reflectivity) and high electrical conductivity.

5.5. Single Junction p-i-n Solar Cells

Amorphous and microcrystalline single junctions have been fabricated in many research laboratories. The thickness of the intrinsic absorber layers are in between 300-500 nm for amorphous device and 1-3 μm for the microcrystalline one. Typical Spectral Responses, which means number of photocarriers collected for each incident photon, measured on amorphous and microcrystalline cells are shown in figure 11: the response in the NIR part of the spectrum observed for the microcrystalline cell is due to the lower optical gap of this material with respect to a-Si:H.

Since the drift lengths of amorphous and microcrystalline materials are higher than 10 μm [76-78], in principle it would not be a problem to fabricate a-Si:H solar cells with i-layer thicknesses of 1 μm or more. In practice, however, a-Si:H cells should be kept thinner than 0.4 μm, otherwise severe recombination losses occur in the stabilized state, thus adversely affecting the fill factor (FF) of the cell.

Microcrystalline silicon solar cells show satisfactory collection [79] for cell thickness up to 4 μm. Here, the reason for using, in practice, much thinner cells (1–2 μm) is not always a collection problem, but is due to economic considerations linked to the correspondingly shorter deposition duration and lower material consumption. For microcrystalline devices the structure of the absorber material is a very important issue. As already discussed (section 3.3) microcrystalline silicon is a highly complex and versatile material, and its structural composition can be adjusted by varying the deposition conditions. The best electrical performances are obtained with the μc-Si:H absorber layer grown near the amorphous-crystalline transition [54,80]. In fact many authors report that cells deposited with high silane concentration in the gas mixture show larger V_{OC} but low FF and J_{SC}, due to the high amorphous-phase volume-fraction present in the absorber material, whereas, when highly crystalline material is used, increased local current drains and problems of carrier extraction limit the device performance [54].

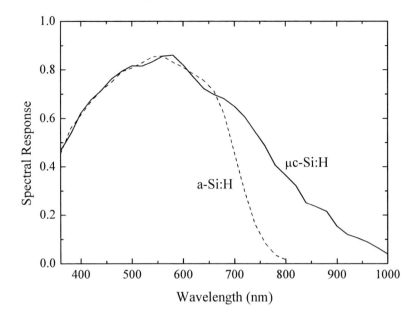

Figure 11. Spectral responses of single-junction amorphous and microcrystalline silicon p-i-n solar cells.

An important issue related to the fabrication of p-i-n junctions is the quality of the window p-layer and of the p/i interface. This layer has to be highly conducting and, furthermore, since optical absorption within the p-layer can limit the photocurrent, material with wide optical gap is desired. In amorphous p-i-n junctions a major improvement in the efficiency has been obtained by incorporating a silicon carbide wide gap p-layer (a-SiC:H), with better window properties [81]. Moreover, the use of an intentionally graded buffer p/i interface enhances the performance in the blue region of the spectrum. Different explanations have been proposed for such effect; among others, the relaxation of the band gap discontinuity by reducing the bond distortion due to lattice strain at the interface, and the prevention of excessive boron diffusion from the p-layer. In general, the improved quantum efficiency in the blue region and the higher open circuit voltage and fill factor have usually been attributed to a reduced density of recombination centers at the junction [82].

The use of p-doped microcrystalline material has been proposed for the superstrate structure because µc-Si:H (p) films show larger doping efficiency and conductivity together with lower optical absorption than a-Si:H. Furthermore, the use of µc-Si:H material is necessary for microcrystalline p-i-n junctions in order to promote the nucleation of the superimposed microcrystalline i layer at the p/i interface [83]. Due to the unavoidable presence of an amorphous incubation layer in the first growth stages of µc-Si:H films, the conductivity and microstructure of the material strongly depend on the film thickness. Great efforts of the researchers are addressed toward the optimization of very thin µc-Si (p) layer (about 20nm) as required in solar devices.

The properties of the n-layer and the i/n interface are less critical to cell performance because the local generation rate is much smaller than at the p/i interface. The n-layer is often made microcrystalline as this has an advantageous influence on the built-in voltage as well as on the contact resistance.

In Table 1 the three key parameters (short-circuit current density J_{SC}, fill factor FF and open-circuit voltage V_{OC}) as well as the overall conversion efficiency η of state-of-the-art single-junction laboratory cells on glass substrates (superstrate configuration) are reported. In section 6.3 the experimental values will be compared with the "basic" limits evaluated within a semi-empirical approach. Essentially, V_{OC} and FF remain rather low, partly due to the p–i–n type structure itself [84]. Furthermore, V_{OC} (and, hence, also FF) is additionally limited in a-Si:H, with its pronounced bandtail states, because of the difficulty of obtaining efficient doping. The highest additional gain is to be obtained, both for amorphous and for microcrystalline silicon solar cells, by a further increase in J_{SC}. Practical values for J_{SC} are in both cases so far still relatively low, because thin i-layers need to be used. J_{SC} could be increased by further improving the light trapping (see Section 4) and, thus, by increasing the optical absorption within the cell.

Table 1. Comparison between the photovoltaic parameters of state-of-art amorphous and microcrystalline single-junction laboratory solar cells

	a-Si:H [85]	μc-Si:H [86]
J_{SC} (mA/cm^2)	17.3	24.4
V_{OC} (mV)	877	539
FF (%)	66.6	76.6
η (%)	10.1	10.1

5.6. Light Soaking Effects

The light induced degradation is one of the main problems for a-Si:H solar cells: their initial efficiency significantly decreases during the first few hundred hours of illumination. The single junction amorphous cell loses about 30% of its initial efficiency after about 1000 h. The initial performance of the device can be largely restored by annealing the cell at moderate temperatures (150°C) for several hours. The reversibility of the device performance shows that the initial loss is not due to diffusion of ions or dopants, or other irreversible processes. All the phenomena in solar cells are consistent with metastable defect creation due to the Staebler-Wronski effect already discussed for individual thin films. The additional defects created in the intrinsic layer will trap photo-generated charge carriers as they drift towards the electrodes. As a result of the trapping, a space charge region develops in the intrinsic a-Si:H layer and the internal electric field can be strongly distorted, which in turn leads to lower drift and thus to a lower collection efficiency (see Section 6). The largest relative changes occur in the fill factor. The relative changes in the short circuit current are significantly smaller, whereas the open circuit voltage usually does not degrade.

Intrinsic layer properties such as hydrogen configuration void structure and Si network order can sometimes be correlated with the device stability. For example, cells with intrinsic layers made from highly hydrogen-diluted silane discharges were shown to stabilize at a higher efficiency than those with an intrinsic layer made from pure silane [87].

The Sbaebler-Wronsky effect contributes to noticeable seasonal variations in the conversion efficiency of a-Si:H-based modules, due to partial recovery of the performance when cells operate at elevated temperature.

Microcrystalline silicon solar cells have, at first, been considered not to be affected by light induced degradation. But, µc-Si:H is a mixed material that exhibits a wide range of microstructures depending on both deposition conditions and substrate material (see section 3.3). One could expect some light-induced degradation in µc-Si:H solar cells, especially when deposited near or within the crystalline to amorphous transition, that has been recognized as the region where the best electrical performances can been obtained. As a matter of fact the presence of a mild form of light-induced degradation in µc-Si:H solar cells has been revealed: Cells with a crystallinity comprised between 50 and 60 % have shown after light-soaking a relative efficiency reduction lower than 10 % [88,89].

The light-induced degradation can be partially compensated by using thinner absorber layers, in which case the internal electric field is higher and therefore less sensitive to distortion. However, the total light absorption in the solar cell will then decrease. To compensate the consequent current loss, good light trapping by a hazed TCO and reflection from a highly reflective back contact is desirable. Another way to increase optical absorption and stability with respect to light-induced degradation is the use of tandem or multi-junction cells.

5.7. Multijunction Solar Cells

Thin film silicon solar cells can be fabricated in a stacked structure to form multijunction solar cells. If two or more cells with different optical gap are arranged in series on top of each other the conversion efficiency can be much improved compared to a single cell. Figure 12 shows the structure of a triple-junction substrate cell grown on stainless steel foil.

The fundamental concept underlying multijunction solar cells is "spectrum splitting." The multijunction approach requires that incident photons are directed onto a junction that is tuned to the photon energy. Consider what happens if a second p-i-n junction structure is deposited on top of the first one. The second structure "filters" the sunlight: photons absorbed in the top junction are of course removed from the light that reaches the bottom cell. In other words, the junctions themselves act as optical elements to distribute the spectrum to the appropriate junctions for multijunction photoconversion. In practice, for a tandem structure, the top component is adjusted so that it filters out about half of the photons that would otherwise have been absorbed in the bottom cell. The band gaps must decrease from the top to the bottom of the stack. This is the "spectrum-splitting" effect. Such effect is evidenced in figure 13 where the quantum efficiency curves of the components cells of a triple-junction device are shown [90].

The use of multijunctions allows also to improve carrier collection and stability. This is due to the fact that each component cell has a smaller thickness than that of a single junction. On the other hand, it is more challenging to fabricate multiple-junction solar cells than single-junction cells. The performance of a multijunction cell is more sensitive to the spectrum of the incident light due to the spectrum-splitting feature. The control of the band gaps and thicknesses of the individual layers is critical. This is a very important issue because, since the component cells are connected in series to form a two-terminal device, the cell with minimum current density during operation will limit the total current of the device. Therefore the current densities of each component need to be matched at the maximum power point in sunlight.

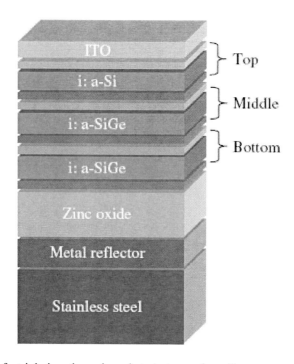

Figure 12. Structure of a triple-junction n-i-p substrate-type solar cell.

Another area that needs attention in fabricating a multijunction solar cell is the tunnel junction connecting adjacent p-i-n cells. For example, a tandem cell incorporates one internal tunnel junction at the interface between the n and p layers of the two adjacent component cells. Photogenerated electrons in the top cell and photogenerated holes in the bottom cell are driven to the interconnect junction where all carriers must recombine. If the tunnel junction does not provide enough recombination centers, then a charge will accumulate at this junction, where the field distribution will change resulting in a reverse dipole layer.

A good interconnect junction must allow for very high recombination rates, show an Ohmic characteristic with low series resistance, have negligible optical absorption, and should be easily integrated in the deposition process. The properties of the doped layers can have significant effect on this critical recombination junction. Any parasitic junction loss at the tunnel junction, electrical or optical, leads to deleterious effects on the overall characteristics of the device. Increasing the conductivity of the n and p layers is one way to improve the Ohmic character of the tunnel junction, but this is in contrast with the required high transparency [91]. The use of microcrystalline Si doped layers is an efficient approach [92] because of the reduced optical losses.

Tandem a-Si:H/a-Si:H cells can be fabricated using intrinsic amorphous layers characterized by different energy gap. As already pointed out, the optical gap of intrinsic a-Si:H depends on deposition temperature and gas mixture: An increase of optical gap is obtained when the temperature is lowered and/or hydrogen dilution is added to the gas source [93].

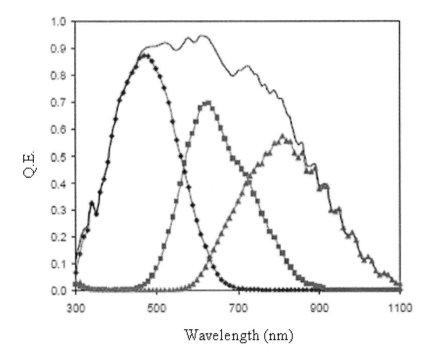

Figure 13. Quantum efficiency curves of the component cells of a triple-junction solar device [90]. The thin black line corresponds to the sum of the three contributions.

Tandem or multijunction devices can be also fabricated using alloys. As reported in section 3.4, when a-Si:H is alloyed with other elements such as Ge or C, amorphous alloy materials with different gaps can be obtained. This allows the selection of appropriate band gap combinations. Since the band gap of the a-SiGe:H alloy can be continuously adjusted between 1.7 and 1.4 eV when different amounts of Ge are incorporated in the alloy, such material can be used as middle and bottom cell absorber layer for a triple junction solar cell.

On the other side, the band gap of a-SiC:H can be adjusted between 1.7 and 2.2 eV, depending mainly on the C concentration [94]. After extensive research, most workers decided that a-SiC:H is not suitable for use as the i-layer of the uppermost cell in a multijunction structure. After light soaking, a-SiC:H material, that has an appreciable band gap increase over a-Si:H, is fairly defective and must be used in very thin layers: these layers would not absorb enough sunlight to allow for the realization of a device.

Tandem a-Si:H/a-SiGe:H and triple a-Si:H/a-SiGe:H/a-SiGe:H junctions with good performance have been fabricated. Among these, a-Si(1.8eV)/a-SiGe(1.6 eV)/a-SiGe(1.4 eV) triple-junction solar cells are at present time among the most efficient thin film silicon based cells (see table 2), and using the a-Si:H/a-SiGe:H/nc-Si:H cell structure, where nc-Si stands for nanocrystalline silicon, at UNI-SOLAR an initial efficiency of 15.4% has been measured [90].

5.8. Micromorph Tandem Solar Cells

A recent multibandgap cell concept is the so called micromorph concept [95], in which an amorphous top cell ($E_G \sim 1.75$ eV) and a microcrystalline silicon bottom cell ($E_G \sim 1.1$ eV) are combined. Apart from the fact that this band-gap combination is almost optimal for the terrestrial solar spectrum [96] (see section 6.3), the advantages of this structure are that it circumvents the difficulties in achieving high quality a-SiGe:H with an optical band gap lower than 1.4 eV and that the microcrystalline component is more stable than a-Si:H based alloys. Due to the low optical energy gap of the microcrystalline material, the bottom cell component enlarges the spectral sensitivity of the tandem cell in the near-infrared region of the AM1.5 solar spectrum (figure 14).

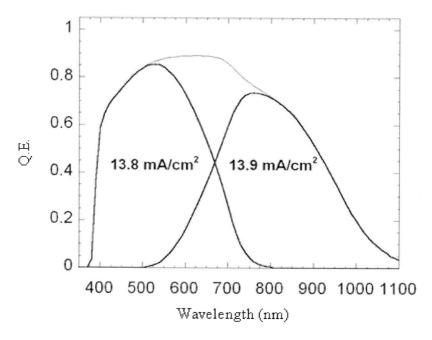

Figure 14. Spectral response curves for the two component cells of a micromorph device [97]. The grey line corresponds to the sum of the two contributions. The short circuit current density for each component cell is indicated.

From the point of view of commercialization, the micromorph tandem structure could allow for a reduction of material costs with respect to the multijunctions that use the a-SiGe alloy, because it does not require the use of the costly GeH_4 gas in the deposition process. On the other side, µc-Si cells require much thicker *i*-layers (2-3 µm) to absorb the sunlight: This is an effect of the lower interband absorption coefficients in (indirect band gap) crystals compared to amorphous semiconductors (see section 3.3).

Due to the high current potential of the microcrystalline bottom cell (> 27 mA/cm^2), particular care should be used to design the two components in order to obtain a good current matching. For example, the use of amorphous silicon deposited without hydrogen dilution could increase the short circuit current by decreasing the optical gap of the material [98].

For the bottom microcrystalline cell, as for the single junction, the structure of the absorber material needs to be optimized. Also in this case the better electrical parameters have been found for intrinsic material grown within the transition region from crystalline to amorphous phase. It has been recently shown that growth regimes characterized by different power to pressure ratio can influence the performance of the devices: low values of this ratio seem to favor a better quality of the p/i interface, and then higher V_{OC} values (see section 6.2), while larger values produce a weaker dependence of J_{SC} and FF on silane concentration, making the deposition regime very interesting for industrial application [99].

A further requirement for the top cell component is to keep the thickness of the a-Si:H intrinsic material reasonably thin to minimize the impact of light-induced degradation. This requirement, however, goes in the opposite direction with respect to the necessity to get high current from the top cell. Therefore, to conserve sufficient current generation capability in the top absorber layer, the insertion of an intermediate reflector between the two cells has been proposed [100]. For an intermediate layer to act as a reflector, its refractive index n must be lower (typically about 2) than that of silicon (n_{Si} = 3.8 at 600 nm). The refractive index jump, then, would cause the reflection of light at the material interface. At the same time, the layer has to be sufficiently conductive to avoid blocking current and as transparent as possible to minimize the current losses due to absorption of light outside the active layers. Several groups have used zinc oxide (ZnO) as intermediate reflector [100-102]. The advantage of ZnO is its low optical absorption and sufficiently low refraction index; the drawback is the rather high conductivity which asks for an additional insulating laser scribe for the monolithic series interconnection, when fabricating modules. Moreover, the deposition of such intermediate layer by sputtering or MOCVD requires the interruption of the PECVD deposition sequence to transfer the devices to a different vacuum chamber. Alternatively, the use of doped silicon oxide has been recently proposed and good results have been shown [103,104]. By applying this material as intermediate reflector in micromorph tandem cell, an initial conversion efficiency of 12.6% (stabilized to 11.1%) has been obtained. The use of silicon oxide combined with the use of anti-reflective coating for the air/glass interface have permitted to reach initial efficiency values higher than 13% [97].

Table 2. Record efficiencies for laboratory cells

Cell type	Organization	Area (cm^2)	V_{OC} (V)	J_{SC} (mA/cm^2)	FF	Efficiency (%)
a-Si/a-SiGe/a-SiGe	United Solar [67]	0.25	2.30	7.56	0.697	12.1*
a-Si/nc-Si/nc-Si	United Solar [14]	0.25	2.01	9.11	0.684	12.5*
a-Si/µc-Si	U. Neuchatel [97]	1.2	1.32	12.5	0.672	11.1*
a-Si/µc-Si	U. Neuchatel [97]	1.2	1.36	13.8	0.708	13.3**
a-Si/ µc-Si$_{1-x}$Ge$_x$	AIST [106]	1.0	1.38	11.1	0.727	11.2**

* Stabilized;
** Stability not investigated.

As it has been already reported, due to the weak infrared absorption in the microcrystalline material, a very thick bottom cell layer has to be formed (2-3 µm) to obtain the photocurrent matching between the two component cells. To deposit these thick microcrystalline layers high deposition rate methods need to be used such us VHF PECVD or HWCVD (illustrated in section 4). Recently, the replacement of µc-Si:H with hydrogenated microcrystalline $Si_{1-x}Ge_x$ (µc-$Si_{1-x}Ge_x$:H) alloys in the bottom cell has been proposed. As a matter of fact, it has been demonstrated that µc-$Si_{1-x}Ge_x$:H (x ~ 0.1-0.2) single junction solar cells provide a significant enhancement in the infrared response with reduced absorber thickness [105]. Using this material as an alternative bottom cell absorber, the bottom cell thickness can be reduced by more than a half while preserving the photocurrent matching with the top cell. An initial efficiency of 11.2% has been obtained using such tandem structure with thickness lower than 1 µm for the µc-$Si_{1-x}Ge_x$:H bottom absorber layer [106]. Finally in table 2 the record efficiencies of multi-junction laboratory cells are reported.

6. THE PHYSICS OF THIN FILM SI-BASED SOLAR CELLS

6.1 Mathematical Description

Device modelling can furnish information that is not available from experiments, such as, to name a part, electric field, concentration of free charge carriers, and recombination rate as a function of the position in the solar cells, thus helping to understand the physics governing device operation and eventually give hints for improvements. The procedure involves the solution of a set of equations which form a mathematical description for device operation, making also use of models that describe the material properties. Only under simple approximations analytical solutions can be furnished. Generally, the solution of the mathematical equations is carried out numerically. A number of sophisticated device program packages have been developed. Schropp and Zeman give a good overview in Part II of their monograph on a-Si:H and µc-Si:H solar cells [107]. The codes are mostly designed for modelling crystalline semiconductor devices, but are continuously updated to adapt to a-Si:H and µc-Si:H based devices such as solar cells and thin film transistors. For these materials special attention has to be paid to modelling the continuous density of states distribution within the bandgap and the recombination-generation statistics of these states. The standard model consists of parabolic conduction and valence bands, exponentially decaying band tails and two Gaussian distributions of states around midgap representing the defect states related to dangling bonds (Figure 2). Mostly, the available codes are limited to one dimensional modelling, which is, in principle, suited for a-Si:H solar cells on flat substrates. A more rigorous description of devices grown on textured substrates and/or employing spatially inhomogeneous materials, like µc-Si:H, would require two dimensional modelling.

The basic semiconductor equations include Poisson's equation (1), the continuity equations for electrons and holes (2 and 3) and the equations for electron and hole current densities (4 and 5). Under steady-state conditions, in one dimension, the equations are:

$$\frac{d}{dx}\left(\varepsilon\frac{d\varphi}{dx}\right) = -\rho \qquad (1)$$

$$\frac{1}{q}\frac{dJ_n}{dx} + G(x) - R(x) = 0 \qquad (2)$$

$$-\frac{1}{q}\frac{dJ_p}{dx} + G(x) - R(x) = 0 \qquad (3)$$

$$J_n = n\,\mu_n\frac{dE_{Fn}}{dx} \qquad (4)$$

$$J_p = p\,\mu_p\frac{dE_{Fp}}{dx} \qquad (5)$$

where ε is the material permittivity, φ is the electrostatic potential referred to the vacuum level[19], ρ is the charge density (including free and localized charge carriers, and ionized donors and acceptors), q is the electric charge, J_n and J_p are electron and hole current densities, G(x) is the optical generation rate, R(x) the net recombination rate resulting from band-to-band recombination and recombination traffic through gap states[20], n and p are the free electron and hole concentrations, μ_n and μ_p are the electron and hole mobilities, and finally E_{Fn} and E_{Fp} are the quasi-Fermi energy levels. The expressions relating the quasi-Fermi energies to the density of mobile electrons and holes are:

$$E_{Fn} = E_C + KT\ln\left(\frac{n}{N_C}\right),\ E_{Fp} = E_V - KT\ln\left(\frac{p}{N_V}\right)$$

With N_C and N_V effective density of states in the conduction and valence bands respectively, and K the Boltzmann constant. We want to remind that the conduction and valence band edges E_C, E_V (mobility edges) correspond to potential energies for electrons and holes respectively, being:

$$E_C = -q\varphi + C, \qquad E_V = -q\varphi - E_G + C \qquad (6)$$

Where E_G is the mobility gap already defined in Section 2 and $C = E_0 - \chi$ is a constant. Here E_0 is the vacuum energy, and χ is the electron affinity, the minimum energy needed to free an electron from the bottom of the conduction band and take it to the vacuum level.

The primary function of a simulator is to solve this set of coupled differential equations accompanied by the appropriate boundary conditions (the two physical boundaries in one dimension are the front and back contacts). Different sets of independent variables can be chosen: most commonly (φ, n, p) or (φ, E_{Fn}, E_{Fp}). One of the main difficulties in a-Si:H device modelling is the large number of parameters required for a detailed description of the material properties to be used as input data. Moreover, due to the inherent nature of the material, its properties may vary from one laboratory to another, and no commonly accepted

---

[19] The vacuum energy E_0 serves as a convenient reference point and is universally constant with position. An electron at the vacuum energy is, by definition, completely free of influence from all external forces.

[20] The traffic back and forth between the delocalized bands and localized gap states is usually described in terms of the Shockley-Read-Hall (S-R-H) capture and emission mechanisms, in more or less sophisticated versions [107].

default values can be defined. The parameter values are usually based on laboratory-dependent measurements complemented with data from literature.

The first comprehensive computer simulation of amorphous silicon based p-i-n solar cells is reported in Ref. [108]. The authors showed that carrier collection is mainly reduced by bulk recombination, and the performance is mostly determined by the transport properties of the limiting carrier, i.e. holes, due to the asymmetry in the density of localized states. The hole being the limiting carrier, they quantitatively demonstrated that solar cells illuminated through the p-layer have a higher efficiency than cells illuminated through the n-layer. Moreover, with illumination through the n-layer the device performance is found to be very sensitive to the cell thickness because the blue response depends on the ability of holes to travel across the entire i-layer without recombining. On the other hand, when illuminating the p side, for strongly absorbed radiation (blue light) the collection of carriers is controlled by electron transport, depending on the ability of the electrons to reach the n-layer without recombining along the way, while for radiation absorbed in the middle of the device the collection is again hole transport controlled.

6.2. Understanding a-Si:H Solar Cell Operation

To discuss the main effects happening within a-Si:H devices[21], the behaviour of the amorphous p-i-n structure schematized in Figure 15(a), illuminated through the p-layer, is here simulated, making use of one of the various available codes[22]. The code solves equations (1-5) in terms of the unknown state variables φ, E_{Fn}, E_{Fp} at each user-defined mesh grid-point in the device. Once these three state variables are obtained as a function of the position within the device, the band edges, electric field, recombination profiles, current densities, and any other transport information can be obtained.

A set of idealized but reasonable parameters has been considered, including appropriate absorption coefficient spectra. In particular, for the mid-gap states only one Gaussian level has been considered, assuming uniform density throughout the entire layer. The device current-voltage characteristics, calculated for two different values of total dangling bond density in the material, shown in Figure 15(b), are very close to real behaviour, thus validating the input values. Note that the short circuit current is rather low, since no back reflector or texturing effects have been included in the simulation. It is evident that an increase of the density of defect states in the intrinsic layer deteriorates the performance of the cell, clearly explaining the continuous effort at producing a material with low defect level.

The most frequently used tools for understanding semiconductor device physics are the bandedge plus Fermi level/quasi-Fermi level profiles. In figure 16 the band diagrams at open-circuit in equilibrium conditions (left panel) and under illumination (right panel) are illustrated. In isolation, *p*-type and *n*-type materials have very different Fermi energies. When the *pin* device is assembled, these Fermi energies must be equalized to create thermal equilibrium. Electrons are donated from the *n*-layer to the *p*-layer, generating a built-in electric field. The space-charge distribution is determined by ionized dopants and trapped

---

[21] The discussion roughly holds also for microcrystalline silicon devices.
[22] Here we used AMPS-1D©, a free general computer simulation tool for Analysis of Microelectronic and Photonic Structures (AMPS) developed at The Pennsylvania State University.

charges, and in the intrinsic layer mostly lies close to the p-i and i-n interfaces. While the Fermi energy itself is constant, the electric field within the device causes all electron energy level such as E_C and E_V, to vary in space, with bandedge gradients directly related to the internal electric field (eq. 6). The original difference in Fermi energies becomes the "built-in energy potential" eV_{BI} across the device illustrated in figure 16 (left panel). In a first approximation an average field, equal to V_{BI}/d_i, with d_i the i-layer thickness, is present between the doped layers. In this field, electrons and holes will drift in the directions shown in the figure.

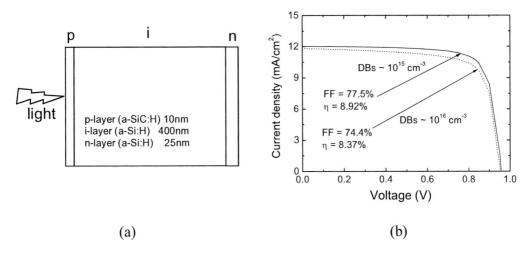

Figure 15. (a) Schematic structure of the analyzed p-i-n device. (b) Illuminated current-voltage characteristic numerically calculated by using the procedure described in the text, adopting two possible values of defect state density (solid line: DBs ~ 10^{15} cm^{-3}, dashed line: DBs ~ 10^{16} cm^{-3}).

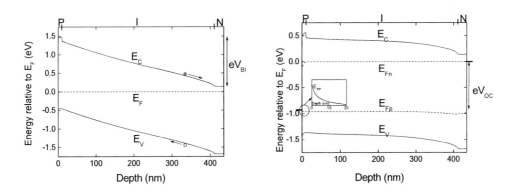

Figure 16. Bandedge and Fermi-level/quasi-Fermi-level profiles at open-circuit at thermodynamic equilibrium (left panel) and under illumination (right panel). The built-in potential V_{BI} and open-circuit voltage V_{OC} are also illustrated. Note that the *p*-layer has a slightly larger band gap than the *i*-layer.

Note that the profiles were calculated assuming that the *p*-layer has a larger band gap than both intrinsic layer and *n*-layer. The use of a wider band gap *p*-layer allows both to increase V_{BI}, improving the collection efficiency, and to reduce the optical absorption in this layer. Note also that the *i*-layer thickness affects the photocurrent collection negatively in two ways: by increasing the recombination probability, since the carriers have to travel over longer distances, and by reducing the driving force of charge carrier separation. For this reason the use of light trapping schemes in thin film Si-based solar cells is crucial.

When illuminating the device, the generation of electron-hole pairs produced by the photon absorption does not allow to use the equilibrium Fermi distribution. Due to the increased electron density, the Fermi energy describing their distribution in the conduction band must be closer to the conduction band than in the dark. Similarly, due to the increased hole density, the Fermi energy describing their distribution in the valence band must be closer to the valence band. The quasi-Fermi energies, E_{Fn} and E_{Fp}, are then used to describe the occupation of states in conduction and valence bands.

The spatial gradient of the quasi-Fermi levels is the driving force of electron and hole current flows (eqs. 4 and 5). At thermodynamic equilibrium, $E_{Fe} = E_{Fp} = E_F = constant$ and no current flows in the device. The situation changes under illumination. The right panel of figure 16 shows the band diagram of the illuminated device at open circuit. In this condition the quasi-Fermi levels are nearly constant across the cell, showing variations only close to the edges, where the electron quasi-Fermi level catches down to E_{Fp}, on the p-side, and the hole quasi-Fermi level catches up to E_{Fe}, on the n-side. As a result of surface recombination, the Fermi energies for the conduction and valence bands, which are different inside the semiconductor, merge into a single Fermi energy at the surface (one value on the left and a different value on the right, evidenced by the thick segments in Figure 16 – right panel), fixing the carrier concentrations at the surfaces at different values. Therefore, despite the presence of light, an ordinary Fermi level can be defined at these edges, and the difference between the two Fermi levels defines the product eV_{OC}. In this sense it is said that solar cells act like ordinary batteries, which also maintain different Fermi levels at the two terminals. The difference in quasi-Fermi levels in the middle of the cell defines the maximum V_{OC} allowed by the absorber layer (its intrinsic limit). However, the difference in Fermi energies at the terminals is generally lower than the quasi-Fermi level separation within the i-layer (inset to Figure 16 – right panel). In real devices the non ideal p-i interface, where a hole quasi-Fermi level gradient can be observed, often reduces V_{OC} below its intrinsic limit. The quality of such interface region (through which light enters) is a critical factor that needs particular attention.

The built-in voltage is distributed non-uniformly over the intrinsic layer. While for rough estimates an effective electric field $E_{eff} = (V_{BI}-V)/d_i$ can be used, the electric field profile is highly nonuniform, as shown in figure 17 under illumination at different voltages. The field strength is obviously dependent on the doping level in the n- and p-layers, with major effect especially close to the p-i and i-n interfaces (dashed line in Figure 17). In general, the high space-charge densities in the doped layers cause large interface fields, the order of $10^4 - 10^5$ V/cm, while values around $10^3 - 10^4$ V/cm are found in the bulk, ensuring that carrier transport is a drift process in the bulk of the device. At V>0, or equivalently in presence of forward bias at the terminals, such voltage is virtually all dropped across the bulk of the intrinsic layer: the bulk field strength decreases, while the interface field strengths are more o less independent of bias.

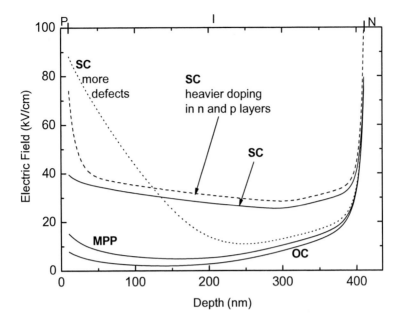

Figure 17. Electric field strength profile under illumination at different voltages (solid lines). Here SC, MPP, and OC stand for short circuit, maximum power point, and open circuit, respectively. At short circuit, the electric field profile in case of doped layers characterized by heavier doping (dashed line), and in case of increased defect state density in the intrinsic material is also shown (dotted line).

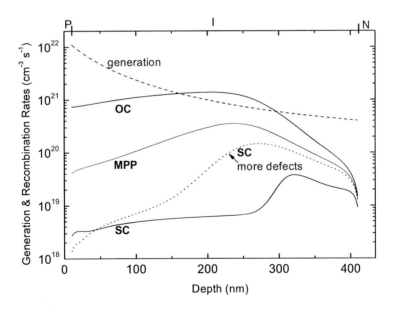

Figure 18. Generation rate profile in case of flat interfaces and no back reflector (dashed line), and recombination rate profiles at different voltages (solid lines). Here SC, MPP, and OC stand for short circuit, maximum power point, and open circuit, respectively At short circuit, the recombination rate profile in case of increased defect state density (one order of magnitude more) in the intrinsic material is also shown (dotted line).

It should be noted that an increased defect density (as in the light induced degraded state) leads to a reduced electric field (dotted line in Figure 17), which results in poorer carrier collection. The reduced collection is associated with an increased recombination rate, as shown in figure 18. Here the generation rate profile (in this example flat interfaces and no back reflector are considered) is shown together with the recombination rate profiles for different voltage values. The regions of increased recombination roughly correspond to the regions with lower field strength. It should be pointed out that at low voltage the highest-recombination/lowest-field region is positioned close to the i-n interface. With light entering through the p-layer, and generation being stronger close to the p-i interface, the collection depends on the ability of electrons (in large number generated close to the p-side) to travel across the entire i-layer and finally reach the n-layer. In presence of rather strong electric field a significant amount of such carriers will recombine only towards the end of the ride. Thus, a proper design of the i-n interface region, and not only of the p-i interface mentioned above, should be adopted to improve collection.

At V > 0 an increase of the recombination traffic, with the photocarriers drifting in a less intense electric field, is observed. In particular at V_{OC} the electric field is very weak and a strong recombination is active. For cells that have attained the intrinsic limit, these fundamental processes determine V_{OC}. The loss in current due to increased recombination results in the collected photocurrent in a-Si:H solar cells being bias dependent. Therefore an enhanced curvature of the illuminated current-voltage characteristic is observed with respect to c-Si solar cells. This is the intrinsic reason explaining why fill factors of a-Si:H devices are always lower than their crystalline counterpart.

6.3. Semi-empirical Limits for the Device Parameters

The presence of a bias-dependent recombination channel asks for a revision of the equivalent circuit generally used for photovoltaic solar cells, essentially consisting of a current source shunted by a diode plus parallel and series "parasitic" resistances, introduced to describe the behaviour of real solar cells with their technical limitations. To describe thin film silicon-based cells an additional current sink (J_{rec}), describing the recombination losses, needs to be included (Figure 19). Recombination losses, besides being voltage dependent, are also in a first approximation proportional to the carrier concentrations, and thus to the photogenerated current, as theoretically demonstrated in ref. [109].

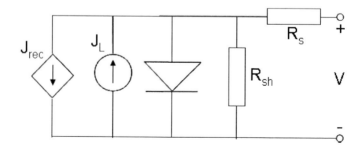

Figure 19. Equivalent circuit for thin film silicon-based photovoltaic solar cells.

Based on this equivalent circuit, the illuminated J(V) characteristic can be written as:

$$J(J_L,V) = J_L - J_{rec}(J_L,V) - J_0\left(e^{\frac{qV}{nKT}} - 1\right)$$

where J_L is the photogenerated current, J_0 is the reverse saturation current, q is the elementary charge, K the Boltzmann constant, T the absolute temperature and n the ideality factor (usually, $n \approx 2$ for p-i-n junctions). It should be kept in mind that, due to the presence of dangling bonds, the thermal generation from mid-gap states contributes as an additional term to the dark current and then to J_0.

Starting from semi-empirical considerations, fundamental limits for the device parameters have been evaluated in ref. [96] as a function of the energy gap for thin film silicon-based p-i-n solar cells. The upper limit on J_{SC} is computed by considering the normalized AM 1.5 spectrum and assuming that all photons with hv>E_G are absorbed, converted into electron–hole pairs, and collected (which is, in principle, possible at short circuit conditions). Lower limiting values have been estimated regarding V_{OC}, FF and η, when compared to the usually reported limits valid for ordered semiconductors. The extrapolations from the curves in case of a-Si:H (with $E_G \sim 1.75$ eV) and μc-Si:H ($E_G \sim 1.1$ eV) are reported in Table 3 together with the limits on J_{SC}. Comparing these values with the experimental results reported in Table 1, it appears that major improvements should be attempted in order to reach the limits on J_{SC}, mainly by working on the light trapping schemes.

Table 3. Semi-empirical limits for the photovoltaic parameters of amorphous silicon and microcrystalline silicon solar cells [96]

	a-Si:H	μc-Si:H
J_{SC} (mA/cm^2)	21.1	43.6
V_{OC} (V)	1.25	0.65
FF (%)	82	78
η (%)	24	25

In the same study, efficiency limits for tandem solar cells were also computed considering:
(1) a perfect balance between top and bottom cell current densities,
(2) the idealized case where the bottom cell absorbs all the light transmitted by the top cell.

Such limits are shown in figure 20 as contour plot of constant solar conversion efficiency for varying bandgaps of top and bottom cell. From this estimate the "micromorph" combination of a-Si:H in the top cell and μc-Si:H in the bottom cell is found to be very close to ideal, with an efficiency limit of 35%.

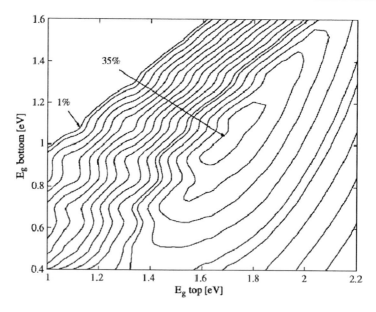

Figure 20. Upper limit of the efficiency of p–i–n/p–i–n tandem solar cells as a function of the energy gaps of top and bottom cells (E_g top and E_g bottom). [Reprinted from Solar Energy Materials & Solar Cells 90, F. Meillaud et al., "Efficiency limits for single-junction and tandem solar cells", pp 2952–2959, Copyright (2006), with permission from Elsevier].

7. MODULES AND COMMERCIAL PRODUCTS

7.1 Fabrication of Large Area Modules

In order to produce feasible currents and voltages and reduce the series resistance losses that occur upon scaling up the cell area, most commercial silicon thin film modules are constructed with monolithically interconnected cells. The series connection of the individual cells on the same substrate is achieved including three laser patterning steps in order to obtain the structure shown in figure 21, where the electrode coupling region between two subsequent cells is schematised. For scribing the TCO front electrode, the appropriate wavelength is $\lambda = 1064$ nm. For removing the silicon single junction or tandem structure, and the metal back electrode, the wavelength $\lambda = 523$ nm is used. The laser cuttings selectively remove narrow regions (50 - 150 μm) of a specific layer, ideally without affecting the previously deposited layers. An accurate alignment of the laser and, in order to prevent edge damage along the patterned zone, a tight control of the laser parameters are required to obtain good module performance. Furthermore, a high precision of the laser system allows for reducing the dead area losses down to the basic limitations of the individual scribe width, and, thus, simply improves the efficiency of the modules by a better structuration.

It is interesting to note that the output power of modules is increased in multijunction technology with respect to single junctions even if the cell efficiency is the same, merely by reducing the interconnection area losses. This is because the current density generated in multijunction cells is lower than in single junction cells of the same efficiency. This allows for the module to be constructed from wider subcells, so that the number of interconnection zones, and thus the inactive area, is reduced.

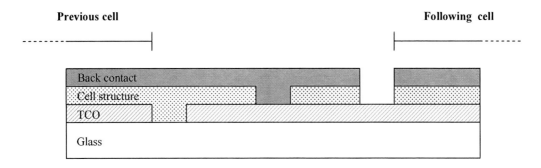

Figure 21. Cell interconnection of superstrate-type solar cells to form a module.

Comparing typical small area laboratory cell efficiency and equivalent large area monolithically interconnected prototype panel efficiencies, a gap in performance can be observed. Different issues can contribute to this gap. Together with the interconnection area losses due to the module fabrication itself, the laser scribing could cause interconnect contact series resistance losses due to incomplete material removal or re-deposited debris, or local shunting due to edge damage along the patterned zone. The gap of module performance could be also due to nonuniformity of different deposition steps. As a matter of fact, any local reduction of photovoltaic parameters (V_{OC}, J_{SC}, or FF) in the string of subcells will affect the entire panel. For example the built-in voltage (and thus V_{OC} and FF) is very sensitive to p-layer nonuniformities, and also J_{SC} can be reduced if the p layer becomes too thick. Specifically in multijunction cells, thickness non-uniformity in the intrinsic layers can adversely affect the current matching properties and thus reduce the overall current output. Finally, both in single and multijunction devices, nonuniformity in the TCO layer can locally alter the transmission or the light trapping capability. Irregularities in the morphology of the TCO can also generate local shunting paths in the device.

Considering the different issues that can cause potential losses when scaling up, improving the quality of the laser patterning is only one of the necessary approaches to decrease the performance gap between small area cells and large area modules.

In substrate modules realized on conducting carriers, by definition, monolithic interconnection cannot be implemented. For example, when a stainless steel roll is used as a substrate, the TCO plus junction coated roll is first cut into slabs of selected sizes, then different numbers of strip cells are connected together in series.

Finally, module encapsulation is a critical step in order to avoid water penetration that could cause corrosion or delamination of contact layers. Superstrate modules fabricated on glass are protected on the window side of the panel. Water vapour or even ions originating from salt do not pass through a few mm thick glass and the glass protects from environmental effects. A second piece of glass is also the most effective way to incapsulate the rear side. It can be attached to the module using a resin, generally EVA (ethyl-vinyl-acetate copolymer). The two glass panel edges are hermetically joined in order to protect the interior of the module.

Substrate modules on conducting carriers as well as on plastic foils have the drawback that they need to be encapsulated at both faces. The use of quite expensive Tefzel (combined with EVA) is needed on the window side of the modules, because this polymer is transparent and resistant to UV exposition.

Table 4 shows some of the highest stabilized efficiencies obtained for R&D prototype modules.

Table 4. Stabilized efficiencies of R&D prototype modules

Cell type	Organization	Area (cm^2)	Efficiency (%)
a-Si/a-SiGe	Sanyo [110]	1200	9.5
a-Si/-a-SiGe/a-SiGe	United Solar [111]	905	10.4
a-Si/µc-Si	Oerlikon [112]	62.3	9.94

7.2. Production and Market for Thin Film Silicon Solar Cells

In 2009, the photovoltaic industry continued its growth and delivered worldwide about 9340 MWp of photovoltaic generators growing by almost 35% in comparison with the previous year. The thin film segment, starting from a very low level, grew by almost 80% and reached 196 MW or 8% of total PV production in 2006, with about 100 MW constituted of thin film silicon modules. During 2009 thin film production has reached 18% of the total module production [1].

The high growth rate of thin film production and the increase of the total production share indicate that the thin film technology is gaining more and more acceptance. A thin film market share of 25% in 2012 seems not to be unrealistic. This takes into account the fact that more and more PV manufacturers, among them the current market leader Q-Cells, are diversifying their production portfolio. Moreover, significant expansion of thin film production capacity of existing manufacturers is under way and a large number of new manufacturers is entering the market to answer to the growing demand for PV modules.

The huge investments launched on thin film PV based on CdTe, CIS/CIGS, and silicon find their basis on the higher cost reduction potential of these technologies with respect to conventional wafer-based PV [113,114]. Such cost reduction potential, however, will be demonstrated only when the production volume increases significantly. Up to now the main advantage of the wafer-based crystalline silicon technology was that complete production lines could be bought, installed and be up and producing within a relatively short time-frame. This predictable production start-up scenario constitutes a low-risk placement with high expectations for return on investments. However, the market entry of companies offering turn-key production lines for thin film solar cells is now leading to a massive expansion of investments into the thin film sector.

In 2006, WTC, a consulting company based in Munich, Germany, has identified more than 130 companies involved in the thin film solar cell production process, ranging from R&D activities to major manufacturing plants. The majority of these companies are silicon based. The reason is probably that, in the meantime, several companies started offering complete turn-key production lines for amorphous and/or micromorph silicon solar cells.

Oerlikon Solar and Applied Material, the leaders among such companies, in particular, are developing the technology to produce thin film silicon-based large area modules. Both companies have realized micromorph prototype modules on 1.4 m^2 area with efficiency of ~10% [115,116]. Applied Material is also scaling up the technology to much larger area (5.7 m^2) and prototype modules have recently been shown [117]. The present status and future goal for thin film silicon technology performance has been summarized in the roadmap presented by the U.S. Department of Energy in 2008 [118]. Table 5 highlights relevant items.

Table 5. Thin-Film Silicon Technology Performance [118]

Parameter	Present Status (2007)	Future Goal (2015)
Production volume	100 MW/yr	>5 GW/yr
Equipment and material cost	1–2 $/W	0.7 $/W
Substrate cost	12–20 $/m^2	4 $/m^2
Module manufacturing cost for a-Si	125–200 $/m^2	70 $/m^2
Commercial a-Si modules stabilized efficiency	5%–8%	10%–13%

8. CONCLUSION

Over the last quarter of a century, significant progress has been made in the understanding of properties and of deposition processes for amorphous and microcrystalline silicon materials and solar cells. There have been impressive achievements both in increasing the conversion efficiency of solar cells and in reducing the fabrication costs. In 1997, a-Si:H-based multijunction solar cells with 12.1% stable efficiency were demonstrated [14], and an initial efficiency of 15.4% has been recently reported [90]. Additionally, the use of microcrystalline silicon as the narrow band gap absorber layer in tandem solar cells has been demonstrated, and micromorph cells, consisting of an amorphous top cell and a microcrystalline bottom cell, with initial efficiency higher than 13% have been realized [97]. Due to the low optical energy gap of microcrystalline silicon, the use of this material in combination with a-Si:H allows to enlarge the spectral sensitivity of the tandem cell in the near-infrared region of the AM1.5 solar spectrum. Moreover, cells incorporating microcrystalline material show superior light stability over extended light soaking.

The advantages of the thin film Si PV technology are significant. The thin film silicon product is made through a low-cost process: amorphous and microcrystalline silicon are deposited using the PECVD technique at low temperature (150-250°C) on inexpensive substrates, such as glass, plastics, and stainless steel. Significant progress has been made in the development of rapid deposition processes (VHF-PECV, HWCVD) aiming at obtaining device quality material with growth rate above 5 Å/s. The use of low temperature processes allows to reduce both the production costs and the energy payback time (the time required for a module to generate the energy used in its production) with respect to the wafer-based crystalline silicon technology. When deposited on selected substrates, the product can be made lightweight and flexible, which is important for many applications. Due to the positive effect of higher working temperatures on a-Si:H modules, they may even outperform modules based on crystalline silicon in terms of annual energy yield divided by the module nominal power.

At production level, the thin film segment, starting from a very low level, grew by almost 80% and reached 196 MW (100 MW silicon based) or 8% of total PV production in 2006. During 2008, the thin film production has exceeded 10% of the total module production [113]. Compared with other thin-film PV technologies, such as CdTe and CIGS PV technologies that have demonstrated higher efficiency in small-area R&D type cells, a-Si:H photovoltaics looks attractive because (1) it has been developed for approximately 20 years and the production process is more mature and proven and (2) the product does not contain any hazardous material, such as Cd in CdTe-based photovoltaics, or materials with availability problems, and then costly, like In in CIGS-based PV, or costly in perspective, like Te in the CdTe case (the materials in thin film silicon–based cells originate, instead, in raw materials that are abundant on earth).

To increase the application of the thin film silicon-based PV significantly beyond today's level and to further decrease the costs of the technology, better stable efficiencies should be obtained. A stable efficiency of 14% using multi-junction solar cells can be reasonably envisioned in the near future. In order to give solid grounds to the thin film silicon-based PV, the following issues are critical and must be addressed:
- critical assessment of all tandem component layers to identify the most important limitations and derive optimal designs;
- development of advanced light trapping schemes including improved back and intermediate reflectors, optimized TCO, and possible antireflection coatings;
- identification of optimal growth regimes for a-Si:H and µc-Si:H, with faster deposition rates (>10Å/s) and high gas utilization that (at least) preserve the conversion efficiencies achieved by present processes (this is critical for low-cost and high throughput manufacturing);
- technical or cost-related improvements in module design, framing, and encapsulation;
- exploration of new applications for thin film silicon PV products in all of its present markets, including building-integrated PV, space power, and consumer electronics as well as grid-connected large-scale power generation.

As these critical issues are successfully addressed, with the growing demand of PV products, we expect that thin film Si-based solar cells will have appreciably lower costs (even below 0.5 €/Wp), the right figures to boost the PV market.

REFERENCES

[1] Marketbuzz 2010: Annual World Solar Photovoltaic *Industry Report*, (http://www.solarbuzz.com).
[2] Shah, A.V.; Schade, H.; Vanecek, M.; et al. *Prog Photovolt. Res. Appl.* 2004, 12, 113–42.
[3] Kroll, U.; Meier, J.; Benagli, S.; et al. *Proceedings of the 22nd European photovoltaic solar energy conference, Milan,* 2007; pp 1795–1800.

[4] Chae, Y.; Won, T.K.; Li, L.; et al. *Proceedings of the 22nd European photovoltaic solar energy conference, Milan*, 2007; pp 1807–1809.
[5] Takatsuka, H.; Yamauchi, Y.; Fukagawa, M.; et al. *Proceedings of the 21st European photovoltaic solar energy conference, Dresden*, 2006; pp 1531–1534.
[6] Chittick, R.; Sterling, H.; in Adler, D., Fritzsche, H. Eds. *Tetrahedrally Bonded Amorphous Semiconductors*, Plenum Press: New York, NY, 1985; pp 1–11.
[7] Fritzsche, H. *Mater. Res. Soc. Symp. Proc.* 2001, 609, A17.1.1–12.
[8] Spear,W.; LeComber, P. *Solid State Commun.* 1975, 17, 1193.
[9] Carlson, D.; Wronski, C. *Appl. Phys. Lett.* 1976, 28, 671.
[10] Catalano, A.; D'Aiello, R.; Dresner, J.; Faughnan, B.; Firester, A.; Kane, J.; Schade, H.; Smith, Z.E.; Swartz, G.; Triano, A. *Proceedings of the 16th IEEE Photovoltaic Specialists Conference, San Diego*, 1982; pp 1421–1422.
[11] Staebler, D.L.; Wronski, C.R. *Applied Physics Letters* 1977, 31, 292–294.
[12] Meier, J.; Dubail, S.; Fluckiger, R.; Fischer, D.; Keppner, H.; Shah, A. *Proceedings of the 1st World Conference on Photovoltaic Energy Conversion (WCPEC1), Hawaii*, 1994; pp 409.
[13] Veprek, S.; Marecek, V.; Anna Selvan, J.A. *Solid State Electronics* 1968, 11, 683–684.
[14] Yan, B.; Yue, G.; Guha, S. *Mater. Res. Soc. Symp. Proc.* 2007, 0989-A15-01.
[15] Brodsky, M. H.; Title, R. S. *Phys. Rev. Lett.* 1969, 23, 581–585.
[16] Hishikawa, Y. *J. Appl. Phys.* 1987, 62, 3150–3155.
[17] Molenbroek, E. C.; Mahan, A. H.; Gallagher, A. *J. Appl. Phys.* 1997, 82, 1909–1917.
[18] Acco, S.; Williamson, D. L.; Stolk, P. A.; et al. *Phys. Rev. B* 1996, 53, 4415–4427.
[19] Acco, S.; Sark, W. G. J. H. M. V.; der Weg, W. F. V.; Williamson, D. L.; Roorda, S.; Polman, A.; *J. Non-Cryst. Solids* 1998, 227-230, 128–132.
[20] Vanecek, M.; Kocka, J.; Stuchlık, J.; Trıska, A. *Solid State Commun.* 1981, 39, 1199–1202.
[21] Jackson, W. B.; Amer, N. M. *Phys. Rev. B* 1982, 25, 5559–5562.
[22] R.A. Street, *Hydrogenated amorphous silicon*, Cambridge Univ. Press, 1991.
[23] Ley, L. *J. Non-Cryst. Solids* 1989, 114, 238.
[24] Jackson, W.; Kelso, S.; Tsai, C.; Allen, J.; Oh, S. *Phys. Rev. B* 1985, 31, 5187.
[25] Cody, G.; Tiedje, T.; Abeles, B.; Brooks, B.; Goldstein, Y. *Phys. Rev. Lett.* 1981, 47, 1480.
[26] Tiedje, T. In Joannopoulos, J.; Lucovsky, G. Eds. *Hydrogenated Amorphous Silicon II*, Springer-Verlag: New York, 1984; pp 261–300.
[27] Tauc, J. In Abeles, F. Ed. *Optical Properties of Solids*, North Holland: Amsterdam, 1972; pp 277–313.
[28] Mott, N. *Conduction in Non-Crystalline Solids*, Oxford University Press: Oxford, 1987.
[29] Chen, I.; Wronski, C. *J. Non-Cryst. Solids* 1995, 190, 58.
[30] Antoniadis, H. ; Schiff, E. *Phys. Rev. B* 1992, 46, 9482–9492.
[31] Han, D.; Melcher, D.; Schiff, E.; Silver, M. *Phys. Rev. B* 1993, 48, 8658.
[32] Lee, J.; Schiff, E. *Phys. Rev. Lett.* 1992, 68, 2972.
[33] Mott, N.F.; Davis, E.A. In Marshall, W.; Wilkinson, D.H. Ed. *Electronic process in non-crystalline materials*, 2nd ed, The international series of monographs on Physics, Clarendon Press: Oxford, 1979.
[34] Dersch, H.; Stuke, J.; Beichler, J. *Appl. Phys. Lett.* 1981, 38, 456–458.
[35] Stutzmann, M.; Jackson, W. B.; Tsai, C. C. *Phys. Rev. B* 1985, 32, 23–47.

[36] Santos, P. V.; Johnson, N. M.; Street, R. A. *Phys. Rev. Lett.* 1991, 67, 2686–2689.
[37] Carlson, D. E.; Rajan, K. J. *Appl. Phys.* 1998, 83, 1726–1729.
[38] Godet, C. J. *Non–Cryst. Solids* 1998, 227-230, 272–275.
[39] Kroll, U.; et al. *J. Appl. Phys.* 1996, 80, 4971.
[40] Bailat, J. ; et al. *J. Non-Cryst. Solids* 2002, 299-302, 1219-1223.
[41] Delli Veneri, P.; Mercaldo, L.V.; Tassini, P.; Privato, C. *Thin Solid Films* 2005, 487, 174–178.
[42] Droz, C.; Vallat-Sauvain, E.; Bailat, J.; Feitknecht, L.; Shah, A. *Proc. 7th European Photovoltaic Solar Energy Conference, WIP, Munich*, 2001, pp 2917.
[43] Poruba, A.; Fejfar, A.; Remes, Z.; et al. *J. Appl. Phys.* 2000, 88, 148 - 159.
[44] Bass, M. Ed. *Handbook of Optics*, 2nd ed., McGraw-Hill: New York, 1995.
[45] Guha, S.; Payson, J.; Agarwal, S.; Ovshinsky, S. *J. Non-Cryst. Solids* 1987, 97–98, 1455.
[46] Chittick, R.; Alexander, J.; Sterling, H. *J. Electrochem. Soc.* 1969, 116, 77–81.
[47] Spear, W.; LeComber, P. *J. Non-Cryst. Solids* 1972, 8–10, 727–738.
[48] Gallagher, A. *J. Appl. Phys.* 1988, 63, 2406–2413.
[49] Brodsky, M.H. *Thin Solid Films* 1977, 40, L23-L25.
[50] Hama, S.; Okamoto, H.; Hamakawa, Y.; Matsubara, T. *J. Non-Cryst. Solids* 1983, 59–60, 333.
[51] Ueda, M.; Imura, T.; Osaka, Y. *Proc. 10th Symp. on Ion Sources and Ion-Assisted Technology* (1986).
[52] Platz, R.; Hof, C.; Fischer, D.; Meier, J.; Shah A. *Solar Energy Mater. Solar Cells* 1998, 53, 1.
[53] Guha, S.; Narasimhan, K. L.; Pietruszko, S. M. *J. Appl.Phys.* 1981, 52, 859.
[54] Vetterl, O.; et al. *Sol. Energy Mater. Sol. Cells* 2000, 62 ,97.
[55] Rath, J.K. *Solar Energy Mater. Solar Cells* 2003, 76, 431–487.
[56] Roca i Cabarrocas, P.; Nguyen-Tran, Th.; Djeridane, Y.; et al. J. Phys. D: *Appl. Phys.* 2007, 40, 2258–2266.
[57] Curtins, H.; Wyrsch, N.; Shah, A. *Electron. Lett.* 1987, 23, 228–230.
[58] Meier, J.; Kroll, U.; Vallat-Sauvain, E.; Spitznagel, J.; Graf, U.; Shah, A. *Solar Energy* 2004, 77, 983–993.
[59] Heintze, M.; Zedlitz, R. *J. Non-Cryst. Solids* 1996, 198-200, 1038-1041.
[60] Wiesmann, H.; Gosh A. K.; McMahon T.; Strongin M. *J. Appl. Phys.* 1979, 50, 3752.
[61] Matsumura, H. *J. Appl. Phys.* 1989, 65, 4396-4402.
[62] Mahan, H. ; Nelson, B.P. ; Salamon, S. ; Crandall, R.S. *J. Non-Cryst. Solids* 1991, 137-138, 657-660.
[63] Schropp, R.E.I. *Thin Solid Films* 2002, 403 –404, 17–25.
[64] Takeuchi, Y.; Takano, A.; Mashima, H.; Kawamura, K.; Yamauchi, Y.; Takatsuka, H. *Proceedings of the 19th European Photovoltaic Solar Energy Conference and Exhibition, Paris* 2004; pp 1378.
[65] I. Balber and Y. Lubianiker, *Phys. Rev. B48*(12) (1993), 8709-8714.
[66] Carasco, F.; Spear, W. Philos. Mag. B 1983, 47, 495; Schiff, E. *J. Non-Cryst. Solids* 1995, 190, 1.
[67] Yang, J.; Banerjee, A. ; Guha, S. *Appl. Phys. Lett.* 1997, 70, 2975.
[68] Gordon, R.; Proscia, J.; Ellis, F.B.; Delaoy, A.E. *Solar Energy Mater.* 1989, 18, 263.
[69] Mizuhashi, M.; Gotoh, Y.; Adachi, K. *Jpn. J. Appl. Phys.* 1988, 27, 2053.

[70] Drevillon, B. *J. Non-Cryst. Solids* 1989, 114, 139.
[71] Kluth, O.; Löffl, A.; Wieder, S.; et al. *Proceedings of the 26th IEEE Photovoltaic Specialists Conference, Anaheim,* CA, 1997; pp 715–718.
[72] Rech, B.; Wieder, S.; Beneking, C.; Löffl, A.; Kluth, O.; Reetz, W.; Wagner, H. *Proceedings of the 26th IEEE Photovoltaic Specialists Conference, Anaheim*, CA, 1997; pp 619–622.
[73] Benagli, S.; Hoetzel, J.; Borrelo, D.; et al. *Proceedings of 23rd European photovoltaic solar energy conference, Valencia,* 2008, pp 2414.
[74] Ballif, C.; Bailat, J.; Dominé, D.; Steinhauser, J.; Faÿ, S.; Python, M.; Feitknecht, L. *Proceedings of 21st European photovoltaic solar energy conference, Dresden,* 2006; pp 1552.
[75] Addonizio, M.L.; Manoj, R.; Usatii, I. *Proceedings of 22nd European photovoltaic solar energy conference, Milan,* 2007; pp 2129.
[76] Beck, N.; Wyrsch, N.; Hof, C.; Shah A. *J. Appl. Phys.* 1996, 79, 9361.
[77] Daudrix, V.; Droz, C.; Wyrsch, N.; Ziegler, Y.; Niquille, X.; Shah, A. *Proceedings of the 16th European photovoltaic solar energy conference,* 2000; pp 385.
[78] Droz, C.; Goerlitzer, M.; Wyrsch, N.; Shah, A. *J. Non-Cryst. Solids* 2000, 266–269, 319.
[79] Meier, J.; Torres, P.; Platz, R.; et al. *Proceedings of the Materials Research Society Symposium, 420,* 1996; pp 3.
[80] Delli Veneri, P.; Mercaldo, L. V.; Bobeico, E.; Spinillo, P.; Privato, C. *Proceedings of 19th European Photovoltaic Solar Energy Conference, Paris,* 2004; pp 1469.
[81] Tawada, Y.; Okamoto, H.; Hamakawa, Y. *Appl. Phys. Lett.* 1981, 39, 237.
[82] Arya, R.R.; Catalano, A.; Oswald, R.S. *Appl. Phys. Lett.* 1986, 49, 1089.
[83] Hamma, S.; Roca i Cabarrocas, P. *Sol. Energy Mater. Sol. Cells* 2001, 69, 217.
[84] Shah, A.V.; Meier, J.; Vallat-Sauvain, E.; Droz, C.; Kroll, U.; Wyrsch, N.; Guillet, J.; Graf, U. *Thin Solid Films* 2002, 403–404, 179–187.
[85] Benagli, S.; et al. *Proceedings of 24th European Photovoltaic Solar Energy Conference, Hamburg,* 2009; pp 2293.
[86] Yamamoto, K.; Toshimi, M.; Suzuki, T.; Tawada, Y.; Okamoto, T.; Nakajima, A. *Proceedings MRS Spring Meeting, San Francisco,* April, 1998.
[87] Guha, S.; Narasimhan, K. L.; Pietruszko, S. M. *J. Appl.Phys.* 1981, 52, 859.
[88] Meillaud, F.; Shah, A.; Vallat-Sauvain, E.; Niquille, X.; Dubey, M.; Ballif, C. *Proceedings of the 20th European Photovoltaic Solar Energy Conference, Barcelona,* 2005; pp 1509.
[89] Yan, B.; Yue, G.; Owens, J. M.; Yang, J.; Guha, S. *Appl. Phys. Lett.* 2004, 85, 1925.
[90] Yan, B.; Yue, G.; Yang, J.; Guha, S. *Proc. of 33rd IEEE Photovoltaic Specialists Conf., San Diego,* 2008.
[91] Shen, D. S.; Schropp, R. E. I.; Chatham, H.; Hollingsworth, R. E.; Bhat, P. K.; Xi, J. *Appl. Phys. Lett.* 1990, 56, 1871.
[92] Hegedus, S. S.; Kampas, F.; Xi, J. *Appl. Phys. Lett.*1995, 67, 813.
[93] Platz, R.; Fischer, D.; Dubail, S.; Shah, A. *Solar Energy Mater. Solar Cells* 1997, 46, 157.
[94] Hamakawa, Y.; Tawada, Y.; Nishimura, K.; Tsuge, K.; Kondo, M.; Fujimoto, K.; Nonomura, S.; Okamoto, H. *Conference Record of the 16th IEEE Photovoltaic Specialists Conference,* 1982; pp 679–684.

[95] Meier, J.; Dubail, S.; Flückiger, R.; Fischer, D.; Keppner, H.; Shah, A. *Proceedings of 1st World Conference on Photovoltaic Energy Conversion*, 1994; pp 409.

[96] Meillaud, F.; Shah, A.; Droz, C.; Vallat-Sauvain, E.; Miazza, C. *Solar Energy Mater. Solar Cells* 2006, 90, 2952–2959.

[97] Dominé, D.; Buehlmann, P.; Bailat, J.; Billet, A.; Feltrin, A.; Ballif, C. *Proceedings of 23rd European photovoltaic solar energy conference*, 2008; pp 2091.

[98] Delli Veneri, P.; Mercaldo, L. V.; Privato, C. *Renewable Energy* 2008, 33, 42.

[99] Delli Veneri, P.; Mercaldo, L. V.; Caputo, D.; Usatii, I.; Privato, C. *J. Non-Cryst. Solids* 2008, 354, 2478.

[100] Fischer D. *Proceedings of the 25th IEEE Photovoltaic Specialists Conference*, 1996; pp 1053.

[101] Yamamoto, K.; et al. *Sol. Energy Mater. Sol. Cells* 2002, 74, 449.

[102] Myong, S. Y.; Sriprapha, K.; Miyajima, S.; Konagai, M. *Appl. Phys. Lett.* 2007, 90, 263509.

[103] Buehlmann, P.; Bailat, J.; Dominé, D.; Billet, A.; Meillaud, F.; Feltrin, A.; Ballif, C. *Appl. Phys. Lett.* 2007, 91, 143505.

[104] Lambertz, A.; Dasgupta, A.; Reetz, W.; Gordijn, A.; Carius, R.; Finger, F. *Proceedings of 22th European Photovoltaic Solar Energy Conference, Milan*, 2007; pp 1839.

[105] Matsui, T.; Chang, C. W.; Takada, T.; Isomura, M.; Fujiwara, H.; Kondo, M. *Appl. Phys. Express* 2008, 1, 031501.

[106] Matsui, T.; Jia, H.; Fujiwara, H.; Kondo, M. *Proceedings of 23rd European photovoltaic solar energy conference, Valencia*, 2008; pp 2113.

[107] Schropp, R.E.I.; Zeman, M. *Amorphous and microcrystalline silicon solar cells*, Kluwer Academic Publishers; Boston, Dordrecht, London, 1998.

[108] Hack, M.; Shur, M. *J. Appl. Phys.* 1985, 58, 997-1020.

[109] Hubin, J.; Shah, A. V. *Philos. Mag. B* 1995, 72, 589–599.

[110] Terakawa, A.; Shima, M.; Kinoshita, T.; Isomura, M.; Tanaka, M.; Kiyama, S.; Tsuda, S.; Matsunami, H. *Proceedings of 14th European photovoltaic solar energy conference*, 1997; pp 2359.

[111] Yang, J.; Banerjee, A.; Glatfelter, T.; Hoffman, K.; Xu, X.; Guha, S. *Conference Record, 1st World Conference on Photovoltaic Energy Conversion, Hawaii*, 1994; pp 380.

[112] Meier, J.; Kroll, U.; Benagli, S.; et al. *Proceedings of 23rd European Photovoltaic Solar Energy Conference, Valencia*, 2008; pp 2057.

[113] Woodcock, J.M.; Schade, H.; Maurus, H.; Dimmler, B.; Springer, J.; Ricaud, A. *Proceedings of 14th European Photovoltaic Solar Energy Conference*, 1997; pp 857.

[114] Zweibel, K.; van Roedern, B.; Ullal, H. *Photon International*, October 2004, pp 48.

[115] Kluth, O.; et al. *Proceedings of 24th European Photovoltaic Solar Energy Conference, Hamburg*, 2009; pp 2715.

[116] Obermeyer, P.; et al. *Proceedings of 24th European photovoltaic solar energy conference, Hamburg*, 2009; pp 2313.

[117] Vetter, M.; Mata, C.; Andreu, J. *Proceedings of 23rd European photovoltaic solar energy conference, Valencia*, 2008; pp 2075.

[118] U.S. Department of Energy, April 2008, Solar Energy Technologies Programme (Solar Programme): *2008-2012 Multi-Year Programme Plan*.

In: Thin Film Solar Cells: Current Status and Future Trends ISBN 978-1-61668-326-9
Editors: Alessio Bosio and Alessandro Romeo © 2010 Nova Science Publishers, Inc.

Chapter 9

ORGANIC AND HYBRID SOLAR CELLS

Thomas M. Brown, Andrea Reale and Aldo Di Carlo
CHOSE, Dept of Electronic Engineering, University of Rome – Tor Vergata, Italy.

ABSTRACT

Interest in novel types of thin film technology, in which the active absorbing media are not conventional inorganic semiconductors but carbon based organic molecules, has been rising steadily in the last years. The electronic properties of these organic materials, including light absorption and electronic transport, can be tailored via chemical synthesis of a variety of molecular designs and via a careful control of deposition parameters and device architectures. Special strategies have been developed to improve efficiencies since photoexcitation in organic compounds give rise to tightly bound excitons, not free charge carriers as in conventional semiconductors. One of the biggest appeals of molecular based photovoltaic devices is that the active layers can be deposited either through simple evaporation processes or in the form of liquid solutions or pastes employing low-cost, low-temperature casting techniques such as those from the established printing industry. In this chapter we will review the status of fully organic solar cell technology, namely small molecular weight solar cells and conjugated semiconductor polymer solar cells and also hybrid devices, in particular dye solar cell technology.

1. INTRODUCTION

In the last 5 years, world-wide production of PV cells has increased steadily by an average of more than 30% per year [1,2]. Renewable energy sources have come into the spotlight in recent times due to the increase in oil prices and for political and environmental considerations. In order for photovoltaics (PVs) to become a widespread technology, fabrication costs have to be driven down. In inorganic crystalline semiconductor PVs, the ratio between fabrication costs and energy delivered can be high. This is due to processing parameters that require elevated temperatures, energetically expensive purification and doping steps and separate rigid framed substrates for panel construction [3].

Thin film solar cells [4-6] promise a reduction in costs by being able to reduce materials costs, use less of these per square metre, and also by utilising coating techniques that permit deposition of photovoltaic cells over large areas rather than construction of single cells of limited area and their subsequent lamination and series connection as occurs in crystalline and polycrystalline semiconductor photovoltaic panels [7].

Interest in novel types of thin film technology [8-11] in which the active absorbing media are not conventional inorganic semiconductors but carbon based organic (or metallo-organic) molecules has been rising steadily and rapidly in the last 15 to 20 years; first being developed and researched mainly in academic and research institutions in the nineties and with industrial outfits starting to develop a strong interest and allocating resources in this latter decade. The basic architecture of these cells is remarkably simple: a sandwich structure comprising of a transparent substrate, generally glass but also flexible plastic foils, covered by thin transparent conducting oxides (TCOs) and one or more layers of thin photoactive films including at least a molecular absorber and a top electrode. Molecular materials have a delocalized π electron system along the molecular chain or ring (via an alternation of single and double bonds- i.e. conjugation) which give rise to their peculiar optoelectronic properties including conferring real semiconducting behaviour to films consisting of these organic materials. The electronic properties of these materials, including light absorption and electronic transport, can be tailored via chemical synthesis of a variety of molecular designs and a careful control of deposition parameters.

The PV process in organic devices can be divided into its essential components: photoexcitation of an electron/hole pair, charge separation, transport and finally collection at the contacts. Primary photoexitations in organic materials do not directly lead to free carriers as in conventional inorganic semiconductors, but to excitons, i.e. coulombically bound electron-hole pairs. Because the binding energy is of the order of 0.1 eV - 1 eV $\gg$ kT, [12-17] only a small fraction of the photoexcitations in a homogenous material lead to free charge carriers, making the PV process an inefficient one. A successful way to break excitons is to do so at the interface between two different materials, utilizing the local electric fields present due to differences in the electron affinities and ionization potentials: a donor material in which often the bulk of the absorption of the solar spectrum usually occurs and through which holes are transported and an acceptor material into which electrons are injected and transported to the opposite contact. In fully organic solar cells the acceptor materials are carbon based. In hybrid devices such as dye solar cells these are inorganic high band gap semiconductors.

The range of types of cells that is being developed internationally is broad, each of which can be found in different stadia of research and technological maturity. In short, it comprises of fully Organic Solar Cells (OSCs – also reffered to as "plastic") where both the absorbing media and charge carrier transporting media are organic molecules [8,11], and hybrid devices where the absorbing medium is an organic or metallo-organic molecule and the electron transporting medium is a high band semiconducting oxide such as titanium dioxide as in Dye (sensitized) Solar Cells (DSCs) [9] or where the function of absorbing incident light and/or carrier transport is also taken up by nanocrystals such as in hybrid nanocrystal/polymer based solar cells (NCSCs) [18].

Recently, solar power efficiencies between 4% and 7%, [19-29] approaching those of amorphous silicon, have been reported for fully organic solar cells, and over 11% for dye solar cells [30] intensifying the interest in these systems as a promising route towards a low cost flexible renewable energy source.

Further R&D is required to increase the conversion efficiency and lifetime of organic solar cells, but the potential for low-cost PV modules is huge [20,26]. The fact that certified solar power efficiencies of up to 7.9 % (Solarmer) and 7.7 % (Heliatek-IAPP) have been reported for fully organic cells and over 11% for dye solar cells [30], and the possibility of applying these cells onto flexible and transparent substrates, has induced an ever increasing number of academic groups and companies, both established (e.g. Sharp, Sony, Aisin Seki, General Electric) and those just dedicated to this type of technology (e.g. Solarmer, Dyesol, Konarka, Heliatek, Plextronics, Solaronix, Peccell, 3GSolar), to declare an interest in the full development of organic PV devices.

In this chapter we will review the status of fully organic solar cell technology which we will divide in two sections - small molecular weight solar cells and conjugated semiconductor polymer solar cells (where we will also briefly mention nanocrystal based solar cells) - and conclude the chapter by illustrating dye solar cell technology.

2. SMALL MOLECULE SOLAR CELLS

Small molecular weight organic PV cells [28,31-33] have attracted attention since a 0.95% power efficient thin film organic cell was reported by Tang in 1986 in the Eastman Kodak Company research laboratories [10]. The novelty of this kind of device, which led to a dramatic jump in device efficiencies, was the introduction of a double layer composed of a Donor–Acceptor (D-A) heterojunction. The solar cell consisted in an Indium Tin Oxide (ITO) coated glass substrate on which a thin 30nm layer of copper phthalocyanine (CuPc) was deposited by conventional vacuum evaporation. CuPc functions as the main absorbing and donor material and has semiconducting properties, i.e. is optically active with high peak absorption coefficients ($\alpha > 10^5$ cm^{-1}) [34] in the visible range and also charge carrier transport mobilities of the order of (~0.02 cm^2/Vs) [35,36]. A second layer, the acceptor layer, (a perylene tetracarboxylic derivative) also tens of nm thick, was deposited on top of the CuPc layer by vacuum evaporation. The cell was completed by evaporating a top Ag electrode.

The DA interface allows for efficient dissociation of excitons in comparison to that possible with a single organic layer. For the latter, power conversion efficiencies (η) are low, $\eta \approx 0.01\%$ [32,37].

For a single organic layer (eg CuPc) sandwiched between ITO and a low work function metal such as Al or Ca, the internal built-in fields across the layers due to the differences in the work functions of the opposite contacts are not large enough to dissociate the excitons (see Figure 1). The exciton may diffuse to one of the contacts before dissociating [26]. Since exciton diffusion lengths L_D are of the order of 1–10 nm, [38] which is much less than the optical absorption length ($1/\alpha$) of most organic semiconductors, the fraction of photogenerated excitons that is able to dissociate is small. Other important loss factors for single layer devices are non-radiative recombination at the interfaces and non-geminate recombination at impurities or trapped charges [26].

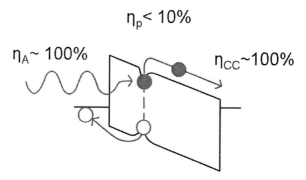

Figure 1. Schematic illustration of the three consecutive steps in the generation of photocurrent from incident light in a single layer organic PV cell: (1) photon absorption with efficiency η_A, (2) exciton dissociation, where the fraction of excitons dissociating is given by η_D; and (3) collection of the carriers at the electrodes, with efficiency η_{CC}. The dips in the energy level diagram represent the exciton binding energy. Redrawn from Reference [28].

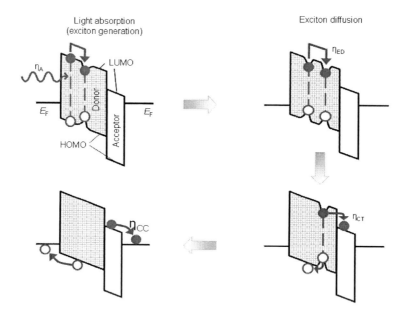

Figure 2. Schematic energy level diagrams of the photovoltaic process in a double layer donor-acceptor (D-A) organic solar cell: (1) photon absorption and exciton generation, (2) exciton diffusion to D-A junction, (3) exciton dissociation and (4) charge transport through the D-A layers and collection at the electrodes. Redrawn from reference [31].

The D-A heterojunction strategy yields a greatly improved efficiency when the difference in electron affinities/ionization potentials (or Lowest Unoccupied or Highest Occupied Molecular Orbitals - LUMOs and HOMOs) between the donor and acceptor materials at the heterojunction generate high enough interfacial electric fields which enable an efficient dissociation of the excitons [14,28,32] (which also has the advantage of occuring away from the metal/organic-semiconductor interface) as shown in Figure 2. This process is intrinsically much more efficient than dissociation of excitons in the bulk of an organic material. Holes are then transported in the donor material and electrons in the acceptor later to the opposite contacts by drift diffusion thus generating the photovoltaic effect. The maximum open circuit voltage achievable in these type of structures is roughly given by the difference between the ionization potential (HOMO level) of the donor and the electron affinity (LUMO level) of the acceptor.

To produce thin films of small molecular weight materials for organic PV cells, there are a number of options available which include organic molecular beam deposition (OMBD), vacuum thermal evaporation (VTE), and organic vapour phase deposition (OVPD) [32,39,40]. These processes are solvent-free and thus permit the facile fabrication of multilayer devices.

Since Tang's work [10], copper phthalocyanine (CuPc) and derivatives remain amongst the donor materials of choice as these have high (even if not very broad) absorption coefficients and sufficient hole carrier mobilities for the very thin layer utilised in fabrication of devices. As an acceptor molecular material, fullerene, C_{60}, has become the common choice because of its relatively high electron affinity, large exciton diffusion length and good electron transport properties. A third layer, consisting of an exciton blocking layer (EBL), is also commonly inserted between the acceptor-type molecular layer and the cathode.

The EBL plays a variety of roles, the most important of which is to prevent damage to the active layers during cathode deposition: this has beneficial effects in reducing exciton quenching at the acceptor/cathode interface and also avoids formation of defect states within the organic layers which can lead to a high resistance between the electrode and C60 [28,32]. Bathocuproine (BCP) works particularly well as EBL and dramatically improves electron transport out of the C60 film into the Al electrode [41].

One of the devices described above developed by Xue et al. at Princeton University, with the highest efficiencies reported, is shown together with its energy level schematics in Figure 3 [42,43]. In this device another strategy to improve the fraction of photogenerated excitons that diffuse to the D-A interface is implemented. This consists of a bulk heterojunction BHJ. The same scheme is also (and even more so) utilised in the construction of polymer solar cells as detailed in the next section of this chapter. Since the diffusion length L_D of excitons is $\ll$ than the optical absorption length ($1/\alpha$) of most organic semiconductors, the fraction of photogenerated excitons that is able to reach the D-A interface in bilayer devices with thick films is small. The alternative is to use very thin bilayer D-A films of the order of L_D so that most photogenerated excitons reach the junction which also keeps the transport paths of the separate carriers small. However in this latter case the optical densities of the films are limited so that a substantial fraction of the incident sunlight is not absorbed. To try and increase the optical densities utilising thicker films whilst maintaining the fraction of excitons dissociating high, the BHJ concept is introduced. In small molecular weight devices donor and acceptor materials are codeposited to form an interpenetrating network resulting in a spatially distributed D-A interface in the bulk of the film (i.e. a BHJ) which ideally lies within a diffusion length of every photoexcited exciton. In this way excitons can diffuse to the interface and dissociate before recombining. A bulk HJ expands the photocurrent generation region of the device, allowing excitons a higher probability to reach a nearby DA interface where they can dissociate. In the example of figure 3, this mixed layer is placed between the CuPc and C60 layers each of which have a thickness equal to the exciton diffusion length. A maximum power conversion efficiency of $\eta = (5.0 \pm 0.3)$ % under 1 – 4 suns, AM1.5, solar illumination was achieved [42].

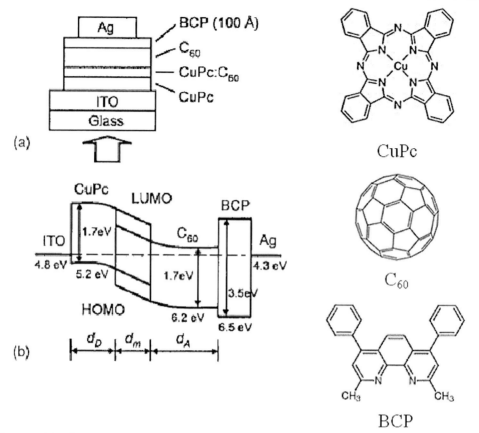

Figure 3. (a) Schematic device structure of an organic hybrid planar-mixed heterojunction PM-HJ photovoltaic cell employing the donor, copper phthalocyanine CuPc, and the acceptor, C60, molecules. Indium tin oxide ITO was used as the anode, and bathocuproine (BCP) as the exciton-blocking layer. The arrow indicates the light incident direction. (b) Schematic energy-level diagram of the hybrid PM-HJ cell under illumination, where HOMO and LUMO are the highest occupied and lowest unoccupied molecular orbitals, respectively. The thicknesses of the donor, mixed, and acceptor layers for the most efficient cells are d_D =15 nm, d_m =10 nm and d_A,= 35nm respectively. Reproduced from reference [42].

Tandem solar cells have been constructed to optimise the amount of photogenerated AM1.5 illumination by stacking two devices on top of each other. Stacking cells overcomes the limited bandwidths that are typically not broad enough to absorb the entire spectrum. The tandem strategy represents also a possible solution for devices that are too thin to absorb all the light because of short carrier collection lengths or short exciton diffusion lengths [28,44]. Making a tandem device is relatively straightforward utilising the deposition techniques of small molecular organic material technology. Record power conversion efficiencies reached are $\eta = (5.7 \pm 0.3)\%$ [45] for this type of devices.

In order for a PV technology to become appealing at the commercial level both conversion efficiencies and lifetimes need to hit the particular targets associated with the applications which are sought. For roof top applications lifetimes often need to be in excess of 20 years but for charging portable electronic devices the requirement can be just a few years. Nevertheless, for small molecular devices systematic lifetimes studies in the literature are relatively few [46-49]. Recent studies show degradation of between 3% and 6% after 1400 h [46] and 6100 h [48] respectively.

Some of the technological and scientific challenges organic devices need to face include continuing to increase lifetimes and efficiencies (by for example broadening the spectra of absorbed light, creating an ordered vertical heterojunction) and to transfer the small single cell technology from the lab scale to large areas for module production using low-cost very high-throughput deposition and processing techniques.

3. CONJUGATED POLYMER SOLAR CELLS

Since discovery of their surprising electronic [50] and optoelectronic [51] properties at the Cavendish Laboratory, conjugated polymer semiconductors have rapidly come into the international R&D focus for the huge potential they have for large area, low-cost optoelectronics. Polymer semiconductors are attractive because they can be processed over large areas from liquid solution, using low-cost, low-temperature casting techniques such as those from the printing industry and also because of the possibility of tuning their colour and energy levels through chemical structure. Furthermore, the capability for monolithic integration of different electronic devices onto one substrate [52,53] makes them very appealing for a variety of applications such as flat panel displays, smart cards, e-paper, electronic tags, RF identification, photodetector arrays and solar cells [19-21,26,54-60].

Figure 4 shows the typical structure of polymer solar cell. On the TCO covered glass or flexible substrate a film of poly(3,4-ethylene dioxythiophene) doped with poly styrene sulphonate (PEDOT:PSS) is deposited from a water based dispersion. PEDOT:PSS was initially developed as a hole injecting layer for polymer light-emitting diodes [61,62] and has been found to be beneficial also in solar cells by smoothening the anode surface, increasing its workfunction as well as facilitating hole extraction. The photoactive layer is deposited over the PEDOT:PSS film and the basic device finally completed by evaporating a top metallic contact, usually Aluminium, but also other multilayer contacts also "borrowed" from PLED technology such as Ca/Al and LiF/Al [63-66] which guarantee a smaller work function which can lead to higher built-in fields [64] depending on electron affinities of the photoactive semiconducting polymers in the active layer.

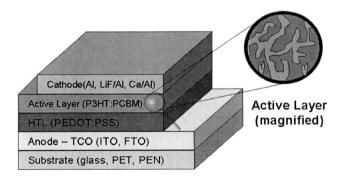

Figure 4. Schematic structure of a polymer solar cell. The active layer is sandwiched between an ITO/PEDOT:PSS anode and an Al (Ca/Al or LiF/Al) cathode. The anode is transparent and faces the light source.

The same donor acceptor- strategy described for small molecular weight devices (see Figure 2) for efficient charge separation of the photogenerated exciton is utilised for conjugated polymers devices as well. The photoactive layer is composed of at least two different semiconducting polymers. A wide range of donor materials have been utilised for the construction of devices, comprising derivatives of phenylene vinylenes [8,11,67] such as MDMO-PPV [66], polyfluorene- and triarylamine-based polymers [68-71], poly(3-alkylthiophenes) [72,73] and poly(3-hexylthiophene) P3HT [74-78]. Different acceptor materials have been used from polyfluorene, triarylamine and phenylene vinylenes derivatives, but the most commonly used polymer is a soluble derivative of fullerene, i.e. [6,6]-phenyl C 61 butyric acid methyl ester (PCBM) as it maintains a relatively high electron affinity and good electron mobility as well as good solution processability. The chemical structures of PEDOT:PSS, P3HT, MDMO-PPV and PCBM are shown in figure 5. Typical thicknesses are 40nm -100 nm for PEDOT:PSS, 70nm – 200nm for the active blend layer and 100-200nm for the top Al layer (sometimes including a thin LiF or Ca interlayer between the active semiconducting film and the top Al electrode).

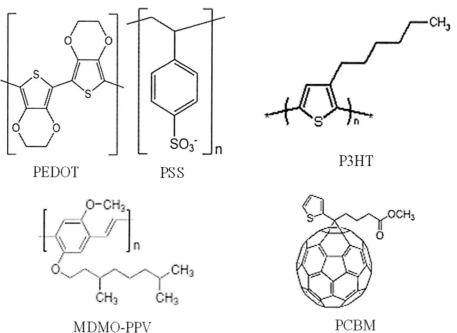

Figure 5. Chemical structure of PEDOT:PSS (anodic layer), P3HT, MDMO-PPV (Donor layers) and PCBM (acceptor layer) which are a selection of commonly utilised polymers for the construction of the conjugated polymer photovoltaic devices schematized in Figure 4.

Another PV technology stems from replacing one of the components of the active layer with an inorganic semiconductor material. These cells are often referred to as hybrid devices. PCBM has been replaced with a mesostructured n-type high band gap inorganic semiconductor such as TiO_2 or ZnO [44,79]. Chemically synthesised semiconductor nanocrystals (NC) and nanorods such as CdSe, CdTe, PbS, PbSe can be used to fabricate solution processed hybrid solar cells together with polymers [18,80-84]. By controlling nanocrystal dimensions it is possible to change the distance on which electrons are transported and to tune the band gap and thus the absorption spectrum of the devices. It is not the scope of this chapter to go into the details of these types of solar cells but it is interesting to note that recently all-inorganic conductor/NC/metal photovoltaic devices with the NC layer deposited via solution processing have been reported with efficiencies ranging between 2.1% and 3.3% [85-87].

One of the biggest appeals of conjugated polymer and molecular semiconductors is that they can be deposited in the form of liquid solutions employing low-cost, low-temperature casting techniques such as those from the printing industry. Instead of using the bilayer structures of Figure 2, which are difficult to fabricate since orthogonal solvents would have to be sought for the two materials, it is simple and easy to dissolve both the donor and acceptor polymers in a common solvent (or solvent mixture). This bi-component "ink" is then used to deposit the active layer. In fact, the highest polymer solar cell efficiencies can be achieved maximizing the interfacial area by blending the two different components, creating an extended interpenetrating network throughout the bulk of the device [8,11,19,21,58,60,69,70,88] as shown in the inset of Figure 4. Roughly a two order of magnitude increase in efficiency was observed when passing from a single layer device to a blend device [11]. This type of PV device is also known as bulk heterojunction (BHJ) conjugated polymer solar cell for which power efficiencies up to 5% - 7% have now been reported [22,25].

At the heart of the BHJ concept is to mix the donor and acceptor components closely so that on average there exists an D-A interface which is no further than the exciton diffusion length so that each photogenerated exciton can diffuse to it and dissociate. With the hole being transported to the anode through the donor phase (eg P3HT) and the electron collected at the metallic cathode after being transported through the acceptor material (eg. PCBM) via drift of the carriers arising from built-in fields generated by the asymmetric electrode contacts and diffusion due to concentration gradients of the respective charge carriers [58].

The estimates for the exciton diffusion length are around 10nm [38,59]. Thus for efficient charge transfer the domains of the individual components of the bulk heterojunction ought to be close to that length scale. However, once the charges are separated these need to be collected at the opposite electrodes so a continuous path in each polymer component along the whole thickness of the device (of the order of 0.1 µm) would facilitate charge extraction. Larger phase separated domains would hence be desirable. These two sometimes opposing requirements makes the control of the morphology of the blend at the nanophase scales crucial in determining device performance [68-70,89-91].

Phase separation length scales have been varied through several different processing routes. Shaheen at al.[66] reported a strong dependence of the performance on the solvent used for the active blend: whereas toluene cast devices yielded power conversion efficiencies of 0.9%, switching to chlorobenzene almost tripled the efficiency ($\eta = 2.5\%$). Post-production annealing of polymer solar cells [25,72] has been shown to increase performance significantly as shown in Figure 6.

Varying the solvent, the relative ratio of the two components of the blend, the substrate temperature, the saturation of the atmosphere over the transforming film and the time allowed for solvent-enhanced self-organization one is able to vary the characteristic phase separation length scales from tens of nanometers or less up to tens of microns [11,24,25,66,68-70,72,75,77,89-91].

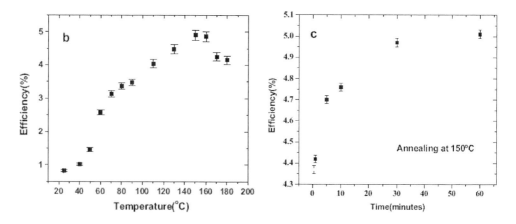

Figure 6. Evolution of power conversion efficiencies (AM1.5, 80mWcm^{-2}) with thermal annealing temperature at constant annealing time of 15 minutes (left) and annealing time at a constant annealing temperature of 150C (right). The device structure is ITO/PEDOT/P3HT:PCBM/Al. The characteristics of the best device were JSC = 9.5 mAcm^{-2}, FF = 68%, V_{OC} = 0.63V, $\eta \cong$ 5%. Reproduced from reference [25].

Post treatments like annealing not only cause a reorganization of the domain sizes but may also improve the intimacy of the contact of the active layers to the electrodes and also cause an increase in the crystallinity of the polymers with an associated lowering of the series resistance of the cells. The benefits could be manifold such as carrier mobilities being enhanced and morphological stability also improved [25,89]. Polymer blends may be engineered at several length scales at once to optimize, for example, charge separation and transport to the contacts [69].

Figure 7 shows the schematic structure of a desirable model organic solar cell in which the heterojunction is ordered in such a way that a D-A interface exists within an exciton diffusion length (i.e. ~ 10nm) of each point in the device whilst maintaining an overall thickness comparable to the absorption length of the radiation (hundred(s) of nm) and a clear homogenous transport path to the electrodes within each of the interdigitated phases.

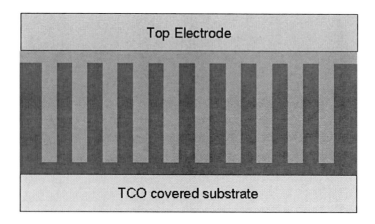

Figure 7. Schematic structure of a desirable model organic solar cell.

Such a well-organized nanostructure is not easy to obtain in classical polymer mixtures due to disorder [58]. The trick is to obtain an ordered structure without utilising conventional techniques such as photolithography, which at the length scales required would make any device prohibitively expensive, but use the self-ordering and assembly processes developed by the chemical and materials sciences.

Surface treatments and solvent evaporation control have been used to promote vertical segregation in polyfluorene-blend thin films thereby improving charge transport in these blend structures [68]. Solution processing also opens up the possibility of employing high resolution "soft" lithography techniques to deposit and pattern the materials down to the 10s nm-100nm scales which start to approach the exciton diffusion length of the photoexcited carriers [23]. Complex structures can be engineered from novel materials such a discotic semiconductors [92]. Peet et al. [93] have shown that by incorporating a few volume per cent of alkanedithiols in the solution used to spin-cast films comprising a low-bandgap polymer and a fullerene derivative, the power-conversion efficiency of photovoltaic cells can be increased from 2.8% to 5.5% through altering the bulk heterojunction morphology. The above examples illustrate a few of the strategies that are being employed and researched in order to attempt the control the shapes of the blend phases for maximising charge separation and carrier transport throughout the thickness of the photovoltaic devices.

The solar cells that exhibit the best performance of any bulk heterojunction system studied to date, apart from tandem structures, were based on a PCDTBT/PC$_{70}$BM active layer with J_{SC} = 10.6 mA cm^{-2}, V_{OC} = 0.88 V, FF = 0.66 and η = 6.1% under air mass 1.5 global (AM 1.5 G) irradiation of 100 mW cm-2 [94]. The internal quantum efficiency was close to 100%, implying that essentially every absorbed photon resulted in a separated pair of charge carriers and that all photogenerated carriers were collected at the electrodes [94].

If the internal quantum efficiencies are indeed approaching unity, further improvement in performance maybe be achieved by engineering devices with smaller band gaps and broader absorption spectra to harvest a greater fraction of the incident sunlight whilst also minimising voltage drops due to high contact/series resistances and unnecessarily large interfacial energy level offsets. Theoretical work by Scharber et al. [95] who calculated the expected efficiency of BHJ cells as a function of the band gap and the LUMO level of the donor (assuming a fill factor of 65%) has shown that it is in principle possible to achieve 10% efficiencies by synthesising appropriate donor molecules whose energy levels are tailored to the most commonly acceptor molecule used today, i.e. PCBM (see figure 8). It follows that further gains in power conversion efficiencies could be achieved by not only concentrating on the synthesis of the donor materials but also of novel acceptor materials, thus finding the ideal donor-accept pair combinations.

Increasing the conversion efficiencies is not only a matter of tailoring the optoelectronic properties of the ensemble of constituent materials but also optical effects come into play since the typical layer thicknesses are of the order of the wavelength of the light that is being absorbed.

The electric field of the incident radiation needs to become zero at the reflecting top metal reflecting electrode. Thus, a vanishing radiation intensity may be present throughout a significant fraction of the active layer [96,97]. Kim et al. [74] have introduced an optically transparent TiO$_2$ spacer, whose electronics properties are also compatible with maintaining smooth electron transport, between the active layer and the Al electrode in order to push the active layer away from the top metal electrode. The observed increase in photocurrent when inserting the TiO$_2$ film resulted in an approximately 50% increase in device efficiency showing that optical and interference effects need to be accounted for.

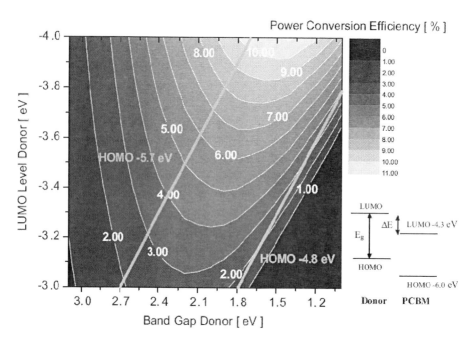

Figure 8. Contour plot showing the calculated energy-conversion efficiency (contour lines and shades of grey) versus the bandgap and the LUMO level of the donor polymer according to the model described in reference [95]. Straight lines starting at 2.7 eV and 1.8 eV indicate HOMO levels of –5.7 eV and –4.8 eV, respectively. A schematic energy diagram of a donor PCBM system with the bandgap energy (Eg) and the energy difference (DE) is also shown. Reproduced from reference [95].

Two major losses that occur in all types of solar cells are due to the sub-band gap transparency of semiconductors and the thermalisation of hot charge carriers which mean that a large fraction of photons and photon energy are not exploited. One way to circumvent both effects simultaneously is the realization of a tandem solar cell. The light which is not absorbed in the first device can be absorbed by the second stacked cell instead. Moreover, the thermalization losses are lowered due to the use of materials having different band gaps [54]. Via this approach it is possible to utilise different materials to cover a broader fraction of the spectrum of the sun (since organic semiconductors have relatively narrow absorption spectra) and also maintain the thicknesses of the films small so as to facilitate charge transport. Tandem device structures have been designed and developed by various groups [54,98-100]. State of the art today is: Jsc = 7.8mA/cm^2, Voc = 1.24 V, FF = 0.67, and η = 6.5% [22]. Calculations suggest that tandem cells showing efficiencies of 15% can be envisaged provided that some necessary optoelectronic properties required from the materials are achieved [54,101].

Success of the conjugated polymer solar cells will not only lie in a continual improvement of device power conversion efficiencies but also of lifetimes and stability [102]. The process of transferring the technology from the lab scale to large modules utilizing low cost high throughput techniques is also fundamental.

The requirements on encapsulation for most organic semiconductor devices are currently extremely high in order to achieve long shelf life and operation lifetimes required for commercialization. A straightforward solution is to cover the solar cells with glass. Inverted P3HT:PCBM solar cells sealed with glass plates maintained 90% of the initial efficiency after 1500 h of continuous illumination, corresponding to a light dose equivalent to approximately 1.5 years exposure to sunlight [103]. Though glass ensures optimum encapsulation, one of the most important advantages of the plastic solar cell, i.e. its mechanical flexibility, is sacrificed [104]. Efficient encapsulation with barriers that have very low oxygen and water permeabilities will be needed before future commercialisation can be anticipated [105]. The effects of multilayer transparent and flexible ultra high barrier foils on the lifetime of organic photovoltaic cells has been investigated [104,106]. Accelerated testing showed that lifetimes exceeding 1250 h, even at high temperature/high humidity conditions, may be reached for flexible polymer solar cells [107]. Flexible modules survived over 1 year of outdoor exposure without performance losses [108].

While a simple PET film as encapsulation provided a shelf lifetime of less than 6 h, ultra-high barrier material yielded lifetimes exceeding 3000 h for MDMO-PPV:PCBM based devices, comparable to those reported for the same type of cells encapsulated between glass plates [104]. These results suggest the importance of developing materials or material combinations with high intrinsic stability rather than relying on encapsulation alone.

In order for organic semiconductor technology to enter the applications market it is necessary to upscale organic solar cells from laboratory design to larger modules [104,105,109,110]. Series connection of solar cells is desirable to increase the output voltage of the module, and also allows the effective use of the surface area [104].

Tipnis et al. [109] fabricated modules by organizing individual cells in six parallel columns, each containing nine cells that are connected in series by internal wiring as shown in figure 9a. Another strategy consists in are more integrated approach as that shown in figure 9b. This is the one used by Lungenschmied et al. [104] who structured the ITO bottom electrode by deposition through a shadow mask in the case of foils as substrates and by chemical wet etching in the case of ITO on glass. The active layer was structured by etching. The devices were finished with a thermally evaporated Al top electrode, using a shadow mask. The highest efficiencies on modules certified by NREL were 1.1% over total area (233 cm^2) and 2.4% over active area (108cm^2). The decay rate of these modules was also monitored corresponding to a lifetime of 5000h [109].

As a technology, polymer optoelectronics biggest manufacturing advantage resides in being able to be mass produced via printing techniques. Established printing technologies can be employed for production. Complete PV devices have been constructed by spin coating, screen printing, blade coating and pad printing of the active conjugated semiconductor blend layer [25,26,76,105,111-117]. Inkjet printing of films (which does not require additional patterning) yielding 3% efficient devices has also been demonstrated [116,118,119]. Development of very high throughput printing technologies and processes like flexographic and gravure printing [55,71,120,121] and the up-scaling of these to reel to reel manufacturing [122] are very important drivers for delivering the promise of organic photovoltaic elements on low–cost, thin plastic substrates with future substantial cost reductions [20], thus making polymer solar cells an established and important technology in the field of photovoltaics.

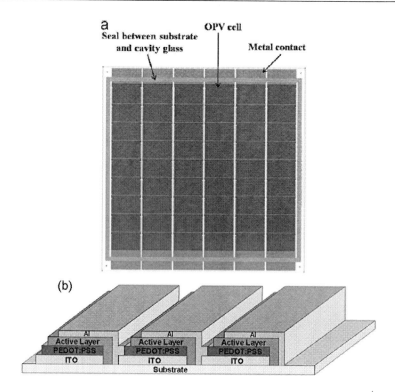

Figure 9. (a) Design of a PV module on 15.2cm x15.2 cm glass substrate described by Tipnis et al. [109]. 54 identical cells were organized in six parallel columns of nine cells each. Reproduced from reference [109]. (b) schematics of an integrated approach to series connection where the layers are patterned and the top electrode of one cell is made to contact the bottom ITO electrode of one neighbouring cell.

4. DYE SOLAR CELLS

Dye Solar Cells (DSCs) [123-128], first developed at EPFL by O'Regan and Graetzel [9], are interesting both from the academic and the commercial point of view, thanks to their ability to convert light into electricity at low cost, with a relatively high efficiency [123,129] and using relatively simple fabrication methods, similar to those used in printing, over a variety of substrates (glass, metal, plastic). The basic structure (see figure 10) is that of a conducting transparent substrate onto which a mesoporous nanostructured high band gap semiconducting material is deposited (usually TiO_2) and subsequently sintered. A monolayer of dye is anchored to the TiO2 film. The device is closed by another conducting substrate onto which a catalytic layer is deposited beforehand, encapsulated and filled with an electrolyte solution. The kernel of the system is represented by the TiO_2 working electrode on which a layer of sensitizing dye (a polypiridinic Ru-complex) is adsorbed [130,131]. The function of the sensitizer is to convert the light energy into excited electrons, which are transferred to the semiconducting oxide: light excites an electron from an occupied to an empty molecular orbital and the electron is transferred from the excited state to the conduction band of the semiconductor. A Red-Ox couple, typically I^-/I_3^-, is the core component of an electrolyte solution that refurnishes the dye with the electron it has just lost to the TiO$_2$. The reduced form of the redox couple is regenerated at the counter-electrode, through the interaction with a catalyst layer (typically a nanometric Pt film). No permanent chemical transformation occurs during this basic process. In this way, the dye solar cell manages to continually transform light from the sun into electric current and energy once a load (e.g. an electrical appliance) is applied between the two terminals.

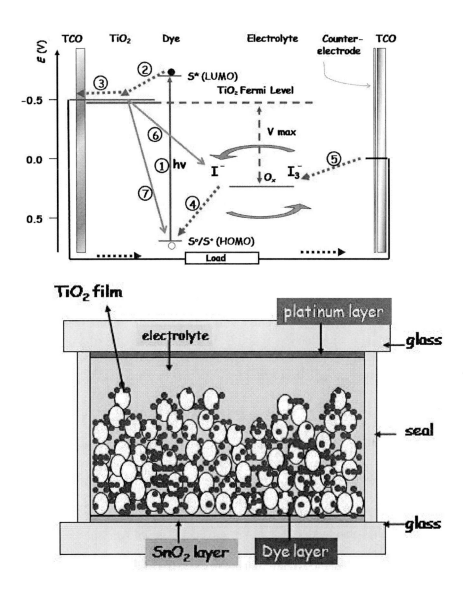

Figure 10. Working principle of a DSC (top); schematic picture of a DSC (bottom). The different processes are: (1): dye excitation; (2): electron injection and (3) electron diffusion into the conduction band of TiO$_2$; (4): reduction of the dye; (5): regeneration of the electrolyte ions by the catalytic later; (6),(7): main electron recombination losses.

4.1. Materials and Architectures for DSCs

4.1.1. Substrates

DSC construction typically makes use of printing technology for the deposition of device layers. The fabrication processes are thus low cost and easy to implement. Among the different techniques used, the most common today are screen printing and blade coating. A great advantage of techniques such as screen printing resides in the fact that these are additive processes, which allow the deposition of the needed amount of material only, also reducing the impact on the environment. Furthermore, these methods are compatible with both batch (mainly glass substrates) and roll-to-roll production methods (metal and plastic foils). Transparent conducting electrodes are typically realized by deposition of transparent conducting oxides (TCO). Among the various choices, $F:SnO_2$ (FTO) offers better performance for glass substrates, because this type of material can withstand the high temperatures necessary for subsequent sintering processes carried out on the TiO_2 layer without degradation of its conductivity. Indium Tin Oxide (ITO) is used for PET and PEN plastic substrates. A number of methods have been suggested to decrease the series resistance arising from the relatively high surface resistivity of TCOs, mainly by the introduction of metal grids [132].

4.1.2. Titania Layer

At the centre of DSC functionality lies a highly porous nanocrystalline TiO_2 film (producing a large adsorptive surface area), enhanced in recent designs by a light-scattering topcoat, with a high molar extinction coefficient dye as sensitizer to form the working electrode of the solar cell [133]. Techniques in TiO_2 film formulation and fabrication are, thus, very important in the production of highly efficient DSCs. TiO_2 film preparation utilizes the techniques of screen-printing for the nanocrystalline and submicron-crystalline- TiO_2 film layers as well as a chemical bath deposition for the $TiCl_4$ treatment [134]. A photon scattering and trapping effect is induced by the use of a the combination of a transparent nanocrystalline TiO_2 film with a microcrystalline light-scattering layer [135,136], in conjunction with an anti-reflecting film (ARF) [133] and other techniques [137-139], thus enhancing the incident photon to current collection efficiency (IPCE). $TiCl_4$ treatment is usually adopted as a pre and post treatment for TiO_2 deposition, since the pre-treatment influences positively the bonding strength between the FTO substrate and the porous- TiO_2 layer, and blocks the charge recombination between electrons emanating from the FTO and the I_3^- ions present in the I^-/I_3^- redox couple. The post-treatment enhances the surface roughness factor and necking of the TiO_2 particles thus augmenting dye adsorption and electronic transport resulting in higher photocurrent [133,134]. See figure 11.

After formulation of the TiO_2 paste and its subsequent deposition, the film undergoes a sintering procedure. The latter step is critical since it needs to guarantee good electromechanical bonding between the nanoparticles maintaining a sufficiently large film porosity and surface area. The standard sintering procedure for TiO_2 consists of subjecting the photoanodes to an increasing temperature vs. time profile in an oven or hotplate with a final ~ 30 min step at 450 °C – 500 °C [131,141].To reduce the energy employed in DSC fabrication, and for compatibility with plastic substrates, several techniques have been suggested for achieving low temperature sintering including pressure [142], electrophoresis [143] and microwave [144]. Laser raster scanning processing is also very promising, and recent results demonstrate equivalent performances to standard oven sintering as shown in Figure 12 [145].

In order to optimize charge collection efficiency, several attempts have been made to introduce ordered nanostructures in the working electrode of the DSC. The most common material is nanostructured ZnO [146], that presents band gap energy and electron affinity similar to TiO$_2$. ZnO and TiO$_2$ nanofibers can be deposited at relatively low temperatures [146-150] for continuous, roll to roll production schemes, even on flexible substrates.

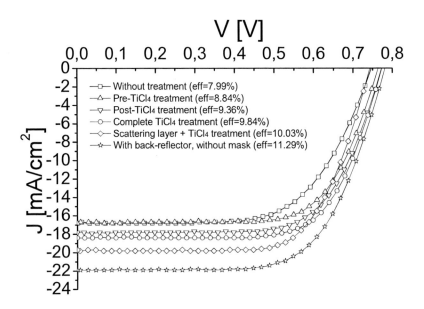

Figure 11. Effects of TiO$_2$ optimization. (a), (b), (c), (d), (e), (f). Details of cell preparation: the substrates are 8 Ω/□ FTO-glass (2 x 2 cm^2, Pilkington ®) cleaned in an ultrasonic bath 10 min in acetone and 5 min in ethanol. 0.5 x 0.5 cm TiO$_2$ films (thickness ~ 15μm) deposited via blade coating using Dyesol® DSL 18 NR-T TiO$_2$ paste, dried for 5 minutes at 65 °C and then sintered at 525°C for 30 min. After sintering the substrates were soaked in a 0.5 mM N719 (Solaronix®) dye solution in ethanol for 15 h, and then rinsed in ethanol. Counterelectrodes were prepared by depositing a Pt precursor solution (Platisol, Solaronix®) onto the FTO-coated substrates and then annealed at 400°C for 5 minutes. The cells were completed by sealing together the two electrodes via a 60 μm thick Surlyn™ gasket. Dyesol® HPE electrolyte was inserted into the cell via vacuum backfilling. The "pre-treated" cells had the first TiCl$_4$ treatment after the cleaning step. The "post-treated" cells had TiCl$_4$ treatment after transparent TiO$_2$ sintering step. Pre and post treatment is the complete treatment. The scattering layer is deposited after TiCl$_4$ "post-treatment". Cell are measured without masking (leading to an estimated increase in cell performance of ~15% in relative terms).[140].

4.1.3. Dye

Traditional dye systems as the one shown in Figure 13 are based on polypiridinic Ru-complexes [130]. The lowest excited state is of charge-transfer nature and involves the direct transfer of an electron mainly localized on the Ru atom to a molecular orbital of the carboxy-bipiridine ligand [151]. Therefore, the absorption bands are due to metal to ligand charge transfer (MLCT) transitions and a careful choice of the ligands can give the desired energy levels and proper redox properties. Indeed, the possibility of influencing the relative position of the orbitals involved allows to engineer the properties of the complex; for instance the introduction of electron-donors ligand groups enhances the charge on the Ru atom, with the consequence of moving the MLCT transition towards the blue end of the spectrum. Also the non-chromoforic ligands play an important role in changing the relative energy levels and in influencing the solvatochromatic properties of the complex.

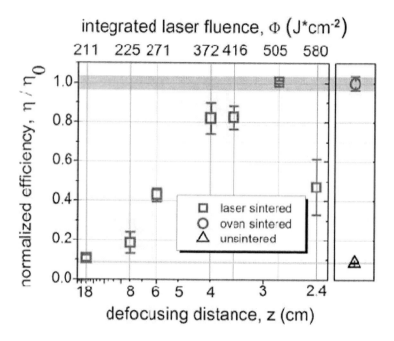

Figure 12. Power conversion efficiencies η (squares) of DSCs as a function of defocusing distance z (and estimated integrated laser fluence Φ, top axis) employed for the laser sintering of the TiO$_2$ layers measured under AM1.5, 10^3 W/m^2 conditions. η are normalized to the average efficiency (η$_0$) of DSCs fabricated with oven sintered TiO$_2$ layers (circles). Also reported are efficiencies values (triangles) of unsintered devices. From reference [145].

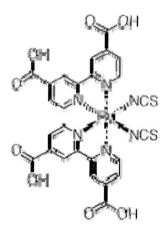

Figure 13. Molecular structure of the N719"red" dye.

With this type of sensitizer it has been possible to reach efficiencies of 11% [152] and noticeable long-term stabilities [153]. Improved performances in terms of stability have been also demonstrated by using hydrophobic dyes. Recently the interest on fully organic dye sensitizers is growing, particularly driven by their lower cost with respect to Ru-complexes. Moreover, the organic dyes show i) very high extinction coefficients, ii) simple synthesis and purification processes, iii) considerable potential in terms of molecular engineering. To date the best photovoltaic conversion efficiencies obtained with organic dyes in DSCs have reached a promising 9.8% [154] (see Figure 14) with encouraging long-term lifetime performance.

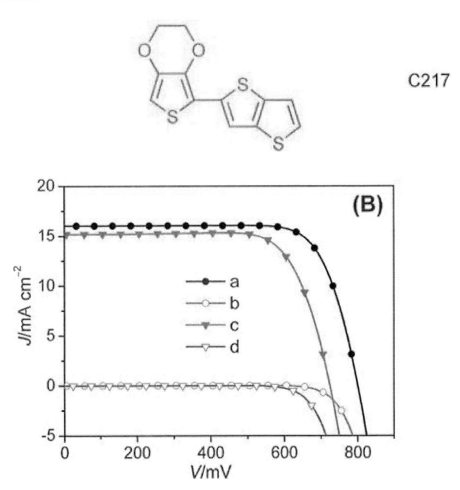

Figure 14. (above) Molecular structure of the organic dye C217; (below) J–V characteristics of cell with 9.8 % efficiency in standard liquid volatile electrolyte (a and b) and cell with 8.1 % efficiency in ionic liquid electrolyte (c and d). Curves (a), (c) measured under AM 1.5, (b) and (d) measured in the dark. Cell tested with mask having active area 0.158 cm^2 and antireflection film. Reproduced from reference[154].

Tailoring of the optoelectronic properties of a dye is possible by chemical synthesis. This, together with controlling the degree of transparency of the cell via TiO$_2$ paste formulation and the patterning via printing techniques allows a great flexibility in the design of the aesthetic properties of PV dye based devices as shown in figure 15.

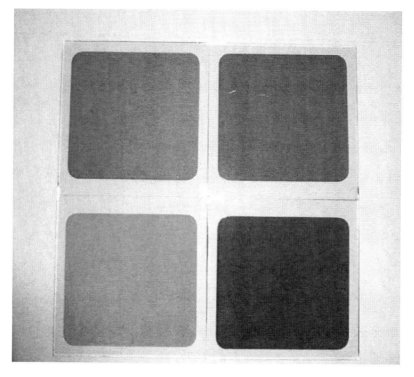

Figure 15. Coloured, transparent large area DSCs realized at CHOSE., Department of Electronic Engineering, University of Tor Vergata, Rome, Italy.

4.1.4. Electrolyte

The electrolyte is one of key components for dye solar cells and its properties have a great bearing on the conversion efficiency and stability of the solar cells [124,126,128,155].

The requirements for the electrolyte include good charge transfer characteristics to the dye after electron injection:

$$Dye^+ + 3/2\ I^- \rightarrow Dye + \tfrac{1}{2}\ I_3^-$$

and good charge transfer characteristics at the CE (low charge transfer resistance):

$$\tfrac{1}{2}\ I_3^- + e^- \rightarrow 3/2\ I^-$$

There are also some "loss processes" in DSCs. An excited dye molecule may directly relax into its ground state without injection of an electron into the TiO_2. This process is negligible, as injection is about 1000 times faster. Also, electrons from the conduction band of the TiO_2 may recombine with the oxidized dye molecule, before the dye is reduced by the electrolyte. However, the latter rate is about 100 times greater. The most significant loss mechanism in the DSC is the recombination of TiO_2 conduction band electrons with the holes in the electrolyte, i.e. I_3^-. The transport of electrons by diffusion in the TiO_2, and their recombination with the electrolyte are the two competing processes in DSCs.

The I^-/I_3^- redox couple is very efficient in these cells because of a fortunate confluence of the right kinetics for at least four different heterogeneous electron-transfer reactions [156]:

- The photoexcited dye must inject an electron faster than it reacts with the mediator;

- The oxidized dye must be reduced by the mediator more rapidly than it recombines with the photoinjected electron;
- The oxidized mediator must, itself, react slowly with electrons in both the TiO_2 and the fluorine-doped tin oxide (SnO2:F) contact;
- Finally, the reduction of the oxidized mediator at the cathode (CE) must be rapid.

The electrolyte used in DSC usually contains an organic solvent, a redox couple, and additives. It is divided into three types: liquid electrolyte, quasi-solid state electrolyte, and solid electrolyte. Liquid electrolytes can be divided into organic solvent electrolytes and ionic liquid electrolytes according to the solvent used. Organic solvent electrolytes were widely used and investigated in dye-sensitized solar cells for their low viscosity, fast ion diffusion, high efficiency, easy to be designed, and high pervasion into nanocrystalline film electrode.

Organic solvents used in organic liquid electrolyte include nitriles such as acetonitrile, valeronitrile, 3-methoxypropionitrile, and esters such as ethylene carbonate, propylene carbonate, γ-butyrolactone. A good solvent should have a high boiling point, a high dielectric constant and permittivity to separate the salts. Moreover it needs low viscosity in order to yield higher mobility since the diffusion of the redox couple is inversely related to the viscosity of the electrolyte.

A number of additives have been discovered, which are able to facilitate self-assembly of the dye molecules on to the TiO_2 electrode surface, making it more impermeable. This results in a reduced dark current of the cell. Examples are carboxylic (like hexadecylmalonic acid) and phosphonic acid derivates (like decylphosphonic acid). The additives reduce recombination and maintain dye integrity. Secondary cations were also used to screen electronic charge of electrons injected into TiO_2 nanoparticles to enable higher photocurrents. The commonly used additive in the electrolytes for dye-sensitized solar cells contained 4-*tert*-butylpyridine (TBP) and N-methylbenzimidazole (NMBI). The addition of these additives suppresses the dark current and improves the photoelectric conversion efficiency. TBP reduces the recombination of electrons in the conduction band of the semiconductor and the electron acceptor in the electrolyte through the coordination between N atom and the Ti ion in incomplete coordination state on the surface of TiO_2 film. This results in a large increase of the photovoltage fill factor and conversion efficiencies [157].

Triiodide/iodide is the common choice for the redox couple as it yields the highest efficiencies when used in liquid electrolytes. Despite its good performance, the triiodide/iodide couple has its own disadvantages: the triiodide ion absorbs a significant part of the visible light when employed in high concentrations, its redox potential partially limits the open-circuit voltage available, and its aggressiveness towards silver and the vast majority of metals makes it very difficult to introduce metallic grids that would reduce the series resistances due to the resistive TCO and thus increase cell fill factors [158]. Moreover, some authors report catalyst instabilities in the presence of the I^-/I_3^- couple [132].

Several other redox couples have been considered such Br^-/Br_2 $SCN^-/(SCN)_2$ and $SeCN^-/(SeCN)_2$. Sapp et al.[156] reported substituted bipyridyl cobalt($^{III/II}$) as redox couple in DSCs. However, the performance of this couple is still lower than that with a I_3^-/I^- couple. It also requires a different catalyst on the counterelectrode. Gold and carbon cathodes gives higher J_{sc} than platinum since the reduction of Co(^{III}) is much slower with platinum. Cobalt based electrolytes do however have limited corrosive action on metals for interconnects, allowing an easier engineering and fabrication process for DSC modules [156,159]. These species are inert with respect to metals and plastic materials and show interesting electrochemical properties, among which a noticeable electrochemical inertia on conductive oxides surfaces [160]. However they exhibit lower performances with respect to I^-/I_3^- couples, mainly because of a lower regeneration rate of the oxidized dye. Tuning the cell design in attempting to minimize recombination losses on the working electrode can lead to a significant enhancement of the cell efficiency. Recently Liberatore et al. [161] have shown the great effectiveness of the addition of a alumina blocking layer in cells containing cobalt based electrolyte, by substantially reducing the recombination current. The consequent power conversion efficiency increase is more than double as shown from the IV curves of figure 16.

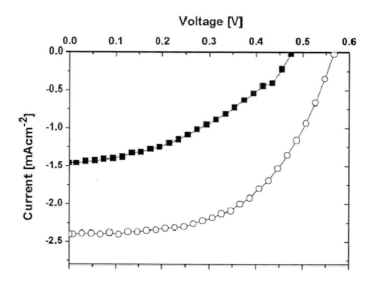

Figure 16. IV curves of cells utilizing cobalt-based electrolyte with standard (squares) or alumina-coated (circles) TiO$_2$ photoanodes under 300 W m^{-2} illumination. Reproduced from reference [161].

Solar cells based on organic electrolytes have some potential disadvantages such as difficulty in maintaining the electrolyte within the solar cells with normal encapsulants due to its volatility. Room-temperature Ionic liquids (RTILs) were developed in recent years and these have good chemical and thermal stability, negligible vapour pressure, non flammability, high ionic conductivity and high solubility for organic or inorganic materials, and a wide electrochemical window, and have been intensively pursued as alternative electrolytes for DSCs and also for other electrochemical devices. When incorporated into DSCs, they can be both the source of iodide and the solvent themselves. Efficient (η = 8.2 %) and stable performance of DSC based on ionic liquids has been recently demonstrated [162].

4.1.5. Counterelectrode

The task of a the counter electrode (CE) in DSCs is to return the charge from the external load to the electrolyte thus keeping the operating cycle of the cell continually running [163]. The charge transfer resistance between the electrode and the electrolyte appears as a voltage loss at the CE and it contributes directly to the series resistance of the cell. Thus, a good CE material must possess low charge transfer resistance (i.e. good electrocatalytic activity), but also good mechanical and chemical stability. The most commonly used CE in DSCs is FTO-coated glass, covered by a thin layer of platinum [130,164] due its high electrocatalytic activity and chemical and mechanical stability. This is either deposited by sputtering or deposited as a precursor which is then converted by a firing step at around 400 °C.

For flexible applications using plastic substrates, alternative methods of platinization are required. In one of these alternative methods, a platinum salt (i.e., $PtCl_4$ or K_2PtCl_6) is diluted in a mixture of organic solvents and by a reduction process carried out at temperatures ranging from 90 to 160° C, Pt is precipitated directly on the supporting material forming a fine, highly nanoparticulate sized layer of Pt. It has been shown that reaction temperatures of 90° C could be high enough to prepare counter electrodes with a sufficiently low RCT (around 0.5 Ωcm^2) and used to fabricate efficient DSCs [30].

There have been efforts to find alternatives to Pt which are low cost (even though the amounts utilized are very small) and low temperature. The most widely investigated materials are carbon based ones such as graphite which has good conducting properties and carbon black which has a very high surface area which is desirable when acting as a catalyst for the redox couple [165-170]. Due to their large scale use, for example as printing toners, carbon blacks (CBs) are already set for cheap industrial mass production and seem thus to be one of the promising alternative to the Pt layers, especially if transparency of the device is not a requirement.

4.1.6. Sealing Materials

In the DSC fabrication process, particular attention has been devoted to the sealing procedure of the cell, to avoid ingress of oxygen and moisture and external evaporation of the electrolyte. Furthermore, encapsulation needs to protect metallic fingers or vertical interconnects in modules from the corrosive I^-/I_3^- ions. Encapsulation of the glass substrates is obtained using thermoplastic foils made out of thermosetting resins such as Suryln® and Bynel® (Dupont) [164]. In order to pass the ageing tests at high temperatures it is necessary that the materials do not degrade. For some applications, the thermo sealing resins alone may not be sufficient to guarantee good hermetic properties since their melting point is around 80 ^0C -120 ^0C. In order to circumvent this problem it is possible to encapsulate DSCs using high melting temperature glass powders (glass frits) that are used today for plasma screen and LCD sealing [171-173]. Glass-frit can also be applied using screen printing techniques since it can be formulated into viscous pastes. It is also thermally and chemically stable. The main disadvantage of this technique is that requires high deposition temperatures (about 400 °C), and both dye and electrolyte needs to be injected after the sealing process.

4.2. DSC Modules

Photovoltaic technology allows the direct conversion of solar energy into electricity. The conversion takes place through photovoltaic cells that must be electrically connected with each other (modules) to achieve the required voltage and current levels. The photovoltaic modules can be used individually (for example 36 silicon cells can load a 12 V battery) or connected in series or in parallel to form strings and fields.

When interconnecting single cells together to form a module, a number of new issues arise from this system. These aspects have partly already been studied in other solar cell technologies, but DSC technology introduces additional problems. Although wide experience gained with previous technologies can be exploited, DSCs are electrochemical devices which require specific solutions as also detailed in reference [155]. For example, the presence and management of a liquid electrolyte is a new specific topic regarding this particular technology.

The connection of two or more cells (or modules) can be performed by connecting cells in parallel to obtain higher currents and in series to obtaining higher output voltages.

Connections can be external (this requires a dedicated post-processing of cells to build them into a module) or integrated. In integrated connection schemes, the cells in a module are built simultaneously, and are contacted with some integrated connection strategy. Integrated connection is generally preferable (when it is technically feasible) as it involves lower production costs.

4.2.1. Series Interconnections

The most simple series connection is the W structure (see figure 17). The single cell is electrically isolated removing TCO by Laser scribing. The removal of TCO occurs every two cells and a catalyst layer (CE) and a TiO_2 layer (photoelectrode) are deposed in alternation on the same glass substrate. In this way the series connection is simply made by the TCO strip that is in common between adjacent counter and photo electrodes of two neighbouring cells. The main advantage of this connection is that it does not necessitate additional conducting elements such as vertical interconnects. The latter are difficult to manage in relation to their possible contact with electrolyte and their stability in temperature. On the other hand, in the W configuration the counter electrode is dipped into dye which is process that can degrade the functioning of cells [174]. Moreover, half of the cells are exposed to light through the counter electrode side. These cells have a performance disadvantage since part of the light is absorbed by the electrolyte and platinum that can lead to electrical mismatch in the module [175]. A proper choice of cell width allows a design where no mismatch occurs. By tuning the electrolyte concentration (i.e. transparency), cell thickness, dye and the optical property of the TiO_2 layer, efficiencies up to 8.2 % have been demonstrated [176] (see figure 17).

The Z connection is a different series connection which includes a vertical connection between identical cells as shown in figure 18. This structure allows placing all counter and photo electrodes on opposite substrates avoiding cell mismatch issues. The onus on this scheme is the development of reliable interconnects that occupy as little space as possible to maintain a high module aperture ratio. The best results with this configuration have been reported at the 3rd DSC-IC Conference on the Industrialization of DSC technology by Sony, reaching conversion efficiencies of 8.2 %. To mitigate the active area disadvantage of Z structures in contrast with W connections, novel strategies are explored to approach the more favorable ratio of the active area on total area that is typical of W design (see schemes in figures 17 and 18).

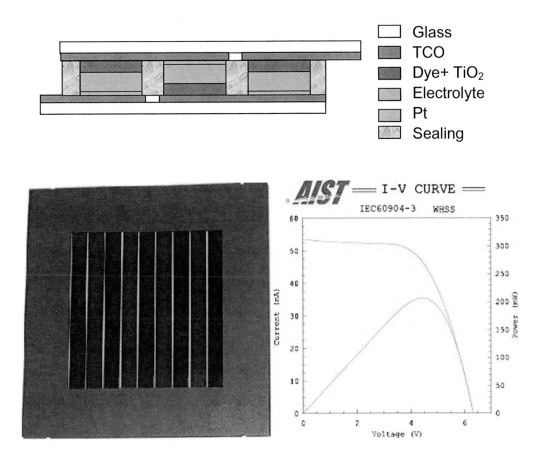

Figure 17. (above) W scheme for the series interconnection of DSC; (below) picture and J-V of a W module realized by SHARP (area 25.45 cm^2, Voc=6.33 V, Isc=53.6 mA, FF=61.2 %, Pmax=207.7 mW. Reproduced from reference[176].

Another important architecture is the so called monolithic design [165,178-181], where only one TCO covered transparent substrate is required (see figure 19). This requires the deposition of an intermediate porous layer that acts as a spacer and contains the electrolyte. The counter-electrode is deposited by screen printing with carbon based materials in a multilayer configuration. The thicker external layer provides an electrical path for carriers with a reasonably low surface resistance connecting two adjacent cells, while the inner layer is optimized for the catalytic activity, and can be Pt-loaded Carbon for example.

The main advantages of the monolithic design for the series interconnection of DSC modules consist in the cost and weight reduction arising from the use of only one TCO substrate, and the possibility of depositing all the internal layers (TiO$_2$, spacer, counter-electrode) by screen printing, and firing them all at once. All the layers are porous, and the modules can be stained in dye after sintering of both electrodes. A possible drawback arising from the use of carbon based counter-electrodes is their opacity although transparent screen printable counter-electrodes are being developed [181].

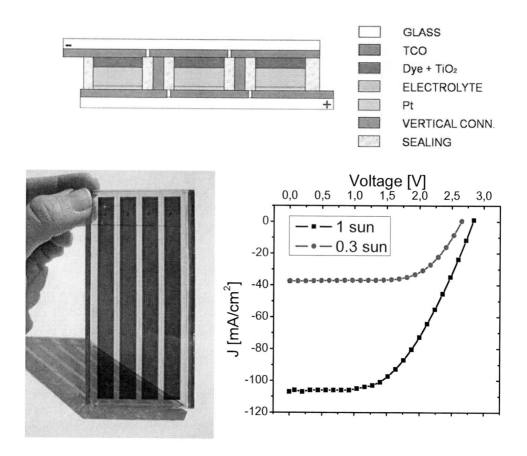

Figure 18. (above) Z scheme for the series interconnection of DSC; (below) picture and J-V of a Z module, with microscopic vertical interconnections fabricated at CHOSE (area = 37 cm^2)
Performances at AM 1.5, 1 sun intensity: V_{oc} = 2.84 V, I_{sc} = 106 mA, FF = 51%, η= 4.15 %.
Performances at AM 1.5, 0.3 sun intensity: V_{oc} = 2.65 V, I_{sc} = 37 mA/cm^2, FF = 64%, η= 5.8 %. [177].

4.2.2. Parallel Interconnections

Parallel interconnections aim at increasing current output delivered by a module (figure 20). A typical scheme requires metal fingers or grids on both electrodes. Cells are built interspersed with conductive fingers, for example, or are externally connected [181]. These maintain the fill factor which would be reduced by the resistive TCO. The encapsulation is necessary to avoid contact between silver and electrolyte, but is also important as it avoids possible short circuits at the expense of active area which is partly occupied by the conductive fingers and encapsulation. Other solutions that can be adopted include the use of conductors that are more resistant to corrosive effect of the electrolyte [132], the protection of the fingers with a layer of TiO_2 [182], TCO [132], glass frit and/or also with an additional polymer protective layer [183] and the development of highly conductive transparent conductive oxides [184].

The best performance of modules with parallel interconnections was shown by H. Arakawa et al. at the 3rd DSC-IC Conference on the Industrialization in Nara, Japan. These have now reached 8.1 % efficiency utilizing a low volatility electrolyte, while more that 10% efficiency has been demonstrated with a high performance volatile electrolyte (the module's aperture area was 82.9 cm^2). Grids for current collection have been also demonstrated on very large area cells by 3G solar, and on plastic substrates by Peccell at the 3rd DSC-IC Conference on the Industrialization in Nara, Japan. Promising performances can be achieved, mainly thanks to the large fill factor guaranteed by the grids.

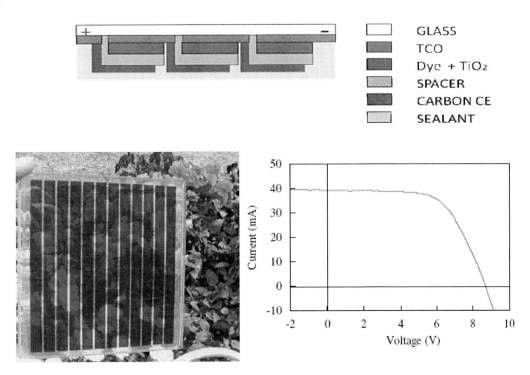

Figure 19. (above) Scheme of the monolithic design for series connection scheme for DSC modules; (below) picture and J-V characteristics of a monolithic module, with transparent counterelectrode, realized by Toyota-AISIN). Reproduced from reference [181].

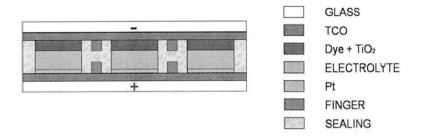

Figure 20. Scheme of the parallel design for high current, low voltage connection scheme for DSC modules.

4.3. Performance of DSC Modules

Several industrial outfits, like Dyesol, G24i, Aisin Seki, Sony, Sharp, Solaronix, Peccell, 3G Solar are investing large resources in developing DSC technology with the intention of a future commercialization. Sharp has the efficiency record of 10.4% on a cell of 1 cm^2 [185]. On smaller cells of 0.219 cm^2 an efficiency above 11% has been demonstrated [124]. Toyota reported that ambient light illumination effectively increases efficiency by 10-20% with respect to indoor simulations. Dyesol has realized a DSC demonstrator covering a building over an area of ~ 200 m^2.

Several strategies have been suggested recently to increase the upper limit of the DSC performance. Among the various techniques, very interesting are the co-sensitized DSCs, where the complementary action of different dyes contribute to increase the overall efficiency. Co-sensitization can occur according to different schemes, also based on tandem cell architecture, where two separate working electrodes are used for selective sensitization [186,187].

4.3.1. Stability

Since DSC technology is still relatively new, a uniformly accepted definition of a set of testing conditions specific for DSC technology is still to be developed and agreed upon. The most commonly used set of reference tests is driven by the test methods for a-Si, such as IEC 68 or JIS C-8938. The most important tests are the following:
- Dry heat cycle test (- 40°C to 90°C, 200 cycles),
- Heat-humidity cycle test (-40°C to 90°C with 85% humidity, 10 cycles),
- Light soaking test (Irradiance: 255W/m^2 (300nm -700nm), 500 hrs)
- Heat-humidity test (85°C with 85% humidity, 1000 hrs)

A typical target for DSC modules is to limit the reduction in module performances to within 10% of initial electrical parameters after tests, on a 10 cm x 10 cm module [188].

Obviously, these tests become meaningful once a reliable encapsulation and hermetic sealing of the cell is guaranteed. Test on cells are useful to study the intrinsic mechanisms which degrade the cell. A number of studies have assessed that DSCs do not degrade significantly under normal illumination [173,175,189,190]. A factor that limits the lifetime of the cells is the exposition to UV light [191]. The degradation mechanism due to UV light has been studied in detail and considerable improvements have been obtained by adding CaI$_2$ or MgI$_2$ to the electrolyte. These compounds have the ability to increase the cell lifetimes up to 3300 hours [189]. In order to go beyond this figure it is possible to integrate a polymeric film in the cell structure that filters the UV radiation [191]. In this way it is possible to reach many years of stability. Another factor that limits lifetime is the presence or ingress of oxygen and moisture. Accelerated lifetime tests have demonstrated a substantial invariance of performance for 27.500 hours at 55 °C and 0.8 sun conditions. Accepting standard acceleration factors leads to the conclusion that DSCs have lifetimes of 44 years under typical sun conditions of central Europe and 26 years in southern Europe [187]. Although these tests have yet to be transferred to modules, these very promising data make envisioning a future commercial use of this technology a reality.

4.4. Outdoor Performance

A very interesting aspect of the evaluation of DSC performances is the behavior of DSC in the different outdoor conditions that are experienced by modules during the year. It is known in fact that DSC technology performs well in low light intensities and diffuse illumination [192,193]. In DSCs the generated electricity does not increase linearly with light intensity [194,195]. The efficiency has a maximum at a radiation intensity of a fraction of a sun (e.g. 0.3 sun) and then decreases with increase in light intensity. A possible reason for this behavior seems to be related to limitation in the mass transport of ions within the cell electrodes.

The abovementioned consideration suggest that for DSC modules, the traditionally method of rating the nominal power capacity of a module based on "Watt-peak" concept can lead to underestimations of the effective potential of DSC technology. While the electrical power for standard crystalline Silicon modules are commonly rated at 1 sun, direct light, this illumination condition is not the most frequent condition for a solar panel working in realistic outdoor conditions. If the power is integrated over the whole year, the energy produced by DSC and c-Si modules rated at the same "Watt-peak" capabilities (c-Si technology will need less area to deliver the same Wp because of higher efficiencies) will produce different amount of electric energy. DSC production can exceed by 10-20 % that of c-Si [192].

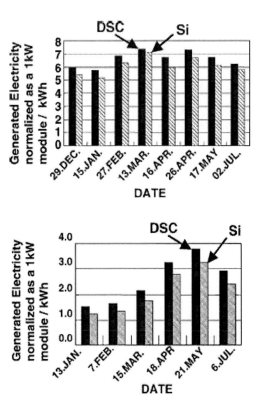

Figure 21. Comparison of DSC and c-Si module performances. (above) sunny day; (below) cloudy day; Modules have same nominal peak electric power, and different areas. Reproduced from reference[192].

CONCLUSIONS

In this chapter we have reviewed small molecule, conjugated polymer and dye solar cell technologies. Light absorption occurs in the molecular components of these devices. The electronic properties of the organic materials, including light absorption and electronic transport, can be tailored via chemical synthesis of a variety of molecular designs and a careful control of deposition parameters and device architectures. One of the biggest appeals of molecular based photovoltaic devices is that the active layers can be deposited either through simple evaporation processes or in the form of liquid solutions or pastes employing low-cost, low-temperature casting techniques such as those from the established printing industry. The potential for low cost flexible manufacturing of photovoltaics is huge.

REFERENCES

[1] A. Jäger-Waldau, *Renewable and Sustainable Energy Reviews 11/7* (2007) 1414.
[2] A. Jäger-Waldau, *Thin Solid Films 451* (2004) 448.
[3] A. Stoppato, *Energy 33/2* (2008) 224.
[4] D. Carlson, S. Div, P. Newtown, *IEEE Transactions on Electron Devices 36/12* (1989) 2775.
[5] K. Chopra, S. Das, *Thin film solar cells*, Plenum Pub Corp, (1983).
[6] A. Shah, P. Torres, R. Tscharner, N. Wyrsch, H. Keppner, *Science 285/5428* (1999) 692.
[7] M. Green, *Silicon solar cells: advanced principles & practice*, Centre for Photovoltaic Devices and Systems Sydney, (1995).
[8] J. Halls, C. Walsh, N. Greenham, E. Marseglia, R. Friend, S. Moratti, A. Holmes, *Nature 376* (1995) 498.
[9] B. O'Regan, M. Grätzel, *Nature 353* (1991) 737.
[10] C. Tang, *Applied Physics Letters 48* (1986) 183.
[11] G. Yu, J. Gao, J. Hummelen, F. Wudl, A. Heeger, *Science 270/5243* (1995) 1789.
[12] S. Alvarado, P. Seidler, D. Lidzey, D. Bradley, *Physical Review Letters 81/5* (1998) 1082.
[13] J.L. Bredas, J. Cornil, A.J. Heeger, *Adv. Mat. 8/5* (1996) 447.
[14] J. Halls, J. Cornil, D. Dos Santos, R. Silbey, D. Hwang, A. Holmes, J. Bredas, R. Friend, *Physical Review B 60/8* (1999) 5721.
[15] I. Hill, A. Kahn, Z. Soos, J. Pascal, RA, *Chemical Physics Letters 327/3-4* (2000) 181.
[16] M. Knupfer, Applied Physics A: *Materials Science & Processing 77/5* (2003) 623.
[17] R. Marks, J. Halls, D. Bradley, R. Friend, A. Holmes, *Journal of Physics: Condensed Matter 6* (1994) 1379.
[18] W. Huynh, J. Dittmer, A. Alivisatos, *Science 295* (2002) 2425.
[19] C. Brabec, *Solar energy materials and solar cells 83/2-3* (2004) 273.
[20] C. Brabec, J. Hauch, P. Schilinsky, C. Waldauf, *MRS bulletin 30* (2005) 51.
[21] H. Hoppe, N. Sariciftci, *J. Mater. Res 19/7* (2004) 1924.
[22] J. Kim, K. Lee, N. Coates, D. Moses, T. Nguyen, M. Dante, A. Heeger, *Science 317* (2007) 222.

[23] M. Kim, J. Kim, J. Cho, M. Shtein, L. Guo, J. Kim, *Applied Physics Letters 90* (2007) 123113.
[24] G. Li, V. Shrotriya, J. Huang, Y. Yao, T. Moriarty, K. Emery, Y. Yang, *Nature materials 4/11* (2005) 864.
[25] W. Ma, C. Yang, X. Gong, K. Lee, A. Heeger, *Advanced Functional Materials 15/10* (2005) 1617.
[26] J. Nelson, *Current Opinion in Solid State & Materials Science 6/1* (2002) 87.
[27] P. Peumans, S. Uchida, S. Forrest, *Nature 425/6954* (2003) 158.
[28] P. Peumans, A. Yakimov, S. Forrest, *Journal of Applied Physics 93* (2003) 3693.
[29] Konarka_Technologies, 2009.
[30] J. Kroon, N. Bakker, H. Smit, P. Liska, K. Thampi, P. Wang, S. Zakeeruddin, M. Gratzel, A. Hinsch, S. Hore, *Progress in photovoltaics: research and applications 15/1* (2007) 1.
[31] S. Forrest, *MRS bulletin 30/1* (2005) 28.
[32] B. Rand, J. Genoe, P. Heremans, J. Poortmans, *Progress in photovoltaics: research and applications 15/8* (2007) **659**.
[33] M. Riede, T. Mueller, W. Tress, R. Schueppel, K. Leo, *Nanotechnology 19* (2008) 424001.
[34] B. Rand, J. Xue, F. Yang, S. Forrest, *Applied Physics Letters 87* (2005) 233508.
[35] Z. Bao, A. Lovinger, A. Dodabalapur, *Applied Physics Letters 69* (1996) 3066.
[36] M. Ofuji, K. Ishikawa, H. Takezoe, K. Inaba, K. Omote, *Applied Physics Letters 86* (2005) 062114.
[37] G. Chamberlain, *Solar Cells 8/1* (1983) 47.
[38] J. Halls, K. Pichler, R. Friend, S. Moratti, A. Holmes, *Synthetic metals 77/1-3* (1996) 277.
[39] M. Baldo, M. Deutsch, P. Burrows, H. Gossenberger, M. Gerstenberg, V. Ban, S. Forrest, *Advanced Materials 10/18* (1998) 1505.
[40] S. Forrest, *Chemical Reviews 97/6* (1997) 1793.
[41] M. Vogel, S. Doka, C. Breyer, M. Lux-Steiner, K. Fostiropoulos, *Applied Physics Letters 89* (2006) 163501.
[42] J. Xue, B. Rand, S. Uchida, S. Forrest, *Journal of Applied Physics 98* (2005) 124903.
[43] J. Xue, B. Rand, S. Uchida, S. Forrest, *Advanced Materials 17/1* (2005).
[44] A. Mayer, S. Scully, B. Hardin, M. Rowell, M. McGehee, *Materials Today 10/11* (2007) 28.
[45] J. Xue, S. Uchida, B. Rand, S. Forrest, *Applied Physics Letters 85* (2004) 5757.
[46] R. Franke, B. Maennig, A. Petrich, M. Pfeiffer, *Solar energy materials and solar cells 92/7* (2008) 732.
[47] H. Hänsel, H. Zettl, G. Krausch, C. Schmitz, R. Kisselev, M. Thelakkat, H. Schmidt, *Applied Physics Letters 81* (2002) 2106.
[48] W. Potscavage, S. Yoo, B. Domercq, B. Kippelen, *Appl. Phys. Lett 90* (2007) 253511.
[49] Q. Song, M. Wang, E. Obbard, X. Sun, X. Ding, X. Hou, C. Li, *Applied Physics Letters 89* (2006) 251118.
[50] J. Burroughes, C. Jones, R. Friend, *Nature 335* (1988) 137.
[51] J.H. Burroughes, D.D.C. Bradley, A.R. Brown, R.N. Marks, K. Mackay, R.H. Friend, P.L. Burn, A.B. Holmes, *Nature 347* (1990) 539.
[52] H. Sirringhaus, N. Tessler, R. Friend, *Science 280* (1998) 1741.

[53] S. Forrest, *Nature 428* (2004) 911.
[54] T. Ameri, G. Dennler, C. Lungenschmied, C. Brabec, *Energy & Environmental Science 2/4* (2009) 347.
[55] C. Brabec, J. Durrant, *MRS bulletin 33* (2008) 670.
[56] C. Brabec, N. Sariciftci, J. Hummelen, *Advanced Functional Materials 11/1* (2001) 15.
[57] K. Coakley, M. McGehee, *Chemistry of materials 16/23* (2004) 4533.
[58] S. Guenes, H. Neugebauer, N. Sariciftci, *Chem. Rev 107* (2007) 1324.
[59] J. Nunzi, *Comptes rendus-Physique 3/4* (2002) 523.
[60] G. Dennler, M. Scharber, C. Brabec, *Advanced Materials 21/13* (2009) 1323.
[61] T.M. Brown, J.S. Kim, R.H. Friend, F. Cacialli, R. Daik, W.J. Feast, *Applied Physics Letters 75/12* (1999) 1679.
[62] S.A. Carter, M. Angelopoulos, S. Karg, P.J. Brock, J.C. Scott, *Applied Physics Letters 70/16* (1997) 2067.
[63] C. Brabec, S. Shaheen, C. Winder, N. Sariciftci, P. Denk, *Applied Physics Letters 80* (2002) 1288.
[64] T.M. Brown, R.H. Friend, I. Millard, D. Lacey, J.H. Burroughes, F. Cacialli, *Applied Physics Letters 77/19* (2000) 3096.
[65] L.S. Hung, C.W. Tang, M.G. Mason, *Applied Physics Letters 70/2* (1997) 152.
[66] S. Shaheen, C. Brabec, N. Sariciftci, F. Padinger, T. Fromherz, J. Hummelen, *Applied Physics Letters 78* (2001) 841.
[67] J.J.M. Halls, K. Pichler, R.H. Friend, S.C. Moratti, A.B. Holmes, *Appl. Phys. Lett. 68/* (1996) 3120.
[68] A. Arias, N. Corcoran, M. Banach, R. Friend, J. MacKenzie, W. Huck, *Applied Physics Letters 80* (2002) 1695.
[69] A. Arias, J. MacKenzie, R. Stevenson, J. Halls, M. Inbasekaran, E. Woo, D. Richards, R. Friend, *Macromolecules 34/17* (2001) 6005.
[70] J. Halls, A. Arias, J. MacKenzie, W. Wu, M. Inbasekaran, E. Woo, R. Friend, *Advanced Materials 12/7* (2000) 498.
[71] C. McNeill, A. Abrusci, J. Zaumseil, R. Wilson, M. McKiernan, J. Burroughes, J. Halls, N. Greenham, R. Friend, *Applied Physics Letters 90* (2007) 193506.
[72] N. Camaioni, G. Ridolfi, G. Casalbore-Miceli, G. Possamai, M. Maggini, *Advanced Materials 14/23* (2002) 1735.
[73] K. Yoshino, X. YIN, S. Morita, T. Kawai, A. Zakhidov, *Solid state communications 85/2* (1993) 85.
[74] J. Kim, S. Kim, H. Lee, K. Lee, W. Ma, X. Gong, A. Heeger, *Advanced Materials 18/5* (2006) 572.
[75] Y. Kim, S. Choulis, J. Nelson, D. Bradley, S. Cook, J. Durrant, *Applied Physics Letters 86* (2005) 063502.
[76] F. Padinger, C. Brabec, T. Fromherz, J. Hummelen, N. Sariciftci, *Optoelectronics Review* (2000) 280.
[77] F. Padinger, R. Rittberger, N. Sariciftci, *Advanced Functional Materials 13/1* (2003) 85.
[78] P. Schilinsky, C. Waldauf, C. Brabec, *Applied Physics Letters 81* (2002) 3885.
[79] K. Coakley, M. McGehee, *Applied Physics Letters 83* (2003) 3380.
[80] A. Alivisatos, *Science 271* (1996) 933.

[81] N. Greenham, X. Peng, A. Alivisatos, *Physical Review-Section B-Condensed Matter 54/24* (1996) 17628.
[82] W. Huynh, J. Dittmer, W. Libby, G. Whiting, A. Alivisatos, *Advanced Functional Materials 13/1* (2003) 73.
[83] W. Huynh, X. Peng, A. Alivisatos, *Proc. Electrochem. Soc* (1999) 99.
[84] L. Li, J. Hu, W. Yang, A. Alivisatos, *Nano Letters 1/7* (2001) 349.
[85] I. Gur, N. Fromer, M. Geier, A. Alivisatos, *Science 310* (2005) 462.
[86] J. Luther, M. Law, M. Beard, Q. Song, M. Reese, R. Ellingson, A. Nozik, *Nano Letters 8/10* (2008) 3488.
[87] W. Ma, J. Luther, H. Zheng, Y. Wu, A. Alivisatos, *Nano Letters* (2009) 12700.
[88] M. Granström, K. Petritsch, A.C. Arias, A. Lux, M.R. Andersson, R.H. Friend, *Nature 395/6699* (1998) 257.
[89] H. Hoppe, N. Sariciftci, *Journal of Materials Chemistry 16/1* (2006) 45.
[90] M. Reyes-Reyes, K. Kim, D. Carroll, *Applied Physics Letters 87* (2005) 083506.
[91] X. Yang, J. Loos, S. Veenstra, W. Verhees, M. Wienk, J. Kroon, M. Michels, R. Janssen, *Nano Letters 5/4* (2005) 579.
[92] L. Schmidt-Mende, A. Fechtenkotter, K. Mullen, E. Moons, R. Friend, J. MacKenzie, *Science 293* (2001) 1119.
[93] J. Peet, J. Kim, N. Coates, W. Ma, D. Moses, A. Heeger, G. Bazan, Nature Materials, 6 (2007) 497.
[94] S. Park, A. Roy, S. Beaupré, S. Cho, N. Coates, J. Moon, D. Moses, M. Leclerc, K. Lee, A. Heeger, *Nature Photonics 3/5* (2009) 297.
[95] M. Scharber, D. Mühlbacher, M. Koppe, P. Denk, C. Waldauf, A. Heeger, C. Brabec, *Advanced Materials 18/6* (2006) 789.
[96] H. Hoppe, N. Arnold, D. Meissner, N. Sariciftci, *Thin Solid Films 451* (2004) 589.
[97] L. Pettersson, L. Roman, O. Inganäs, *Journal of Applied Physics 86* (1999) 487.
[98] G. Dennler, H. Prall, R. Koeppe, M. Egginger, R. Autengruber, N. Sariciftci, *Applied Physics Letters 89* (2006) 073502.
[99] J. Gilot, M. Wienk, R. Janssen, *Applied Physics Letters 90* (2007) 143512.
[100] K. Kawano, N. Ito, T. Nishimori, J. Sakai, *Applied Physics Letters 88* (2006) 073514.
[101] G. Dennler, M. Scharber, T. Ameri, P. Denk, K. Forberich, C. Waldauf, C. Brabec, *Advanced Materials 20/3* (2008) 579.
[102] M. Jørgensen, K. Norrman, F. Krebs, *Solar energy materials and solar cells 92/7* (2008) 686.
[103] B. Zimmermann, U. Würfel, M. Niggemann, *Solar energy materials and solar cells* (2009).
[104] C. Lungenschmied, G. Dennler, H. Neugebauer, S. Sariciftci, M. Glatthaar, T. Meyer, A. Meyer, *Solar energy materials and solar cells 91/5* (2007) 379.
[105] F. Krebs, H. Spanggard, T. Kjær, M. Biancardo, J. Alstrup, *Materials Science & Engineering B 138/2* (2007) 106.
[106] L. Moro, N. Rutherford, R. Visser, J. Hauch, C. Klepek, P. Denk, P. Schilinsky, C. Brabec, *Proceedings of SPIE 2006*, (2006), 63340M.
[107] J. Hauch, P. Schilinsky, S. Choulis, S. Rajoelson, C. Brabec, *Applied Physics Letters 93* (2008) 103306.
[108] J. Hauch, P. Schilinsky, S. Choulis, R. Childers, M. Biele, C. Brabec, *Solar energy materials and solar cells 92/7* (2008) 727.

[109] R. Tipnis, J. Bernkopf, S. Jia, J. Krieg, S. Li, M. Storch, D. Laird, *Solar energy materials and solar cells 93/4* (2009) 442.
[110] S. Yoo, W. Potscavage, B. Domercq, J. Kim, J. Holt, B. Kippelen, *Applied Physics Letters 89* (2006) 233516.
[111] F. Krebs, *Solar energy materials and solar cells 93/4* (2009) 394.
[112] F. Krebs, *Organic Electronics* (2009).
[113] F. Krebs, *Solar energy materials and solar cells 93/4* (2009) 484.
[114] F. Krebs, J. Alstrup, H. Spanggaard, K. Larsen, E. Kold, *Solar energy materials and solar cells 83/2-3* (2004) 293.
[115] F. Krebs, M. Jørgensen, K. Norrman, O. Hagemann, J. Alstrup, T. Nielsen, J. Fyenbo, K. Larsen, J. Kristensen, *Solar energy materials and solar cells 93/4* (2009) 422.
[116] V. Marin, E. Holder, M. Wienk, E. Tekin, D. Kozodaev, U. Schubert, *Macromolecular Rapid Communications 26/4* (2005) 319.
[117] S. Shaheen, R. Radspinner, N. Peyghambarian, G. Jabbour, *Applied Physics Letters 79* (2001) 2996.
[118] C. Hoth, S. Choulis, P. Schilinsky, C. Brabec, *Adv. Mater 19* (2007) 3973.
[119] T. Aernouts, T. Aleksandrov, C. Girotto, J. Genoe, J. Poortmans, *Applied Physics Letters 92* (2008) 033306.
[120] R. Gaudiana, C. Brabec, *Nature Photonics 2/5* (2008) 287.
[121] J. Ding, A. de la Fuente Vornbrock, C. Ting, V. Subramanian, *Solar energy materials and solar cells 93/4* (2009) 459.
[122] L. Blankenburg, K. Schultheis, H. Schache, S. Sensfuss, M. Schrödner, *Solar energy materials and solar cells 93* (2009) 476.
[123] M. Graetzel, *Nature 414* (2001) 338.
[124] M. Graetzel, *Chemistry Letters 34/1* (2005) 8.
[125] M. Gratzel, *Progress in photovoltaics: research and applications 8/1* (2000) 171.
[126] M. Gratzel, *Inorg. Chem 44/20* (2005) 6841.
[127] M. Grätzel, *Journal of Photochemistry & Photobiology, C: Photochemistry Reviews 4/2* (2003) 145.
[128] M. Grätzel, *Journal of Photochemistry & Photobiology, A: Chemistry 164/1-3* (2004) 3.
[129] M. Nazeeruddin, *Coordination Chemistry Reviews 248/13-14* (2004) 1161.
[130] M. Nazeeruddin, A. Kay, I. Rodicio, R. Humphry-Baker, E. Müller, P. Liska, N. Vlachopoulos, M. Grätzel, *Journal of the American Chemical Society 115/14* (1993) 6382.
[131] M. Nazeeruddin, P. Pechy, T. Renouard, S. Zakeeruddin, R. Humphry-Baker, P. Comte, P. Liska, E. Costa, V. Shklover, L. Spiccia, *J. Am. Chem. Soc 123/8* (2001) 1613.
[132] K. Okada, H. Matsui, T. Kawashima, T. Ezure, N. Tanabe, *Journal of Photochemistry & Photobiology, A: Chemistry 164/1-3* (2004) 193.
[133] S. Ito, T. Murakami, P. Comte, P. Liska, C. Grätzel, M. Nazeeruddin, M. Grätzel, *Thin Solid Films 516/14* (2008) 4613.
[134] S. Ito, P. Liska, P. Comte, R. Charvet, P. Péchy, U. Bach, L. Schmidt-Mende, S. Zakeeruddin, A. Kay, M. Nazeeruddin, *Chemical Communications 2005/34* (2005) 4351.
[135] P. Wang, S. Zakeeruddin, P. Comte, R. Charvet, R. Humphry-Baker, M. Gratzel, *J. Phys. Chem. B 107/51* (2003) 14336.

[136] H. Arakawa, T. Yamaguchi, A. Takeuchi, S. Agatsuma, WCPEC-4 Conference Record of the 2006 IEEE 4th World Conference on Photovoltaic Energy Conversion, (2006) 36

[137] S. Colodrero, A. Mihi, J. Anta, M. Ocan~a, H. Míguez, *The Journal of Physical Chemistry C 113/4* (2009) 1150.

[138] S. Colodrero, A. Mihi, L. Haggman, M. Ocana, G. Boschloo, A. Hagfeldt, H. Miguez, *Advanced Materials 21/7* (2009) 764.

[139] C. Hägglund, M. Zäch, B. Kasemo, *Applied Physics Letters 92* (2008) 013113.

[140] L. Vesce, A. Reale, T.M. Brown, A. Di Carlo.

[141] S. Ito, P. Chen, P. Comte, M. Nazeeruddin, P. Liska, P. Pechy, M. Gratzel, *Progress in photovoltaics: research and applications 15/7* (2007) 603.

[142] T. Yamaguchi, N. Tobe, D. Matsumoto, H. Arakawa, *Chemical Communications 2007/45* (2007) 4767.

[143] T. Miyasaka, Y. Kijitori, *Journal of the Electrochemical Society 151* (2004) A1767.

[144] S. Uchida, M. Tomiha, N. Masaki, A. Miyazawa, H. Takizawa, *Solar energy materials and solar cells 81/1* (2004) 135.

[145] G. Mincuzzi, L. Vesce, A. Reale, A. Di Carlo, T.M. Brown, *Applied Physics Letters, 95* (2009) **103312**.

[146] J. Baxter, A. Walker, K. Van Ommering, E. Aydil, *Nanotechnology 17/11* (2006) 304.

[147] S. Albu, A. Ghicov, J. Macak, P. Schmuki, *physica status solidi (RRL)-Rapid Research Letters 1/2* (2007) R65.

[148] L. Greene, M. Law, J. Goldberger, F. Kim, J. Johnson, Y. Zhang, R. Saykally, P. Yang, *Angew. Chem. Int. Ed 42/26* (2003) 3031.

[149] L. Greene, B. Yuhas, M. Law, D. Zitoun, P. Yang, *Nat. Mater 4* (2005) 455.

[150] J. Macak, H. Tsuchiya, L. Taveira, S. Aldabergerova, P. Schmuki, *Angewandte Chemie International Edition 44/45* (2005) 7463.

[151] K. Kalyanasundaram, M. Grätzel, *Coordination Chemistry Reviews 177/1* (1998) 347.

[152] M. Nazeeruddin, F. De Angelis, S. Fantacci, A. Selloni, G. Viscardi, P. Liska, S. Ito, B. Takeru, M. Gratzel, *J. Am. Chem. Soc 127/48* (2005) 16835.

[153] D. Kuang, S. Ito, B. Wenger, C. Klein, J. Moser, R. Humphry-Baker, S. Zakeeruddin, M. Gratzel, *J. Am. Chem. Soc 128/12* (2006) 4146.

[154] G. Zhang, H. Bala, Y. Cheng, D. Shi, X. Lv, Q. Yu, P. Wang, *chem. Comm.* (2009) 2198.

[155] A. Di Carlo, A. Reale, T. Brown, M. Cecchetti, F. Giordano, G. Roma, M. Liberatore, V. Miruzzo, V. Conte, in: I.A. Luk'yanchuk, D. Mezzane (Eds.), *Smart Materials for Energy, Communications and Security*, Springer, (2008), 97.

[156] S. Sapp, C. Elliott, C. Contado, S. Caramori, C. Bignozzi, *J. Am. Chem. Soc 124/37* (2002) 11215.

[157] D. Kuang, C. Klein, H. Snaith, J. Moser, R. Humphry-Baker, P. Comte, S. Zakeeruddin, M. Gratzel, *Nano Letters 6/4* (2006) 769.

[158] H. Nusbaumer, S. Zakeeruddin, J. Moser, M. Gratzel, Chemistry-A *European Journal 9/16* (2003) 3756.

[159] M. Brugnati, S. Caramori, S. Cazzanti, L. Marchini, R. Argazzi, C. Bignozzi, *International Journal of Photoenergy 2007* (2007) 1.

[160] C. Elliott, S. Caramori, C. Bignozzi, *Langmuir 21/7* (2005) 3022.

[161] M. Liberatore, L. Burtone, T. Brown, A. Reale, A. Di Carlo, F. Decker, S. Caramori, C. Bignozzi, *Applied Physics Letters 94* (2009) 173113.

[162] Y. Bai, Y. Cao, J. Zhang, M. Wang, R. Li, P. Wang, S. Zakeeruddin, M. Grätzel, *Nature Materials 7* (2008) 626.
[163] J. Halme, M. Toivola, A. Tolvanen, P. Lund, *Solar energy materials and solar cells 90/7-8* (2006) 872.
[164] S. Ito, M. Nazeeruddin, P. Liska, P. Comte, R. Charvet, P. Pechy, M. Jirousek, A. Kay, S. Zakeeruddin, M. Gratzel, *Progress in photovoltaics: research and applications 14/7* (2006) 589.
[165] S. Burnside, S. Winkel, K. Brooks, V. Shklover, M. Gra¨ tzel, A. Hinsch, R. Kinderman, C. Bradbury, A. Hagfeldt, H. Pettersson, *Journal of Materials Science: Materials in Electronics 11/4* (2000) 355.
[166] K. Imoto, K. Takahashi, T. Yamaguchi, T. Komura, J. Nakamura, K. Murata, *Solar energy materials and solar cells 79/4* (2003) 459.
[167] A. Kay, M. Grätzel, *Solar energy materials and solar cells 44/1* (1996) 99.
[168] T. Kitamura, M. Maitani, M. Matsuda, Y. Wada, S. Yanagida, *Chemistry Letters 30/10* (2001) 1054.
[169] T. Murakami, M. Grätzel, *Inorganica Chimica Acta 361/3* (2008) 572.
[170] K. Suzuki, M. Yamaguchi, M. Kumagai, S. Yanagida, *Chemistry Letters 32/1* (2003) 28.
[171] R. Knechtel, *Microsystem Technologies 12/1* (2005) 63.
[172] W. Lee, E. Ramasamy, D. Lee, J. Song, *Journal of Photochemistry & Photobiology, A: Chemistry 183/1-2* (2006) 133.
[173] R. Sastrawan, J. Beier, U. Belledin, S. Hemming, A. Hinsch, R. Kern, C. Vetter, F. Petrat, A. Prodi-Schwab, P. Lechner, *Progress in photovoltaics: research and applications 14/8* (2006) 697.
[174] G. Tulloch, *Journal of Photochemistry & Photobiology, A: Chemistry 164/1-3* (2004) 209.
[175] R. Sastrawan, J. Renz, C. Prahl, J. Beier, A. Hinsch, R. Kern, *Journal of Photochemistry & Photobiology, A: Chemistry 178/1* (2006) 33.
[176] L. Han, A. Fukui, Y. Chiba, A. Islam, R. Komiya, N. Fuke, N. Koide, R. Yamanaka, M. Shimizu, *Applied Physics Letters 94* (2009) 013305.
[177] F. Giordano, E. Petrolati, S. Mastroianni, T.M. Brown, A. Reale, A. Di Carlo, HOPV Conference, (2009).
[178] T. Meyer, Solaronix, *3rd DSC-IC Conference on the Industrialization of DSC technology, Nara*, Japan, 2009.
[179] K. Noda, Sony, *3rd DSC-IC Conference on the Industrialization of DSC technology, Nara*, Japan, 2009.
[180] H. Pettersson, T. Gruszecki, L. Johansson, P. Johander, *Solar energy materials and solar cells 77/4* (2003) 405.
[181] Y. Takeda, N. Kato, K. Higuchi, A. Takeichi, T. Motohiro, S. Fukumoto, T. Sano, T. Toyoda, *Solar energy materials and solar cells 93* (2008) 808.
[182] K. Mandal, M. Choi, C. Noblitt, R. Rauh, *Materials Research Society Symposium Proceedings*; 836, (2005), 3.
[183] A. Hinsch, I.-. Fraunhofer, *3rd DSC-IC Conference on the Industrialization of DSC technology, Nara*, Japan, 2009.
[184] K. Goto, T. Kawashima, N. Tanabe, *Solar energy materials and solar cells 90/18-19* (2006) 3251.

[185] M. Green, K. Emery, Y. Hishikawa, W. Warta, *Progress in Photovoltaics 17* (2009) 85.
[186] Y. Yoshida, S. Pandey, K. Uzaki, S. Hayase, M. Kono, Y. Yamaguchi, *Applied Physics Letters 94* (2009) 093301.
[187] H. Desilvestro, *SC-IC Conference on the Industrialization of DSC technology*, Nara, Japan, (2009).
[188] H. Arakawa, T. Yamaguchi, S. Agatsuma, T. Sutou, Y. Koishi, *Proc. of Renewable Energy* 2008, Busan, Korea, (2008).
[189] A. Hinsch, J. Kroon, R. Kern, I. Uhlendorf, J. Holzbock, A. Meyer, J. Ferber, *Long-term stability of dye-sensitised solar cells, Progress in photovoltaics: research and applications 9* (2001) 425.
[190] P. Wang, C. Klein, R. Humphry-Baker, S. Zakeeruddin, M. Grätzel, *Applied Physics Letters 86* (2005) 123508.
[191] H. Pettersson, T. Gruszecki, *Solar energy materials and solar cells 70/2* (2001) 203.
[192] T. Toyoda, T. Sano, J. Nakajima, S. Doi, S. Fukumoto, A. Ito, T. Tohyama, M. Yoshida, T. Kanagawa, T. Motohiro, *Journal of Photochemistry & Photobiology, A: Chemistry 164/1-3* (2004) 203.
[193] N. Kato, Y. Takeda, K. Higuchi, A. Takeichi, E. Sudo, H. Tanaka, T. Motohiro, T. Sano, T. Toyoda, *Solar energy materials and solar cells 93* (2009) 893.
[194] K. Hara, T. Horiguchi, T. Kinoshita, K. Sayama, H. Sugihara, H. Arakawa, *Solar energy materials and solar cells 64/2* (2000) 115.
[195] H. Lindström, A. Holmberg, E. Magnusson, L. Malmqvist, A. Hagfeldt, *Journal of photochemistry and photobiology. A, Chemistry 145/1-2* (2001) 107.

INDEX

A

absorption spectra, 263, 264
absorption spectroscopy, 72
acetone, 269
acetonitrile, 273
acid, 177, 260, 273
activation energy, 56, 113, 126, 129, 131, 136, 138, 213, 219
adaptability, 49
additives, 273
adhesion, 41, 45
adsorption, 43, 54, 182, 218, 268
aesthetic, 271
AFM, 150, 175, 176, 184, 215
aggressiveness, 273
aluminium, 59
ammonia, 53, 82, 93, 147
ammonium, 102, 182
amplitude, 117, 127
anisotropy, 41
annealing, 87, 88, 114, 129, 146, 148, 149, 153, 155, 156, 157, 158, 159, 162, 172, 176, 177, 178, 179, 182, 183, 187, 199, 200, 211, 213, 230, 261, 262
annual rate, 34
anodization, 39
antimony, 188
anti-reflective coating, 235
aqueous solutions, 82, 172
argon, 150, 190, 220
assessment, 19, 26, 45, 46, 248
assets, 203
asymmetry, 238
atmosphere, 114, 143, 149, 173, 175, 177, 178, 183, 185, 186, 187, 194, 199, 261
atmospheric pressure, 38, 53, 203
atoms, 42, 51, 52, 69, 76, 146, 147, 148, 149, 172, 175, 180, 183, 185, 186, 189, 191, 192, 193, 194, 198, 208, 209, 210, 213, 214, 217, 221
Auger electron spectroscopy, 101, 102, 103
authorities, 218
automation, 28, 222
awareness, 21

B

backscattering, 153
ban, 94
barriers, 78, 81, 89, 96, 120, 126, 149, 265
base, 31, 36, 79, 81, 82, 92, 178, 218, 223, 242, 243, 248, 274
batteries, 240
bending, 45, 56, 79, 115
beneficial effect, 77, 78, 185, 186, 257
benefits, 9, 27, 81, 170, 203, 262
bias, 41, 112, 113, 114, 115, 116, 117, 118, 120, 121, 122, 123, 124, 125, 126, 129, 131, 133, 136, 138, 240, 242
binding energy, 203, 254, 256
birefringence, 41
blends, 14, 262
Boltzman constant, 129, 237, 243
bonding, 208, 211, 217, 222, 268
bonds, 15, 53, 209, 210, 212, 213, 214, 216, 217, 220, 236, 243, 254
bounds, 128
Brazil, 7, 10, 36
bromine, 154, 156
bulk materials, 167, 188

C

cadmium, 13, 21, 79, 92, 93, 102, 147, 167, 168, 182, 192, 203
candidates, 20, 52, 79, 186
carbon, 19, 212, 217, 253, 254, 273, 275, 277
carcinogen, 149
carcinogenicity, 41
casting, 253, 259, 261, 282
catalyst, 267, 273, 275, 276
catalytic activity, 277

cation, 188
chalcogenides, 182
challenges, 22, 31, 34, 38, 104, 259
charge density, 112, 117, 121, 124, 237
China, 7, 10, 11, 30, 36
chlorine, 149, 150, 172, 176, 177, 178, 180, 181, 194, 199, 200, 202, 203
chlorobenzene, 261
CIS, 17, 31, 78, 83, 90, 92, 93, 95, 109, 246
classes, 56
classification, 13
cleaning, 43, 56, 177, 182, 201, 269
clusters, 40, 42, 147
coatings, 37, 39, 52, 63, 64, 72, 97, 188, 248
cobalt, 273, 274
collaboration, 18
commercial, 16, 20, 21, 22, 34, 35, 42, 44, 49, 51, 58, 87, 116, 144, 146, 159, 162, 172, 193, 227, 244, 258, 266, 280
communication, 141
compatibility, 25, 53, 268
compensation, 15, 128
competition, 12
competitiveness, 31
complexity, 112, 210
composition, 15, 41, 49, 51, 56, 61, 68, 70, 71, 73, 80, 81, 82, 85, 101, 102, 116, 117, 182, 215, 219, 220, 222, 228
compounds, 40, 49, 50, 51, 81, 171, 172, 173, 176, 190, 194, 253, 280
computer, 172, 238
concentration ratios, 18, 135
condensation, 13, 40, 41, 172, 182
conditioning, 47, 57
conductance, 120
conduction, 19, 52, 79, 82, 83, 86, 88, 89, 94, 96, 108, 115, 121, 127, 129, 130, 160, 193, 211, 212, 213, 217, 236, 237, 240, 266, 267, 272, 273
conductivity, 41, 49, 50, 51, 52, 68, 88, 112, 146, 156, 169, 175, 188, 191, 192, 193, 194, 203, 209, 213, 225, 228, 229, 232, 235, 268, 274
conductor, 64, 189, 261
conductors, 51, 278
conference, 63, 91, 92, 95, 248, 249, 251, 252
configuration, 37, 42, 44, 114, 120, 121, 144, 157, 158, 159, 160, 161, 162, 169, 171, 198, 209, 213, 217, 224, 225, 226, 227, 228, 230, 276, 277
confinement, 27, 60
conjugation, 254
conservation, 212
consolidation, 45
constant rate, 68
constituent materials, 169, 263

constituents, 21
construction, 28, 202, 253, 254, 257, 260, 268
consulting, 34, 246
consumers, 12, 32
consumption, 9, 222, 228
contaminant, 217
contamination, 85
contour, 243, 264
contradiction, 116
controversial, 23
convergence, 48
cooling, 48, 61, 121, 202
coordination, 210, 217, 273
copolymer, 245
copper, 39, 68, 71, 72, 73, 75, 76, 77, 92, 93, 120, 155, 156, 157, 159, 167, 188, 194, 255, 257, 258
copyright, 214, 244
correlation, 117, 121, 126, 138, 155, 212
corrosion, 15, 41, 245
cost, 7, 16, 17, 18, 19, 20, 21, 22, 23, 24, 26, 27, 28, 31, 32, 33, 34, 35, 36, 38, 45, 46, 47, 50, 53, 56, 57, 58, 59, 62, 76, 77, 80, 90, 144, 170, 182, 207, 208, 209, 210, 221, 246, 247, 248, 253, 255, 259, 261, 264, 265, 266, 268, 271, 275, 277, 282
covalent bond, 208, 210
covering, 7, 26, 75, 203, 226, 280
cracks, 214
creativity, 59
crises, 9, 10
critical value, 213, 217
CSCs, 254
CSS, 13, 143, 148, 149, 150, 151, 152, 153, 168, 169, 172, 173, 174, 175, 176, 177, 178, 181, 183, 184, 185, 186, 199, 201, 202
customers, 30, 45
CVD, 13, 37, 38, 39, 40, 41, 47, 55, 56, 82, 93, 191, 193, 219, 220, 226
cycles, 36, 82, 159, 280

D

damages, iv, 53
dark conductivity, 213
decay, 216, 265
decomposition, 182, 219, 221, 226
deficiency, 13, 120
degenerate, 188
degradation, 14, 15, 16, 18, 20, 21, 22, 23, 48, 55, 87, 132, 155, 157, 159, 179, 180, 196, 208, 210, 213, 225, 230, 231, 235, 258, 268, 280
degradation mechanism, 21, 23, 280
Department of Energy, 19, 32, 247, 252
depth, 55, 68, 69, 72, 75, 98, 103, 104, 108, 129, 179

Index

derivatives, 257, 260
detection, 74, 98, 99, 100, 113
deviation, 188
dielectric constant, 41, 273
dielectric strength, 41
differential equations, 237
diffraction, 101, 106, 152, 154, 161
diffusion, 15, 16, 23, 41, 42, 43, 45, 48, 54, 55, 70, 72, 77, 78, 85, 103, 146, 149, 153, 155, 158, 171, 175, 179, 180, 190, 191, 192, 194, 198, 203, 213, 217, 219, 223, 228, 229, 230, 255, 256, 257, 258, 261, 262, 263, 267, 272, 273
diffusivity, 69, 72, 155
diluent, 220
dimerization, 54
diode laser, 26
diodes, 49, 259
discharges, 189, 230
discontinuity, 229
disorder, 210, 211, 214, 263
dispersion, 39, 259
dissociation, 39, 54, 183, 218, 255, 256
distortions, 112, 115, 210
distribution, 30, 31, 68, 69, 89, 103, 107, 116, 119, 122, 127, 130, 136, 137, 175, 178, 181, 212, 213, 232, 236, 238, 240
dominance, 37
donors, 120, 122, 123, 124, 127, 128, 180, 237, 270
dopants, 49, 78, 117, 217, 230, 238
doping, 16, 43, 51, 52, 78, 88, 94, 112, 115, 118, 127, 128, 136, 138, 144, 149, 171, 189, 191, 192, 209, 210, 217, 223, 229, 230, 240, 241, 253
double bonds, 254
draft, 20
drying, 173
DSC, 14, 27, 267, 268, 269, 272, 273, 274, 275, 276, 277, 278, 279, 280, 281, 288, 289
ductility, 41
durability, 18, 208
dye, 61, 271, 280, 289

E

efficiency level, 15
engineering, 13, 15, 18, 37, 59, 60, 89, 92, 263, 271, 273
enlargement, 150, 154, 189
entropy, 62
EPR, 210, 213
equilibrium, 43, 120, 126, 136, 149, 175, 238, 239, 240
equipment, 11, 20, 21, 27, 28, 29, 44, 46, 80, 169, 172

erosion, 41
ester, 260
etching, 18, 54, 55, 79, 151, 153, 155, 156, 177, 194, 196, 197, 220, 221, 226, 265
ethanol, 269
ethylene, 37, 259, 273
evaporation, 5, 38, 40, 67, 68, 69, 70, 71, 72, 73, 74, 76, 77, 78, 79, 80, 81, 84, 86, 89, 90, 91, 97, 98, 102, 103, 116, 143, 147, 149, 150, 158, 159, 160, 172, 175, 176, 177, 178, 183, 186, 189, 191, 209, 253, 255, 257, 263, 275, 282
evolution, 48, 150, 180, 211
excitation, 24, 39, 40, 54, 218, 221, 267
exciton, 23, 255, 256, 257, 258, 260, 261, 262, 263
expertise, 193
exploitation, 27
exposure, 18, 49, 54, 87, 171, 265
extinction, 49, 268, 271
extraction, 228, 259, 261

F

fabrication, 27, 37, 46, 47, 48, 51, 52, 53, 55, 56, 59, 89, 95, 144, 145, 157, 158, 159, 167, 168, 169, 177, 183, 189, 203, 207, 208, 209, 225, 226, 229, 244, 245, 247, 253, 257, 266, 268, 273, 275
factories, 28, 31, 202
families, 46
Feast, 284
feedstock, 35, 36, 221
Fermi level, 88, 89, 120, 121, 128, 130, 136, 138, 238, 240
filament, 40, 221, 222
filters, 231, 280
first generation, 57
flammability, 274
flexibility, 217, 265, 271
flexible manufacturing, 282
fluctuations, 15, 73
fluid, 38, 193
fluorine, 181, 185, 186, 188, 190, 191, 192, 193, 198, 273
fluorine atoms, 192
foils, 33, 37, 44, 48, 143, 144, 159, 160, 162, 182, 246, 254, 265, 268, 275
force, 47, 148, 185, 240
formation, 15, 39, 48, 51, 54, 55, 79, 81, 85, 88, 90, 95, 100, 108, 123, 143, 147, 149, 154, 155, 157, 174, 175, 181, 185, 187, 189, 190, 191, 194, 197, 200, 202, 211, 217, 219, 220, 221, 222, 257
formula, 127, 200
fragments, 221
framing, 248

France, 5, 26, 33, 64, 93
freedom, 198
fullerene, 257, 260, 263
funding, 16, 25, 26, 59
funds, 26

G

gallium, 14, 41, 59, 68, 71, 72, 73, 77, 78, 91, 92, 95, 167, 198
gas-surface interactions, 219
geometry, 39, 69, 153, 172, 219
germanium, 59, 190, 212
Germany, 26, 32, 48, 63, 64, 90, 109, 164, 202, 204, 246
glasses, 161
glow discharge, 38, 189, 209, 218, 219, 220, 221
glycol, 173
good behavior, 168
grading, 72, 77, 78, 89, 90
gradings, 77
graph, 122, 135, 201
graphite, 47, 155, 173, 183, 194, 221, 275
greenhouse, 64, 163
grids, 268, 273, 278, 279
guidance, 26
guidelines, 19, 28, 47

H

halogen, 74, 117
hardness, 41, 193
hazards, 172, 227
haze, 43, 226
healing, 115
health, 15, 21, 172, 203
heat transfer, 193
heating rate, 108
height, 51, 114, 127, 128, 129
helium, 220
hemisphere, 42
homes, 20
homogeneity, 68, 147, 222
host, 85, 227
hot pressing, 192
human, 31, 203
humidity, 87, 265, 280
hybrid, 18, 47, 97, 98, 101, 102, 103, 104, 253, 254, 258, 260
hydrides, 54
hydroxide, 102, 182
hypothesis, 51, 199

hysteresis, 88, 113, 125

I

ideal, 13, 62, 80, 116, 158, 217, 240, 243, 263
identification, 248, 259
illumination, 87, 88, 112, 114, 116, 117, 120, 122, 129, 131, 134, 135, 137, 138, 143, 156, 157, 159, 168, 169, 180, 213, 230, 238, 239, 240, 241, 257, 258, 265, 274, 280, 281
image, 181, 210
images, 75, 148, 181
improvements, 20, 35, 56, 79, 87, 88, 236, 243, 248, 280
impurities, 41, 48, 55, 85, 91, 158, 178, 180, 213, 255
incidence, 83, 228
incompatibility, 158
independent variable, 237
India, 7, 10
indium, 14, 16, 21, 39, 46, 49, 59, 68, 70, 71, 72, 73, 77, 78, 85, 86, 88, 91, 92, 93, 94, 95, 146, 167, 169, 188, 189, 190, 191, 198, 226
industrial sectors, 170
industrialization, 47
industries, 7, 9, 28, 32, 38, 97
industry, 9, 10, 20, 26, 27, 28, 31, 32, 33, 35, 36, 37, 45, 84, 193, 246, 253, 259, 261, 282
inertia, 274
infrared spectroscopy, 211
infrastructure, 9
initial state, 122, 134
insertion, 43, 54, 55, 235
institutions, 26, 31, 254
insulators, 53
integrated circuits, 37
integration, 28, 207, 259
integrity, 273
interference, 100, 203, 264
internal field, 223
intervention, 212
intimacy, 262
inversion, 81, 120, 203
investment, 10, 12, 14, 23, 29, 31, 36, 38, 207, 246
investors, 29
ion bombardment, 54, 221
ion implantation, 53
ionicity, 51
ionization, 40, 65, 254, 256
ions, 39, 40, 50, 52, 54, 81, 171, 172, 182, 185, 219, 221, 222, 230, 245, 267, 268, 275, 281
iron, 37
irradiation, 263

ISC, 117
Islam, 288
islands, 54
isolation, 202, 238
issues, 15, 17, 22, 27, 28, 31, 32, 37, 46, 56, 68, 78, 112, 115, 127, 144, 207, 208, 210, 222, 225, 245, 248, 275, 276
Italy, 9, 33, 44, 143, 164, 165, 167, 194, 199, 200, 203, 206, 207, 253, 272

J

Japan, 11, 16, 31, 37, 57, 86, 91, 92, 95, 109, 163, 204, 279, 288, 289

K

kinetics, 41, 42, 54, 55, 56, 272
Korea, 289

L

lamination, 16, 23, 203, 254
lasers, 37, 49, 50
lattice parameters, 152, 153
lattices, 208
lead, 10, 16, 27, 34, 38, 68, 73, 75, 77, 78, 79, 129, 134, 136, 254, 257, 259, 274, 276, 281
leadership, 26
learning, 31
LED, 39
lifetime, 15, 27, 35, 156, 157, 180, 187, 209, 217, 222, 255, 265, 271, 280
ligand, 269
liquid phase, 38
liquids, 274
lithography, 263
localization, 51, 212
low temperatures, 121, 122, 124, 126, 131, 135, 162, 171, 269
luminescence, 178, 180
luminosity, 218
lying, 44

M

magnesium, 86
magnet, 40
magnetic resonance, 211
magnitude, 42, 112, 117, 120, 123, 124, 128, 129, 131, 136, 179, 212, 223, 241, 261

majority, 29, 147, 170, 174, 177, 203, 222, 246, 273
management, 203, 220, 276
manganese, 13
manipulation, 90
manufacturing, 9, 15, 18, 20, 21, 22, 27, 28, 29, 31, 36, 38, 44, 53, 62, 95, 143, 144, 169, 188, 203, 207, 208, 222, 226, 227, 246, 247, 248, 265, 282
mapping, 181
market share, 9, 28, 33, 37, 62, 170, 246
masking, 269
mass, 39, 44, 49, 53, 63, 72, 90, 129, 193, 203, 211, 263, 265, 275, 281
matrix, 177, 180, 188, 190, 192
matter, 144, 146, 149, 220, 224, 231, 236, 245, 263
measurement, 72, 73, 74, 83, 85, 86, 87, 93, 96, 111, 112, 113, 116, 117, 123, 124, 125, 128, 132, 138, 150, 152, 159, 179, 199, 200, 211, 212, 213, 238
media, 40, 216, 253, 254
medical, 37
MEG, 23, 24, 25
metals, 15, 16, 38, 47, 71, 146, 155, 158, 169, 273, 283
meter, 33, 116
methanol, 154, 156, 176, 177, 194
microcrystalline, 13, 44, 45, 47, 207, 208, 210, 214, 216, 218, 219, 220, 221, 223, 225, 227, 228, 229, 230, 232, 234, 235, 236, 238, 243, 247, 252, 268
microelectronics, 36, 222
micrometer, 215
microscopy, 100, 148
microstructure, 15, 41, 90, 154, 159, 161, 162, 214, 229
microstructures, 214, 231
microwaves, 40, 54
migration, 180, 219
misfit dislocations, 149
mixing, 108, 149, 173, 194, 199, 209, 218
MOCVD, 39, 63, 80, 84, 169, 172, 173, 226, 227, 235
modelling, 15, 77, 135, 207, 236, 237
models, 89, 111, 115, 117, 193, 213, 236
modifications, 106
modules, 7, 10, 14, 16, 18, 19, 20, 21, 22, 23, 27, 28, 29, 31, 33, 34, 37, 44, 46, 49, 58, 79, 80, 84, 85, 87, 88, 90, 91, 93, 94, 95, 111, 143, 144, 170, 186, 188, 193, 201, 202, 203, 208, 225, 231, 235, 244, 245, 246, 247, 255, 264, 265, 273, 275, 276, 277, 279, 280, 281
moisture, 54, 275, 280
molecular beam, 41, 257
molecular beam epitaxy, 41
molecular weight, 253, 255, 257, 260
molecules, 69, 222, 253, 254, 258, 263, 273

molybdenum, 160
momentum, 212
monolayer, 182, 266
morphology, 16, 23, 42, 43, 56, 100, 103, 106, 107, 108, 143, 146, 147, 148, 150, 154, 159, 160, 161, 173, 175, 177, 178, 181, 226, 227, 245, 261, 263
Moses, 282, 285
MOVPE, 39, 80

N

NaCl, 78, 158, 159, 160
nanocrystals, 23, 25, 214, 215, 254, 260
nanofibers, 269
nanometers, 198, 261
nanoparticles, 60, 61, 268, 273
nanorods, 260
nanostructures, 269
nanotechnology, 60
nanowires, 25
Netherlands, 27
neutral, 40, 212, 219
New South Wales, 48
next generation, 27
NH2, 53, 102, 182
NIR, 52, 225, 228
NIR spectra, 225
nitrides, 77
nitrogen, 41, 53, 94, 153
nodules, 189, 190, 191
Norway, 36
novel materials, 50, 263
nucleation, 42, 76, 176, 184, 219, 220, 221, 229
nuclei, 42

O

OH, 67, 79, 80, 82, 83, 84, 85, 86, 87, 88, 92, 93, 94, 95, 96, 183
oil, 10, 224, 231, 253
one dimension, 236, 237
opacity, 277
operations, 169
opportunities, 56, 57
optical properties, 21, 27, 44, 49, 50, 51, 52, 56, 60, 73, 74, 91, 111, 189, 208, 216, 226, 228
optimization, 56, 81, 95, 162, 229, 269
optoelectronic properties, 47, 208, 254, 263, 264, 271
optoelectronics, 259, 265
orbit, 51
organic compounds, 173, 253

organic polymers, 61
organic solvents, 275
oxidation, 16, 23, 146, 183
oxygen, 41, 51, 82, 85, 114, 146, 149, 175, 177, 178, 184, 188, 190, 191, 192, 193, 194, 199, 200, 265, 275, 280

P

Pacific, 48
parallel, 28, 60, 72, 147, 160, 187, 218, 242, 265, 266, 275, 276, 279
parallelism, 222
parity, 12, 34
passivation, 15, 35, 43, 52, 53, 54, 56, 57, 64, 79, 180
pathways, 20
payback period, 14, 23
PCT, 204
peer review, 18
periodicity, 214
permeability, 41
permission, 214, 244
permit, 47, 191, 254, 257
permittivity, 41, 237, 273
personal choice, 108
perylene, 255
PES, 116
PET, 224, 225, 265, 268
pH, 182
phase diagram, 50, 73, 149, 163
phase transformation, 220
phonons, 212
phosphorous, 217
phosphorus, 53, 209, 217
photoconductivity, 96, 213
photoelectron spectroscopy, 83, 116
photoemission, 211, 212
photolithography, 263
plants, 28, 36, 144, 202, 246
plastics, 61, 208, 224, 247
platinum, 273, 274, 275, 276
playing, 117
PM, 258
polarization, 41
policy, 9
policy options, 9
pollutants, 198
polyimide, 144, 158, 159, 160, 224, 225
polymer, 14, 46, 61, 143, 158, 182, 188, 219, 220, 221, 224, 246, 253, 254, 255, 257, 259, 260, 261, 262, 263, 264, 265, 278, 282
population, 34

porosity, 41, 268
portfolio, 246
Portugal, 97, 109, 205
power generation, 16, 17, 18, 23, 248
preparation, 19, 35, 90, 104, 106, 144, 145, 158, 162, 165, 169, 171, 184, 190, 192, 268, 269
president, 36
prevention, 229
primary function, 237
principles, 120, 207, 282
probability, 42, 213, 225, 239, 257
producers, 36, 225
profit, 62, 63
project, 18, 26, 27, 28, 45, 63, 109, 167
propylene, 173, 273
protection, 79, 278
prototype, 21, 245, 246, 247
public awareness, 21
public support, 12
publishing, 204
pumps, 60
purification, 36, 253, 271
purity, 40, 171, 173
pyrolysis, 80, 97, 189

Q

quartz, 72, 73, 150

R

radiation, 41, 61, 64, 73, 75, 77, 170, 176, 184, 188, 216, 238, 262, 263, 280, 281
radicals, 43, 54, 55, 56, 57, 219, 220, 221
radio, 89, 185, 218
ramp, 28, 29
raw materials, 47, 49, 170, 248
reactant, 39, 41, 172, 193
reagents, 39
reality, 104, 280
recall, 128
recession, 10
recombination, 24, 45, 52, 57, 58, 67, 68, 72, 76, 77, 79, 80, 82, 85, 91, 100, 103, 122, 135, 137, 156, 168, 178, 179, 180, 187, 203, 212, 223, 225, 228, 229, 232, 236, 237, 238, 239, 240, 241, 242, 255, 267, 268, 272, 273, 274
recommendations, 26
recovery, 10, 134, 149, 231
recrystallization, 75, 78, 145, 146, 149, 150, 152, 153, 154, 155, 162
recycling, 19, 28, 203

redistribution, 120, 130, 131, 138
reflectivity, 77, 188, 193, 228
refraction index, 235
refractive index, 41, 53, 228, 235
regeneration, 267, 274
regression, 129
regression method, 129
rejection, 193
relaxation, 87, 120, 152, 153, 155, 229
relevance, 27, 62
reliability, 16, 19, 21, 22, 35, 62
renewable energy, 9, 10, 34, 170, 255
requirements, 15, 22, 24, 48, 50, 62, 74, 78, 80, 261, 265, 272
research institutions, 26, 31, 254
researchers, 7, 59, 144, 194, 229
residues, 177
resins, 37, 275
resistance, 41, 49, 50, 146, 159, 160, 168, 169, 173, 194, 224, 225, 227, 229, 232, 244, 245, 257, 262, 268, 272, 274, 277
resolution, 263
resources, 9, 31, 254, 280
response, 52, 56, 112, 126, 128, 129, 138, 157, 173, 187, 200, 228, 234, 236, 238
restrictions, 224
retail, 34, 35
risk, 20, 36, 76, 79, 82, 246
risks, 29
rods, 23
room temperature, 57, 88, 102, 116, 126, 138, 143, 160, 173, 176, 180, 181, 195, 216
roughness, 42, 43, 56, 74, 81, 151, 215, 226, 227, 268
routes, 27, 67, 261
rowing, 42
rules, 118, 225

S

safety, 15, 21, 149, 177
salts, 78, 182, 273
saturation, 41, 117, 179, 180, 185, 198, 243, 261
scale system, 73
scaling, 97, 98, 222, 244, 245, 247, 265
scanning process, 269
scatter, 45, 52, 61, 69, 74, 85, 90, 106, 107, 211, 215, 225, 226, 268, 269
schema, 257, 266
science, 15, 21, 34, 37, 52, 62, 89, 94, 193
scientific knowledge, 31
scope, 25, 27, 28, 98, 261
second generation, 57, 167

security, 9, 170
seed, 26, 72
segregation, 73, 86, 149, 171, 190, 203, 263
selenium, 14, 68, 69, 71, 75, 89
self-assembly, 273
self-ordering, 263
self-organization, 261
SEM micrographs, 101, 105, 106
sensitivity, 16, 21, 177, 234, 247
sensitization, 280
sensors, 73
shape, 111, 113, 117, 125, 148, 150, 161, 214, 215
shelf life, 265
shortage, 29, 62
showing, 60, 88, 167, 179, 198, 240, 264
Si3N4, 53
signals, 99, 126
signs, 113
silane, 19, 42, 53, 64, 209, 218, 219, 220, 221, 226, 228, 230, 235
silver, 273, 278
simulation, 70, 117, 238
simulations, 118, 127, 128, 280
single crystals, 13, 112, 178
sintering, 146, 155, 160, 173, 268, 269, 270, 277
SiO2, 52, 53, 190, 191
smoothing, 154
sodium, 67, 78, 85, 91, 109, 161, 191
software, 116
solid phase, 26, 48
solid state, 37, 273
solidification, 47
solubility, 43, 149, 274
solution, 35, 39, 56, 61, 62, 63, 82, 93, 102, 106, 147, 153, 155, 158, 169, 172, 176, 177, 182, 183, 236, 258, 259, 260, 263, 265, 266, 269
solvents, 261, 273, 275
space charge distribution, 112, 125, 132, 133, 138
Spain, 63, 95, 205
specialists, 13
species, 41, 42, 43, 45, 53, 54, 55, 171, 172, 182, 190, 193, 198, 213, 218, 219, 220, 221, 273
spectroscopic techniques, 113
spectroscopy, 53, 72, 83, 101, 102, 103, 111, 112, 113, 116, 211, 215
spending, 48
spin, 48, 158, 159, 263, 265
Spring, 32, 163, 251
stability, 14, 15, 21, 23, 27, 41, 47, 49, 54, 87, 92, 136, 143, 144, 146, 147, 156, 157, 159, 160, 169, 170, 176, 189, 190, 191, 198, 199, 203, 208, 220, 230, 231, 247, 262, 264, 265, 271, 272, 274, 275, 276, 280, 289

stabilization, 157
statistics, 236
steel, 20, 224, 225, 231, 245, 247
stoichiometry, 68, 70, 71, 73, 74, 171, 188, 189
storage, 37, 87
stress, 23, 26, 41, 152, 153, 155
styrene, 259
substitutes, 192, 197
substitution, 108, 188
subtraction, 74
Sun, 283
supplier, 29, 109
suppliers, 10, 28, 29, 36, 203
surplus, 75, 116, 125, 131, 135, 136
Sweden, 67
symmetry, 178, 216
synthesis, 22, 37, 253, 254, 263, 271, 282

T

Taiwan, 11, 30
tantalum, 221
target, 18, 24, 25, 27, 34, 35, 37, 40, 41, 53, 98, 100, 104, 173, 174, 184, 185, 189, 190, 191, 192, 194, 280
tariff, 34
tellurium, 16, 156, 172, 194
TEM, 214, 215
terminals, 240, 267
testing, 22, 46, 88, 211, 265, 280
texture, 48, 72, 149, 155, 228
Thailand, 91, 92, 94
thermodynamic equilibrium, 239, 240
tics, 257
tin, 39, 49, 169, 188, 189, 190, 193, 198, 226, 258, 273
tissue, 214, 215, 216
titanium, 77, 254
toluene, 261
total energy, 42
total internal reflection, 228
total product, 246
toxicity, 38, 41, 109
Toyota, 279, 280
tracks, 35
transducer, 191
transformation, 136, 152, 182, 220, 267
transistor, 52, 210
translation, 68, 118
tungsten, 221

U

uniform, 116, 120, 122, 137, 146, 174, 175, 181, 187, 189, 222, 238
universities, 20

V

vacancies, 96, 120, 171, 180, 188, 190, 191, 192, 193
vacuum, 14, 38, 40, 56, 79, 81, 82, 98, 106, 143, 147, 148, 150, 152, 154, 155, 158, 159, 160, 170, 172, 177, 178, 185, 186, 187, 195, 218, 235, 237, 255, 257, 269
valence, 50, 51, 57, 82, 86, 94, 120, 130, 188, 211, 212, 213, 217, 236, 237, 240
Valencia, 63, 95, 251, 252
vapor, 39, 40, 92, 143, 147, 148, 149, 162, 169, 172, 177, 183, 189, 191, 199, 218
variables, 172, 237, 238
variations, 67, 81, 112, 124, 136, 219, 222, 231, 240
VAT, 35
velocity, 129
versatility, 37
viscosity, 273
vision, 19, 26, 31
voiding, 276
volatility, 274, 279
vote, 127

W

Wales, 48, 57
wall temperature, 69
waste, 183, 193, 227
water, 16, 21, 39, 82, 158, 159, 160, 172, 173, 183, 245, 259, 265
wavelengths, 74, 145, 158, 159, 186, 191, 200
web, 26
weight reduction, 277
wetting, 108
wide band gap, 15, 51, 52
windows, 49, 72, 75, 188, 193
wires, 23
workers, 233
worldwide, 11, 34, 36, 44, 207, 246

Y

yield, 21, 22, 25, 28, 32, 68, 72, 80, 158, 248, 273

Z

zirconium, 77
ZnO, 32, 49, 50, 51, 79, 80, 81, 82, 83, 84, 86, 87, 88, 89, 92, 93, 94, 95, 96, 102, 114, 116, 117, 120, 127, 128, 130, 145, 146, 159, 160, 168, 179, 188, 189, 190, 191, 198, 214, 225, 226, 227, 228, 235, 260, 269